PRINCIPLES OF
COMMUNICATION
FOR SCIENCE
AND TECHNOLOGY

PRINCIPLES
OF COMMUNICATION
FOR SCIENCE
AND TECHNOLOGY

LESLIE A. OLSEN
The University of Michigan

THOMAS N. HUCKIN
The University of Michigan

McGRAW-HILL BOOK COMPANY

New York   St. Louis   San Francisco   Auckland   Bogotá   Hamburg
Johannesburg   London   Madrid   Mexico   Montreal   New Delhi
Panama   Paris   São Paulo   Singapore   Sydney   Tokyo   Toronto

**PRINCIPLES OF
COMMUNICATION
FOR SCIENCE AND TECHNOLOGY**

Copyright © 1983 by McGraw-Hill, Inc.
All rights reserved.
Printed in the United States of America.
Except as permitted under the United States Copyright Act of 1976,
no part of this publication may be reproduced or distributed
in any form or by any means, or stored
in a data base or retrieval system,
without the prior written permission of the publisher.

2 3 4 5 6 7 8 9 0   HALHAL   8 9 8 7 6 5 4 3

ISBN 0-07-047821-X

This book was set in Souvenir Light by Waldman Graphics, Inc.
The editors were James A. Dodd and Claudia Tantillo;
the designer was Nicholas Krenitsky;
the production supervisor was Diane Renda.
The drawings were done by J & R Services, Inc.
Halliday Lithograph Corporation was printer and binder.

Library of Congress Cataloging in Publication Data

Olsen, Leslie A.
     Principles of communication for science and
technology.

     Includes index.
     1. Technical writing.   I. Huckin, Thomas N.
II. Title.
T11.054   1983   808'.0665   82-20404
ISBN 0-07-047821-X

FOR OUR PARENTS AND CHRISTIANE

CONTENTS IN BRIEF

APPENDIXES

CONTENTS

Principles of Communication for Science and Technology is a technical writing textbook for intermediate and advanced students in science and engineering. It focuses on those features which characterize scientific and technical English and which are known to be troublesome for university and industry students. It emphasizes principles and use rather than usage, that is, functional explanations rather than formal rules.

The most important feature of *Principles of Communication for Science and Technology* is its functional-rhetorical approach, which emphasizes the communicative use of language rather than simply its formal aspects. The book does treat formal aspects where appropriate (for example, in discussing report formats, grammar, punctuation, and visual elements), but it places more emphasis on the psychological and rhetorical principles underlying effective communication. As such, it focuses on language in meaningful contexts, not on sentences or words in isolation.

A second important feature of the book—and one which stems from its rhetorical focus—is its treatment of the early stages of writing. How does a writer think up something to say, find and define a topic, find appropriate things to say about a topic? In our experience, even if students or professional engineers have been assigned to do particular studies and to write reports about them (and thus have found topics), they often do not know how to set up and state a report's problem effectively for a given audience or how to define and apply criteria needed to solve the problem. These skills are critical for scientific and technical professionals, and Chapters 4 to 6 of the book deal explicitly with them in an approach strongly indebted to classical argument and to Richard E. Young, Alton Becker, and Kenneth L. Pike's *Rhetoric: Discovery and Change.*

A third important feature of the book is its attempt to provide complete, explicit explanations of the main teaching points. We feel that science and engineering students prefer systematic, step-by-step instruction and that they like to have reasoned explanations for how things function, including how language functions in communication.

This point gives rise to yet another feature of the book, namely, its focus on the *process* by which a student may produce an effective piece of communication for a given audience. Too often, textbooks treat writing merely as a finished product. They present models of good writing for the student to imitate, but they do not provide adequate guidance by which the student can learn to cope with novel situations on his or her own. Thus, we have provided step-by-step procedures that students may follow as guides to the writing process, including a number of flow charts.

Finally, *Principles of Communication for Science and Technology* is designed to be used not only as a course text but as a long-term reference work as well. It thus should be especially useful for (1) university-level students in technical areas who are using English in their studies and who plan to continue into technical careers and (2) practicing scientists and engineers who need a self-instructional reference book in written and oral English for technical communication. To increase the book's usefulness, we have included many exercises, lists of supplementary readings, a reference appendix on punctuation, sample reports and letters, and an index. A teacher's manual is available not only for teachers but also for those using the book for self-instruction.

(Note to teachers of both native-speaker and nonnative-speaker students: a companion book, *English for Science and Technology: A Handbook for Nonnative Speakers,* designed specifically for nonnative speakers but covering much of the same material, is also available.)

ACKNOWLEDGMENTS

The preparation of this book has been a long but exciting project and one which would not have been possible without the help and encouragement of our families, friends, and colleagues, many professional acquaintances, and many students. We would like to thank all those who have contributed in one way or another to the completion of this book, but in particular we would like to thank the following people for their special contributions: Christiane Huckin for her patience and support, as well as for other invaluable behind-the-scenes assistance; our parents, grandparents, and siblings for their unflagging moral support; our teachers and later colleagues—Dwight W. Stevenson, Richard E. Young, Louis Trimble, and J. C. Mathes—for our early training in the field and their later influence and inspiration; our colleagues in the Technical Communication program in the University of Michigan's Department of Humanities—Thomas M. Sawyer, Lisa Barton, Peter Klaver, and James Zappen—and our other Humanities Department colleagues for their ready counsel and support; and the Department of Humanities office staff—Vickie Gordenier, Barbara Cox, and Margaret Cousins—for cheerfully putting up with us even in times of stress.

In addition, we would like to thank many others who have provided special research funds, research materials, intellectual stimulation, and other forms of assistance: The University of Michigan—particularly, the College of Engineering, the Horace H. Rackham School of Graduate Studies, the Center for Research on Learning and Teaching, and the Division of Research Development and Administration—the National Endowment for the Humanities, Hansford W. Farris, Blanchard Hiatt, Julie Wei, W. Ross Winterowd, Francis Christensen, Frederick J. Newmeyer, John M. Swales, Edward P. J. Corbett, Heles Contreras, Virginia J. Tufte, M.A.K. Halliday, Walter Kintsch, Richard C. Anderson, Chaim Perelman, Paul Anderson, Carolyn Miller, Larry Selinker, Richard Enos, David Kaufer, Joan Stigliani, John Lepp, Linda Flower,

David Kieras, Christine Neuwirth, Fay Edwards, Robert Caddell, Hal Schulte, Harold Harger, David V. Ragone, Alfred Sussman, Patricia Sullivan, Burte Waite, and Nancy Elder.

We would also like to thank Karl Drobnic, Quentin Johnson, Louis Holschuh, John Lackstrom, Carol Romett, Christopher Gould, William Woods, Douglas Wixson, Ruth Falor, and O. Allen Gianniny, our reviewers, for conscientiously and expertly examining various drafts of our manuscript and giving us numerous helpful suggestions. We wish to thank Phillip Butcher, Claudia Tantillo, Jo Satloff, Irene Nunes, Nicholas Krenitsky, and Diane Renda of the McGraw Hill staff for their help in producing the book.

Finally, we would like to acknowledge the contribution of many students in our technical communication classes at the University of Michigan. They have been patient and helpful experimental subjects and sources of examples, suggestions, and new ideas. (They would also, no doubt, prefer to remain anonymous.)

Leslie A. Olsen
Thomas N. Huckin

Part I

INTRODUCTION

1

WHY STUDY TECHNICAL COMMUNICATION?

Why should engineers and scientists study technical writing or speaking? Their primary training and interests lie in *technical* areas; most science and engineering students successfully pursue their *technical* subjects without extra writing courses; and practicing engineers and scientists in government and industry work on *technical* projects. It might seem, then, that writing and communication are superfluous to a technical education.

In fact, this is not the case. Scientists and engineers may be technically brilliant, and creative, but unless they can convince coworkers and supervisors of their worth, the technical skills will be unnoticed, unappreciated, and unused. In a word, if technical people cannot communicate to others what they are doing and why it is important, it is they and their excellent *technical* skills that will be superfluous. From this perspective, communications skills are not just handy; they are critical tools for success, even survival, in "real-world" environments.

1.1
THE IMPORTANCE OF TECHNICAL COMMUNICATION

The importance of technical communication has been noted by many technical people, including Russel R. Raney, one-time Chief Engineer, Farm Implement Division, at International Harvester Company. He points out that most industries are composed of internal units that include (at least) a marketing unit, a manufacturing unit, an engineering or design unit, and a management unit. These units, or internal organizations,

> are quite similar to factions in a political party in that, while they all strive toward the same end, they do not necessarily agree among themselves as to the means to be employed in achieving that end. So it is to be expected that the normal operation of the business will be marked by a certain amount of internal conflict as each organization seeks to promote its own particular interest.[1]

This natural conflict, the many details and changes to be managed, and the incredible complexity of a large organization all combine to create a demand for communication skills unforeseen by most science and engineering students:

> In school you can attain your goal of a degree on your own individual efforts with practically no dependence on anyone else. On the other hand, it should

TABLE 1-1 SUBJECTS MOST NEEDED FOR ENGINEERING
CAREERS IN INDUSTRY

RANK	SUBJECT
1.	Management practices
2.	Technical writing
3.	Probability and statistics
4.	Public speaking
5.	Creative thinking
6.	Working with individuals
7.	Working with groups
8.	Speed reading
9.	Talking with people
10.	Business practices (marketing, finance, economics)
11.	Survey of computer uses
12.	Heat transfer
13.	Instrumentation and measurements
14.	Data processing
15.	Systems programming
16.	Economics
17.	Ordinary differential equations
18.	Logic
19.	Economic analysis
20.	Applications programming
21.	Psychology
22.	Reliability
23.	Vector analysis
24.	Electronic systems engineering (circuit design)
25.	Laplace transforms
26.	Solid-state physics
27.	Electromechanical energy transformation
28.	Matrix algebra
29.	Computer systems engineering
30.	Operations research
31.	Law practices (patents, contracts)
32.	Information and control systems
33.	Numerical analysis
34.	Physics of fluids
35.	Thermodynamics
36.	Electromagnetics
37.	Human engineering
38.	Materials engineering

SOURCE: Middendorf, Reference 3. © 1980, American Society for Engineering Education.

be evident from the fore-going description of industrial operation that you will be able to accomplish very little by yourself in an organization. I want to point out two significant results of this situation.

First, it means that for the results of your efforts to be incorporated in any end result you must present them effectively in competition with many conflicting considerations. It will, therefore, inevitably be a part of your responsibility to transmit understandably and persuasively the results of your work to other people, usually both verbally and in writing. Since success in engineering school does not depend much on communication abilities, you may not have developed your skills and may mistakenly resent the absolute dependence you will have on them for accomplishment in an organization.[2]

The importance of communication skills is emphasized in survey after survey. For instance, the American Society for Engineering Education conducted a survey[3] to determine which academic subjects are most needed for engineering careers in industry. Responses were received from 4057 engineers, all having at least several years of on-the-job experience behind them. The results, given in Table 1-1, show that communication skills rank above any other type of skill, capturing five of the nine "most-needed" categories. These include technical writing (2), public speaking (4), working with individuals (6), working with groups (7), and talking with people (9). In contrast, technical skills rank toward the bottom of the list. (It must be noted, however, that this ranking was produced by engineers from a variety of disciplines and that not all of the engineers may have needed all of the engineering subjects in the bottom half of the list. Thus, some of these categories might have received somewhat higher ratings from a more homogeneous group of respondents.)

Another survey of 245 Engineers of Distinction[4] produced an even stronger emphasis on communication skills. For both the distinguished engineers and younger engineers hoping for promotion, writing skills were cited as critical or very important to success, as illustrated in Table 1-2.

1.2

THE FREQUENCY OF TECHNICAL COMMUNICATION

The actual amount of time spent on communication tasks varies from job to job, of course, but an average distribution of communication tasks seems to resemble that presented in Table 1-3 by an industrial consultant, Ed Gilbert. He estimates that a junior engineer spends about 20 percent of the time writing and 20 percent talking, for a total of 40 percent. This escalates to approximately 77 percent for a chief executive, mostly in the verbal medium.[5]

These figures on communication time are supported by a recent survey by the University of Wisconsin's College of Engineering, which discovered that

75 percent of engineering undergraduates take jobs in industry, where at least 25 percent of an engineer's time is spent in the reporting process. As the engineer moves up the managerial ladder, this time can increase to as much as 80 percent.[6]

Obviously, if an employee can make such time-consuming activities more efficient, both the employee and the company will be well served. One

TABLE 1-2 RESPONSES OF PROMINENT ENGINEERS TO QUESTIONS ON THE IMPORTANCE OF TECHNICAL WRITING

How important is the writing that you do, and is the ability to write effectively needed in your present position?

Critically important	110
Very important	124
Somewhat important	9
Minimally important	0
No response	2

When you select or approve someone for advancement, you must, of course, consider many factors. If an individual is otherwise qualified, can inability to write effectively delay or prevent advancement?

Ability to write is usually critical—has strong effect on selection	63
Ability to write is usually important—often affects selection	153
Ability to write is sometimes helpful—may have some effect on selection	25
Ability to write is not usually important—has little or no effect on selection	1
No response	3

SOURCE: Davis, Reference 4. © 1977, American Society for Engineering Education.

TABLE 1-3 TYPICAL COMMUNICATION TIME WITHIN ORGANIZATIONS

HIERARCHY	PORTION OF TOTAL WORK TIME SPENT ON COMMUNICATION, %		
	VERBAL MEDIA	INFORMAL WRITING	FORMAL WRITING
Chief executive	65	10	2
Division manager	60	15	2
Department manager	45	20	5
Section head	40	20	5
Project head	30	20	10
Engineer or scientist	25	15	10
Junior engineer	20	10	10
Technician	15	10	2

SOURCE: Gilbert, Reference 5.

easy way to improve efficiency is to develop effective strategies for analyzing communication tasks and audiences as well as strategies for producing reports and oral presentations. This text, and any technical communication course based on it, should help writers develop such strategies.

1.3
THE INADEQUATE COMMUNICATION SKILLS OF MANY TECHNICAL PROFESSIONALS

Another survey (of civil, electrical, and mechanical engineers) notes the extreme importance of communication skills, but in addition stresses the inadequate preparation of most engineers for communication tasks. For instance, the results for civil engineering presented in Table 1-4 rank writing and speaking as the most important area of competence in civil engineering practice, but rank about two thirds of recent graduates inferior in this area. Similarly, the results for electrical and mechanical engineering presented in Tables 1-5 and 1-6 rank writing and speaking as the first and second most important areas, respectively, in engineering practice, but rank about half of the recent

TABLE 1-4   CAPABILITIES OF RECENT CIVIL ENGINEERING GRADUATES

IMPORTANCE TO CIVIL ENGINEERING PRACTICE				CAPABILITY OF RECENT C.E. GRADUATES (1–5 YEARS)		
MOST IMP.	IMPOR- TANT	LESS IMP.	AREA OF COMPETENCE	SUPER- IOR	ADE- QUATE	INFER- IOR
137	86	9	Writing and speaking	7	69	142
106	127	9	Structural analysis and design	25	160	28
72	156	13	Soil mechanics and foundation	9	157	46
62	125	47	Water and wastewater treatment	13	144	27
55	131	38	Fluid mechanics, hydraulics, and hydrology	10	136	42
53	153	32	Computer and numerical methods	42	139	30
50	132	49	Economics and finance	4	96	106
47	147	54	Construction methods and equipment	9	107	95
46	105	70	Law, labor, and management	4	81	120
20	148	67	Surveying and measurement	4	139	65
19	110	92	Transportation, highways, and traffic	5	161	30
14	124	51	Materials	6	153	35
13	85	132	Social sciences and humanities	12	137	53

SOURCE: Kimel and Monsees, Reference 7. © 1979, American Society for Engineering Education.

TABLE 1-5 CAPABILITIES OF MECHANICAL ENGINEERING GRADUATES

IMPORTANCE TO MECHANICAL ENGINEERING PRACTICE				CAPABILITY OF RECENT M.E. GRADUATES (1–5 YEARS)		
MOST IMP.	IMPOR- TANT	LESS IMP.	AREA OF COMPETENCE	SUPER- IOR	ADE- QUATE	INFER- IOR
27	33	2	Engineering materials	3	37	10
26	27	4	Writing and speaking	1	24	34
23	31	4	Thermodynamics and heat transfer	10	40	1
22	28	6	Fluid mechanics	4	41	6
18	16	13	Dynamics and vibration controls	9	31	6
16	30	11	Machine design	7	31	10
14	24	19	Manufacturing methods	1	24	23
14	20	15	Individual project experience	1	22	24
13	36	10	Engineering graphics	2	31	16
13	33	13	Computer and numerical methods	13	35	1
11	35	11	Measurements and instrumentation	8	24	14
6	28	13	Economics, law	0	28	19
4	20	32	Social science and humanities	3	36	4

SOURCE: Kimel and Monsees, Reference 7. © 1979, American Society for Engineering Education.

TABLE 1-6 CAPABILITIES OF RECENT ELECTRICAL ENGINEERING GRADUATES

IMPORTANCE TO ELECTRICAL ENGINEERING PRACTICE				CAPABILITY OF RECENT E.E. GRADUATES (1–5 YEARS)		
MOST IMP.	IMPOR- TANT	LESS IMP.	AREA OF COMPETENCE	SUPER- IOR	ADE- QUATE	INFER- IOR
86	66	1	Writing and speaking	4	59	62
56	68	22	Power engineering and energy systems	7	68	38
53	71	22	Digital and hybrid computer systems	34	77	20
44	72	26	Discrete and integrated electronics	18	79	13
42	90	10	Solid-state devices	22	88	14
33	77	41	Automatic control	11	86	16
37	81	24	Communication and information systems	19	81	22
35	78	26	Network theory	15	93	15
22	100	24	Economics and finance	3	73	52
17	59	55	Advanced automation	7	90	19
13	52	78	Antennas and wave propagation	7	75	19
11	45	80	Artificial intelligence	4	72	21
9	58	88	Social sciences and humanities	6	100	22

SOURCE: Kimel and Monsees, Reference 7. © 1979, American Society for Engineering Education.

graduates inferior in communication skills. This lack of communication skills has also been noted by all the surveys cited above and is pungently summarized by the Industrial Advisory Board at the University of Tulsa:

> Engineers cannot communicate—cannot spell, cannot make a sketch, have difficulty in all phases of communication with others. The student of today needs more ability than ever and a key need is to increase the ability to communicate both in speech and graphics.[8]

1.4
SKILLS NEEDED IN TECHNICAL COMMUNICATION

Given the situation just described—that communication tasks are important to success in the technical professions, that communication tasks consitute a relatively large part of a job, and that many (if not most) graduating scientists and engineers have inadequate communication skills—what kinds of things must such people learn to become effective and efficient communicators? Most respondents to that question from industrial or organizational settings stress that communicators should understand the situation, uses, and audiences for a given communication and write for those audiences and uses; that they should have clear organization and logical reasoning; that they should write clear and concise sentences; and that they should follow the standard conventions for grammar, punctuation, and other mechanics. In other words, writers should produce sentences that readers can easily understand, and they should place those sentences in contexts—paragraphs and larger units—whose function is clear.

Two surveys provide support for these generalizations as applied to technical writing. An early survey reported by MacIntosh[9] itemizes 18 typical problems noted by senior managers in science and industry (Table 1-7). A more recent survey of distinguished engineers by Davis[10] reports the same concerns, but in more general terms (Table 1-8).

A third survey ranks the types of communication skills needed on the job.[11] The results of this survey, presented in Table 1-9, suggest that people in technical fields should study significant amounts of both oral and written work and learn to communicate in a variety of forms, especially shorter forms.

The organization of this text reflects the surveys just discussed. It moves from a review of basic principles of technical communication to a consideration of a report's audiences and uses to a more detailed consideration of structure and logic to a discussion of sentence construction. It treats both written and oral communication and includes a unit on formatting and visual aids. It includes appendices on punctuation and mechanics and a number of sample reports illustrating various types of communication.

TABLE 1-7 COMPLAINTS ABOUT TECHNICAL WRITING FROM 182 SENIOR OFFICIALS IN SCIENCE AND INDUSTRY

COMPLAINT	NO. RESPONDING
Generally foggy language	182
Inadequate general vocabulary	173
Failure to connect information to point at issue	169
Wordiness	164
Lack of stressing important points	163
All sorts of illogical reasoning	163
Too much "engineering gobbledygook"	160
Poor overall organization	153
No clear overview	143
No clear continuity	142
Very little concept of writing for anyone but fellow specialists	136
Poor grammar	133
General lack of flexibility to suit circumstances	127
Deliberate obscurity	92
Poor punctuation	81
No sense of proper tone for circumstances	72
Poor adaptation of form to purpose	64
Almost meaningless introduction	54

SOURCE: MacIntosh, Reference 9.

TABLE 1-8 DISTINGUISHED ENGINEERS' CHOICE OF TOPICS TO BE COVERED IN A TECHNICAL WRITING COURSE

TOPIC*	DEGREE OF IMPORTANCE			
	ESSENTIAL	OK	NOT IMPORTANT	NO RESPONSE
Clarity of expression	239	4	0	2
Analyzing a situation and producing a communication to fit the reader's needs	209	31	0	5
Organizing reports and other communications	189	50	3	3
Grammar and syntax	165	75	0	5
Writing a draft and completing the finished document	151	79	7	8
Mechanics (punctuation, abbreviations, capitalization, use of numbers, etc.)	120	103	12	10
Style and tone of expression	117	115	7	6
Finding and using published information	90	127	22	6

*Brevity and conciseness were the most often mentioned additions to this list. NOTE: The question asked was, Some courses in technical writing are good and others are pretty poor. If a course in technical writing is included in the curriculum, what topics should be covered?

SOURCE: Davis, Reference 10. © 1977, American Society for Engineering Education.

TABLE 1-9 RANK OF JOB IMPORTANCE RESPONSES

RANK	COMMUNICATION SKILL	MEAN*	STD. DEV.
1	One-to-one talks with technically sophisticated personnel	5.9391	1.3109
2	Writing using graphs, charts, and/or other illustrative aids	5.3955	1.7141
3	Project proposals (written)	5.3324	1.8167
4	Participation in a small group or committee made up of only technically sophisticated members (oral)	5.2694	1.7733
5	Instructions for completing a technical process (written)	5.2444	1.8332
6	One-to-one talks with nontechnical personnel	5.0718	1.9065
7	Project progress reports (written)	5.0609	1.6301
8	Project proposal presentations (oral)	5.0587	1.9787
9	Writing technical information for technical audiences	4.9157	1.9396
10	Oral presentations using graphs, charts, and/or other aids	4.9139	1.9623
11	Technical description of a piece of hardware (written)	4.8687	1.8793
12	Memoranda	4.8169	1.7175
13	Short reports (less than 10 pages typewritten, double-spaced)	4.7967	1.8507
14	Writing technical information for nontechnical audiences	4.7458	2.1377
15	Project feasibility studies (written)	4.7124	1.9305
16	Project progress report presentations (oral)	4.6667	1.9277
17	Participation in a small group or committee including nontechnical members (oral)	4.6639	1.9608
18	Business letters	4.6257	1.8856
19	Telephone reports	4.5838	1.9682
20	Project feasibility study presentations (oral)	4.4258	2.0780
21	Form completion	4.2842	1.8847
22	Formal speeches to technically sophisticated audiences	4.0452	2.2108
23	Writing in collaboration with one or more colleagues	4.0313	2.0175
24	Laboratory reports (written)	4.0184	2.1800
25†	Long reports (10 or more pages typewritten, double-spaced)	3.8249	2.2120
26	Formal speeches to nontechnical audiences	3.5869	2.2020
27	Writing requiring library research	3.2062	2.0723
28	Abstracts/summaries of others' writing (written)	2.8864	1.9910
29	Articles submitted to professional journals	2.8074	2.1499
30	Reports submitted to professional societies (written)	2.7224	2.0827

*Rating scale: 1 = least important; 7 = most important.
†Skills 25 through 30 were ranked "below average to unimportant."
SOURCE: Schiff, Reference 11. © 1980, American Society for Engineering Education.

REFERENCES

1   Russel R. Raney, "Why Cultural Education for the Engineer?" Speech to the Iowa-Illinois Section of the American Society of Agricultural Engineers, April 29, 1953, quoted in John G. Young, "Employment Negotiations," *Placement Manual: Fall 1981,* College of Engineering, The University of Michigan, Ann Arbor, 1981, p. 49.

2   Ibid., p. 50.

3   William H. Middendorf, "Academic Programs and Industrial Needs," *Engineering Education,* May 1980, pp. 835–837.

4   Richard M. Davis, "Technical Writing: Who Needs It?" *Engineering Education,* November 1977, pp. 209–211.

5   Ed Gilbert, "Technical Communication and Writing: Management's Perspective," speech to the Conference on Teaching Technical and Professional Communication, co-chaired by Dwight W. Stevenson and J. C. Mathes, Department of Humanities, College of Engineering, The University of Michigan, 1975. Used with the permission of J. C. Mathes.

6   Nicholas D. Sylvester, "Engineering Education Must Improve the Communication Skills of Its Graduates," *Engineering Education,* April 1980, p. 739.

7   William R. Kimel and Melford E. Monsees, "Engineering Graduates: How Good Are They?" *Engineering Education,* November 1979, pp. 210–212.

8   Sylvester, loc. cit.

9   Fred H. MacIntosh, "How Good Is Our Product?" Speech delivered at the annual meeting of the Conference on College Composition and Communication, April 8, 1967, Louisville, Ky.

10 Davis, op. cit., p. 210.

11 Peter M. Schiff, "Speech: Another Facet of Technical Communication," *Engineering Education,* November 1980, p. 181.

ADDITIONAL READING

Committee on Scientific and Technical Communications of the NAS-NAE, *Scientific and Technical Communication: A Pressing National Problem and Recommendations for Its Solution,* National Academy of Sciences, Washington, D.C., 1969.

Richard M. Davis, "Technical Writing: Who Needs It?" *Engineering Education,* November 1977, pp. 209–211.

Exploration of Oral/Informal Technical Communications Behavior, American
    Institutes for Research, Washington, D.C., 1967.
Eugene Garfield, "Communication, Engineering and Engineers," *IEEE
    Transactions on Professional Communication* **15**(2):123–136 (1972).
H. K. Jenny, "Heavy Readers Are Heavy Hitters," *IEEE Spectrum* **15**(9):66–
    68 (1978).
Patrick Kelly, M. Kranzberg, S. R. Carpenter, and F. A. Rossini, *The Flow
    of Scientific and Technical Information in the Innovation Process: An
    Analytic Study,* Georgia Institute of Technology, Atlanta, 1977.
William Kimel and Melford E. Monsees, "Engineering Graduates: How Good
    Are They?" *Engineering Education,* November 1979, pp. 210–212.
William H. Middendorf, "Academic Programs and Industrial Needs," *Engi-
    neering Education,* November 1977, pp. 209–211.
E. C. Nelson and E. D. Pollack, *Communication Among Scientists and En-
    gineers,* Heath Lexington, Lexington, Mass., 1970.
Peter M. Schiff, "Speech: Another Facet of Technical Communication," *En-
    gineering Education,* November 1980, pp. 180–181.
Hedvah L. Shuchman, *Information Transfer in Engineering,* The Futures
    Group, Glastonbury, Connecticut, 1981.
Nicholas D. Sylvester, "Engineering Education Must Improve the Commu-
    nication Skills of Its Graduates," *Engineering Education,* April 1980,
    pp. 739–740.

Part II

REVIEW OF BASIC WRITING PRINCIPLES

2

WRITING PARAGRAPHS

Unlike readers in the academic world, most readers in the "real world" read *selectively:* rather than thoroughly digesting a piece of writing, they skim-read most of it, skipping from one main idea to another until they come to something that particularly interests them. Such readers are forced to read this way. They are overwhelmed with reading material and simply don't have the time to read *everything* carefully.

As a writer, therefore, you should do everything possible to ease this kind of reader's burden; in particular, you should make your writing *easy to skim-read.* One of the best ways to do this is to write good paragraphs. What makes for a good paragraph in scientific/technical writing? First of all, a good paragraph has *unity:* it focuses on a single idea or theme. Secondly, a good paragraph has *coherence:* one sentence leads to the next in some kind of logical sequence. Finally, a good paragraph has *adequate content:* it has an appropriate selection and number of details to support the main idea of the paragraph. Readers expect to find these qualities in paragraphs, and you as writer should take care not to frustrate their expectations.

There are two principal tools you can use to invest your paragraphs with the qualities just described: (1) a good topic statement and (2) an appropriate pattern of organization.

2.1

WRITE A GOOD TOPIC STATEMENT

The topic of a paragraph is its main idea or theme, i.e., what the paragraph is about. As with a larger piece of writing, readers of a paragraph want to know right away what the topic is. They also like to have some idea of how this topic will be developed. In other words, readers will use whatever cues they can to quickly generate expectations about the paragraph as a whole. This strategy serves two purposes: (1) it allows readers to guess what's coming and thus digest it more easily and (2) it allows them to avoid reading the paragraph altogether if the subject matter holds no interest for them.

You can help your readers, therefore, by providing a good topic statement right at the beginning of the paragraph. It does not have to be confined to a single sentence: often a topic statement is extended over the first two sentences of a paragraph. It should, however, always contain one or more key words directly related to the topic, and it should be as complete a statement of the main idea as possible, without getting into too much detail and making the sentence(s) forbiddingly long. In addition, if possible, it should suggest how the topic will be developed (by comparison and contrast, by cause-and-effect analysis, etc.).

Here is an example of an effective topic statement.

> *Unlike gasohol-powered cars, the fuel cell alternative is virtually pollution-free.* A methanol fuel cell system works through chemical reactions that leave the air clean. A fuel processor breaks the methanol down into carbon dioxide and hydrogen; the hydrogen is then pumped to the cell itself, where it combines with oxygen to form water. Current is then produced when the electrons traded between molecules in this reaction travel through an external circuit. The net products are carbon dioxide, water, and electricity. By contrast, when gasohol is burned in an internal combustion engine, it produces the same nitrous oxides that gasoline does.[1] [italics added]

This topic statement is a good one because it tells the reader immediately what the theme of the paragraph is (fuel cell cars don't pollute) and because it's consistent with how the rest of the paragraph is developed (as a cause-and-effect description of how the fuel cell process works). Notice how the writer has used the key term *fuel cell* in the most important position in the sentence, the main-clause subject position, thus establishing it as the paragraph topic, i.e., what the paragraph is about.

For an example of what *not* to do, here is a paragraph from a student report on whether to use an argon recovery process or a hydrogen recovery process in a proposed ammonia plant:

> **NEGATIVE EXAMPLE**
> Utility costs for the argon process are 75 percent greater than for the proposed hydrogen process. Initial capital cost is $5.4 million, roughly three times the hydrogen process cost. However, annual income from the sale of argon, increased ammonia production, and reduced natural gas requirements elsewhere in the plant is 160 percent higher than that generated by the hydrogen process. Present worth analysis shows that the argon process is the better investment. The present worth of the argon process is $10.25 million. The present worth of the hydrogen process is $4.14 million.

Most readers will quickly conclude, on the basis of the first two sentences of this paragraph, that the argon process is more costly than the hydrogen one and should therefore not be chosen. But this is just the opposite of what the writer wants them to understand! For later on—buried near the bottom of the paragraph—the report states that "the argon process is the better investment." This statement is actually the topic statement of the paragraph. By "burying" it, the writer is running a serious risk of having the readers—especially those busy *important* readers!—completely overlook it.

A few simple changes can easily remedy the situation: (1) promoting the topic statement to initial position in the paragraph, (2) combining the next two sentences and subordinating them to the next one, (3) combining the last two sentences, and (4) adding a few words for emphasis:

> The argon process is clearly a better investment than the hydrogen process. Although it has higher utility costs (by 75%) and a higher initial capital cost (by 300%), it generates annual income—from the sale of argon, from increased ammonia production, and from reduced natural gas requirements elsewhere in the plant—that is 160% greater than that generated by the hydrogen process. Present worth analysis shows that the argon process is valued at $10.25 million whereas the hydrogen process is valued at only $4.14 million.

Notice how much more readable the rewritten version is. The topic statement serves to establish the main point and also to suggest how the rest of the paragraph will be developed (as a comparison-and-contrast pattern). The key term *argon process* (or its pronoun equivalent *it*) is used repeatedly in sentence-subject or clause-subject positions, thus keeping the reader's mind focused on it; all reference to the hydrogen process, by contrast, is deliberately subordinated.

The basic principle behind a well-written topic statement is this: by the time a reader has finished reading the first sentence or two of a paragraph, he or she should be able to predict what the rest of the paragraph is generally about and how it will probably be developed. Suppose, for example, you were reading an article on transport across membranes in a popular-science magazine and began reading this paragraph opening:

> The human body is made up of millions and billions of cells, each of which contains, among other substances, millions and billions of protein molecules. . . .[2]

How do you think the rest of the paragraph will go? What will it be about? Do you expect it to elaborate on other substances, on cells, on the human body generally? Probably not. Instead, there seems to be a narrowing down of focus to the term *protein molecules;* this is probably what the paragraph is about. How will this topic be developed? Well, the pattern of development used in this opening sentence is one of classification-division. Perhaps that pattern will be continued. Or maybe the writer has used some other pattern— a general-to-particular ordering of details, say, or a comparison of protein molecules with other kinds of molecules. Maybe the writer has used two patterns together. Of all the possibilities, though, you'd probably expect the classification-division pattern to be continued. It's a general fact about human nature that once we perceive a pattern in something, we expect it to continue—unless, of course, it's explicitly broken. Let's see what happens with this paragraph:

> ... An average cell contains hundreds of different kinds of proteins, and all of the cells of the human body contain, among them, as many as 100,000 different kinds of proteins. These proteins can perform millions of different functions, a versatility which is largely responsible for the phenomenon called "life."[2]

The writer has indeed continued with the classification-division pattern set up in the first sentence. Notice, too, how the pattern proceeds from general to particular; the grammatical subjects of the three sentences show this progression quite clearly: *The human body. ... An average cell. ... These proteins. ...*

Though paragraphs sometimes exist in isolation, they are usually linked to other paragraphs, forming larger conceptual units. In such cases, either the topic and pattern set up in one paragraph may be carried on to the next or the break between paragraphs can be used to switch to a new topic and/or a new pattern. In any event, it is usually desirable to maintain some kind of continuity when moving from one paragraph to the next. This is most often done by incorporating a key word or term from one paragraph into the first sentence of the next. In the membrane transport article, for example, the first sentence of the next paragraph begins as follows:

> The proteins derive their versatility from their structure—they are made up of chains of molecules of amino acids, substances of which there are 20 different ones in the human organism. ...[2]

Notice how the writer has picked up on the key word *versatility* from the last sentence of the first paragraph and used it as a transition to the subject matter of the second paragraph.

And what *is* the subject matter of this new paragraph? If the topic statement (above) is a good one, you should be able to predict with some assurance what it is. First of all, two new words appear prominently in this statement: *structure* and *amino acids*. We might suppose, therefore, that the theme of this paragraph is the structure of amino acids. Furthermore, since the writer has begun discussing amino acids by saying that there are 20 different types of them, we might expect the rest of the paragraph to be a discussion of the structural variety of amino acids, perhaps according to a classification-division pattern, a comparison-contrast pattern, or a general-to-particular ordering of details. Let's see how it actually does continue:

> An average protein molecule consists of about 500 molecules of amino acids of different kinds (seldom all 20), arranged in some particular sequence. A sequence of 500 amino acids composed of all the 20 different ones would have as many as 1×10^{60} (1 followed by 60 zeros) possible arrangements,

each arrangement having particular chemical properties and therefore chemical capabilities. From these few facts alone, we can readily appreciate how important the study of the amino acids is to our understanding of proteins, of the cell, and of life.[2]

As you can see, the paragraph as a whole *does* satisfy the expectations raised by the topic statement. It *is* about the structural variety of amino acids, and it *does* follow a general-to-particular pattern of development.

EXERCISE 2-1 For each of the following paragraphs, circle an appropriate topic statement from among the three possibilities given. Be prepared to defend your choices.

A   i Many researchers believe that the cost of platinum will level off in the not-too-distant future.

    ii Fuel cell cars may someday be designed to operate with a catalyst different from the one being used today.

    iii Further development of fuel cell cars, despite their advantages, awaits more compelling economics.

While the soaring price of oil might have made the fuel cell car economically attractive, the rising value of precious metals has had the opposite effect. Fuel cells designed to run on methanol use platinum as a catalyst, and the price of that metal quadrupled during 1979. The cost of platinum alone could add a few thousand dollars to the price of a fuel cell car. Researchers have been trying to decrease the amount of platinum needed in the cells or to substitute a different, cheaper catalyst.[1]

B   i The quantity of coal left in the earth is impressive.

    ii Coal is a more viable source of energy than petroleum.

    iii The mining of coal entails a number of difficulties.

There are known to be 198 billion tons of coal at a depth of less than 1000 feet and lying in beds at least 3.5 feet deep for bituminous coal and at least 10 feet thick for beds of lower grade coal. An equal quantity of coal of the same accessibility is identified as "undiscovered recoverable reserves." In addition, there are even larger quantities of marginally available coal resources, amounting to 1 trillion, 400 billion tons. At 35 million Btu's per ton, coal can provide a great deal of energy for many years to come.[3]

C   (i) Many molecular biologists are now convinced that the discovery of movable genetic elements holds the key to the solution of several long-standing mysteries.

    ii It has become evident that in eucaryotes, unlike procaryotes, the genes coding for protein production do not exist as one continuous stretch of DNA.

    iii Wrapping of the DNA on the enzyme with a positive superhelical sense ensures that the reaction will produce negative supercoiling once the wrapped segment is translocated.

It seems to go a long way, for instance, toward explaining how the human body is able to synthesize a million and more different molecular antibody species, each tailor-made to grapple with a specific antigen. Movable elements may help answer the age-old question of differentiation: how a fertilized egg divides and ultimately becomes, in the course of embryonic development, many different kinds of tissue cells. Jumping genes may also provide a mechanism for satisfying scientists who have been arguing that point mutations alone were far from enough to account for the story of evolution.[4]

EXERCISE 2-2   Each of the following sentences has been taken from an original text where it serves as an effective topic statement for a well-written paragraph. See if you can guess roughly how each paragraph is developed beyond the topic statement: what the key words are, what the pattern of organization is.

A   The first modular home to be tested by government engineers for durability exceeded the criteria for the National Bureau of Standards.[5]

B   At the time of its explosion, Mount St. Helens was probably the most closely watched volcano in the world.[6]

C   The production of an important heavy chemical, nitric acid (HNO_3), requires large quantities of ammonia.[7]

D   Until 1922, no one knew how a signal crosses the junction between one nerve cell and another.[8]

E   The basic property of gyroscopic action is that the gyroscope stays spinning in exactly the same direction in space over both short and long periods of time.[9]

2.2

DEVELOP A CLEAR PATTERN OF ORGANIZATION

Once you have written a satisfactory topic statement, you'll want to follow it up with a number of supporting statements. These statements should follow a consistent pattern of organization, one that flows naturally or even predictably from the topic statement. That way, you'll satisfy the reader's expectations and allow her or him to process the paragraph as a unified whole.

Some of the most commonly used patterns of organization in scientific and technical writing are chronological description, cause-and-effect analysis, comparison and contrast, listing, and general-to-particular ordering of details. Each of these patterns has certain characteristic features, and by using these features you can make it easier for the reader to perceive which pattern you're using.

Chronological Description

The use of a time frame to tie sentences together is a well-known pattern of organization which you have no doubt used many times in your writing. It is commonly used, for example, to either describe or prescribe a step-by-step procedure: *First connect the vacuum tube . . . then return the plate . . . finally, close the contact key. . . .* It is used to recount a sequence of past events, as when you want to bring a reader up to date, e.g., in a progress report or in the Review of Literature section of a journal article.

The most characteristic features of chronological description are:

time adverbs and phrases	in 1980, last week, at 10:15, first, second, finally, soon after the project began
verb tense sequencing	Originally we *wanted* to. . . . More recently we *have attempted* to. . . . Now we *are trying* to. . . . In the future we *shall try* to. . . .
grammatical parallelism	*Mount* the grating near the end. . . . *Locate* a rider on the scale. . . . *Adjust* the grating. . . . *Read* the distances on the scale. . . .

Not all of these features are likely to be found in any one type of chronological description. For example, descriptions of standard procedures (e.g., test procedures, experimental procedures, assembly instructions) strongly favor the use of parallelism over the other features. Descriptions of past events, on the other hand, tend to rely more on time adverbs and phrases and on different verb tenses.

The following is a well-written paragraph using chronological ordering as its basic pattern of organization:

> Total U.S. research and development spending *is projected* to reach a current-dollar level of $66.7 billion *in 1981,* an increase of 10 percent over *the 1980 projected level* and nearly double the amount spent on these activities *in 1975.* Even in constant dollars, and despite the reductions in R&D programs contained in the Federal *1980* and *1981* budgets, U.S. R&D spending *in 1981 is expected to follow* the growth trend of *the past five years, when* R&D funding *grew* at an average annual rate of better than 3 percent. That growth *resulted* in large part from increased emphasis on searching for means to resolve energy and environmental problems and a resurgence of defense R&D activity. *Between 1975 and 1978,* the last year for which survey data are available, energy *accounted for* one third of the R&D spending increase while, at the same time, amounting to 10 percent of the national R&D effort.[10] [italics added]

This paragraph contains a great deal of information, and it would probably be quite confusing were it not so well structured. Notice how the topic statement provides a clear overview of the paragraph as a whole: it tells you not only what the main theme is (that U.S. R&D spending is continuing to follow a significant growth trend) but also how this theme will be developed (by chronological order, featuring the years 1981, 1980, and 1975). The remainder of the paragraph is then devoted to fulfilling these expectations. Notice how the writer not only has repeated the key dates but also has taken care to use correct verb tense forms when referring to them.

Cause-and-Effect Analysis

This pattern of organization is used in scientific and technical writing for a number of purposes, including (1) making a logical argument, (2) describing a process, (3) explaining why something happened the way it did, and (4) predicting some future sequence of events. Whenever you are describing causes and effects, it is usually best to describe them in straight chronological order: causes *before* effects. That way, you can minimize the number of "traffic signals" you need. Even when the description is clearly one of causes followed by effects, however, your readers will appreciate it if you occasionally insert such signals. The characteristic signals of cause-and-effect analysis include:

connective words and phrases	therefore, thus, consequently, accordingly, as a result, so
subordinate clauses	since, because (of), due to
causative verbs	causes, results in, gives rise to, affects, requires, produces

conditional constructions	when ozone reacts with nitric oxide, the ozone is destroyed and NO_2 is formed.

In addition, when causes and effects are described in chronological order, the features associated with chronological description can be used.

A good example of cause-and-effect patterning is this paragraph from a physics text explaining what surface tension is:

> One of the most important properties of a liquid is that its surface behaves like an elastic covering that is continually trying to decrease its area. *A result of* this tendency for the surface to contract is the formation of liquids into droplets as spherical as possible considering the constraint of the ever-present gravity force. Surface tension *arises because* the elastic attractive forces between molecules inside a liquid are symmetrical; molecules situated near the surface are attracted from the inside but not the outside. The surface molecules experience a net inward force; and *consequently,* moving a surface molecule out of the surface *requires* energy. The energy E required to remove all surface molecules out of range of the forces of the remaining liquid is proportional to the surface area; *therefore,*
>
> $E = \sigma A$
>
> where σ, the proportionality factor, is called the surface tension,
>
> $\sigma = E/A$
>
> and is measured in joules/m^2. [italics added][11]

Notice how, in addition to using signal words such as *consequently* and *therefore,* the writer has linked the sentences together in a steplike sequence. This has been accomplished mainly through the following technique: after introducing and discussing a new term in one sentence, the writer then uses it in the next sentence as part of the framework for introducing and discussing the next new term:

Surface tension . . . *molecules situated near the surface.* . . .

The surface molecules . . . *requires energy.*

The energy E required . . . is *proportional* to . . .

the *proportionality* factor . . . is measured in joules/m^2.

If this kind of step-by-step linking is done properly, it reduces the need to insert many signal words and markers of subordination. (See Chapter 14 for further discussion of this technique.)

Comparison and Contrast

Often, in technical writing especially, you will find it necessary to compare two or more things that are similar in some ways but different in others. This is particularly common in business and industry, where one is constantly wrestling with cost-benefit trade-offs and other choices that must be made from among various alternatives under various constraints.

In writing a comparison-and-contrast paragraph, *try to avoid jumping back and forth from one alternative to another.* Suppose, for example, that you are comparing items X and Y and are using criteria A, B, C, and D to compare them; suppose further that the first three criteria favor X and the fourth favors Y. In such a case, you should present the comparison in terms of these two criteria groupings: first A through C, then D. This will make it easier for the reader to see that product X wins out over product Y in three of the four criteria. And that's what the reader will most likely want to know—the proverbial bottom line. (If you think the reader will be more interested in the details of the comparison, provide a table.)

Characteristic features of comparison-and-contrast paragraphs include:

connective words and phrases	however, on the other hand, conversely, similarly, likewise, in contrast to
comparative constructions	more than, -er, than, less than, as . . . as, rather than, is different from
modal verbs	program X *will* be easy to implement, whereas program Y *would* entail a number of complications . . .
subordinate clauses	while, whereas, but
parallelism	model X is reliable and efficient, whereas model Y is unreliable and relatively inefficient . . .

One final principle of comparison-and-contrast writing is this: *phrase your words so as to reveal your own preference in the comparison.* In other words, don't just make a simple comparison as if you were a neutral observer. Instead, allow your own interpretation of the facts to color your description. Don't say, "Item X weighs 3.2 pounds, and item Y weighs 2.7 pounds"; say, "Item X weighs 3.2 pounds, whereas item Y weighs only 2.7 pounds." After all, *you* are the one who has made the study, and so *you* know what the facts of the matter are. Using some of the features listed above will enable you to let your reader know what your choice is.

Below is a model comparison-and-contrast paragraph. Note how it uses some of the features listed above.

A one-million-fold increase in speed characterizes the development of machine computation over the past thirty years. The increase results from improvements in computer hardware. In the 1940's ENIAC, an early electronic computer, filled a room with its banks of vacuum tubes and miles of wiring. Today one can hold in the hand a computing device costing about $200 that is twenty times faster than ENIAC, has more components and a larger memory, is thousands of times more reliable, costs 1/10,000 the price, and consumes the power of a light bulb rather than that of a locomotive.[12]

Listing

Scientific and technical writing presents frequent opportunities to put information in the form of lists. If you are describing an experiment, for example, you will probably want to make a list of the equipment used. If you are writing a progress report, you may want to make a list of things already done and another list of things still to do. If you're writing a report recommending the development or purchase of some new product, you may want to list its outstanding features or the reasons you're recommending it.

Lists may be either formatted or unformatted. Formatted lists are set off from the rest of the paragraph by means of indentation and/or numbering or lettering. Unformatted lists do not have such visual cues. In both cases, *all items in a list should be cast in parallel grammatical form.* This principle is especially important in the case of unformatted lists, since it can be quite difficult otherwise for the reader to detect the presence of the list. (Parallelism is important not only in lists but also in comparative constructions, in descriptions of procedures, and in other rhetorical patterns; see Chapter 3 for further discussion.)

A second important principle to follow when constructing lists is this: *if the items in a list are not equally important, they should be arranged in descending order of importance.* A list, by definition, is a set of items all of which have something in common and yet are independent of each other. Thus, in principle, these items can be arranged or rearranged any way at all. You can take advantage of this freedom by arranging them with the most important item in the most prominent position, namely, on top. Here is an example:

> In addition to coal and nuclear energy, a wide variety of other power sources are also frequently discussed in the news and in scientific literature; unfortunately, most are not yet ready for practical use. Geothermal energy is one of the more practical of proposed new sources. It is already in use in Italy, Iceland, and northern California but is not yet meeting all expectations for it. Solar energy seems an elegant idea because it is inexhaustible and adds no net heat or carbon dioxide to the global environment. Yet present methods of exploiting it make solar energy hopelessly inadequate as a major power source in the next few decades. Sophisticated windmills to generate electricity are also under study by some. Biomass conversion is also getting under way. Some of these

sources of energy, which we now generally regard as esoteric, may well prove themselves and make a substantial contribution over the long run if their costs can be brought within reason.[3]

The different power sources discussed in this paragraph constitute an unformatted list:

Geothermal energy . . .

Solar energy . . .

Sophisticated windmills . . .

Biomass conversion . . .

Notice how the effective use of parallelism enables the reader to easily locate the four items making up the list, even though no visual formatting is used.

General-to-Particular Ordering of Details

A final rhetorical pattern commonly used in scientific and technical paragraphs is the ordering of supporting details from the more general to the more particular. Each sentence in this pattern focuses on a smaller frame of reference than the sentence before it.

Here is an example of such a pattern, taken from a discussion of the kinds of bearings needed in flywheel energy-storage systems:

> Magnetic bearings have been developed for aerospace applications, but only recently has their practicality been demonstrated as the heart of energy storage systems. The breakthrough is partly due to the recent development of stronger permanent magnets, such as those made from rare-earth cobalt compounds. Only ten pounds of such magnets could support two tons of rotor. Although the free suspension of a weight with permanent magnets is an unstable condition, an electromagnet servo loop has been used successfully to stabilize the rotor position.[13]

Notice how the topic of the first sentence (and of the paragraph as a whole), *magnetic bearings,* is developed in the second sentence by the writer's focusing on one component of such bearings, the *permanent magnets.* This subtopic is then discussed in even greater detail in the third sentence *(Only ten pounds of such magnets . . .).* Finally, the writer focuses on a subcomponent of these magnets, the *electromagnet servo loop.*

Other Patterns

The patterns of organization described above are not the only ones commonly used in scientific and technical writing. Others—such as classification-division, exemplification, extended definition, and analogy, to mention but a few—are also used. (An example of paragraph development by classification-division, you may recall, was given earlier in this chapter.)

In addition, two or more patterns can often be used together. For example, you might want to give an *extended definition* of a process by describing it in terms of *causes and effects* or in terms of an *analogy* to some more familiar process. You might want to explain something by providing a *list* of *examples*. Or you might want to *divide* some subject into different classes and then *compare and contrast* these classes. Regardless of what kind of combination you elect to use, the important point is to establish at least one pattern of organization that runs through most if not all of the paragraph and thus provides a structural backbone for that paragraph.

EXERCISE 2-3 Each of the following paragraphs follows a particular pattern of organization. Identify the pattern, and point out as many of its features as you can.

A Flywheels appear to be the ideal energy storage element in solar electric or wind power systems. They can smooth the load on the generators by providing the energy to generate electricity when the sun is not shining or the wind is not blowing. And they can also provide peak power to an electrical load during periods when demand exceeds supply, such as during motor start-ups. Indeed, once spinning, flywheels can deliver energy rapidly for transient load conditions, which makes them especially useful in industrial and agricultural applications.[13]

Pattern: _____

Features: 1._____
           2._____
           3._____

B *Scalar and Vector Quantities.* Every quantity requires a statement of at least two things: first, a numerical specification, or a magnitude, and, second, the appropriate unit. For one class of quantities the statement of a numeral and a unit is all that is necessary for a complete specification. For example, the quantity volume is completely specified when we say "25 cubic feet." On the other hand, for another class of quantities one must state, in addition, a direction in order to afford a complete specification. For example, the quantity force is completely specified when we say "10 pounds acting vertically upward." A quantity that involves, other than the statement of a unit, only the idea of magnitude is called a *scalar,* while a quantity that needs for its complete specification a direction as well as magnitude is called a *vector.* Quantities such as displacement, velocity, acceleration, force, weight, and torque are vectors: each involves the idea of direction. Quantities such as speed, mass, and energy are scalars; none of these has associated with it a direction.[14]

Pattern: _____

Features: 1._____
2._____
3._____
4._____

C   *The Vapor-Compression Refrigeration Cycle.* A simple vapor-compression refrigeration cycle is shown schematically in Fig. 2-1. The refrigerant enters the compressor as a slightly superheated vapor at a low pressure. It then leaves the compressor and enters the condenser as a vapor at some elevated pressure, where the refrigerant is condensed as a result of heat transfer to cooling water or to the surroundings. The refrigerant then leaves the condenser as a high-pressure liquid. The pressure of the liquid is decreased as it flows through the expansion valve, and, as a result, some of the liquid flashes into vapor. The remaining liquid, now at a low pressure, is vaporized in the evaporator as a result of heat transfer from the refrigerated space. This vapor then enters the compressor.[15]

Pattern: _____

Features: 1._____
2._____
3._____
4._____

D   One of the newest branches of number theory is *analytic number theory.* A vast and intricate subject, it is largely a creation of the twentieth century. It has been called the science of approximation, for it is concerned mainly with determining the order (relative size) of the errors made when a calculation is approximate rather than exact. Its techniques have had an important impact on many departments of applied mathematics, including statistical mechanics and the kinetic theory of gases, where exact results are sometimes humanly unattainable.[16]

Pattern: _____

Features: 1._____
2._____

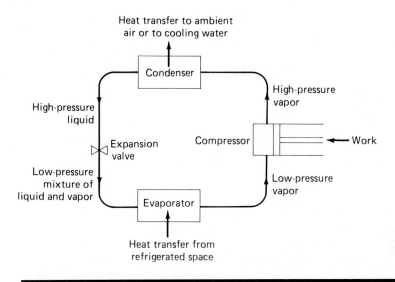

FIGURE 2-1   Schematic diagram of a simple refrigeration cycle.

EXERCISE 2-4   The following passage is taken from a popular-science journal article, where it was divided into seven paragraphs. Using your knowledge of topic statements and patterns of paragraph organization, see if you can determine where the six paragraph divisions should be.

We generally think of volcanoes as sizable mountains that belch lava and smoke from craters at their summits. But the essential element of a volcano is not its aboveground structure but the underground conduit that brings molten rock to the surface. The molten rock is known as magma as long as it is under the ground; after it erupts it is called lava. As the lava flows or explodes from a vent at the top of the conduit, it starts to build a volcanic edifice that may grow into a Vesuvius, a Mauna Loa, or a Mount St. Helens. Few people have witnessed the birth of a volcano, but Dionisio Pulido was present on February 20, 1943, when the Mexican volcano Paricutin made its first appearance in a cornfield on his small farm in the valley of Rancho Tepacua. Pulido, his wife and son, and a neighbor watched as smoke and ash began to rise from a small hole in the middle of the field. The smoke and ash were accompanied by rumbles, hisses, and hot particles of rock that set fire to nearby pine trees. By the next day, ash and rock debris had built a cone 10 meters high; in a week it had grown to 140 meters. By the time Paricutin finally quieted down in 1952, the cone towered 410 meters above the original level of Pulido's cornfield. Paricutin is an unusually large cinder cone, one of four basic types of volcanic edifices. Cinder cones, made up entirely of lava fragments, rarely rise more than 300 meters above their surroundings. They are the simplest type of volcano and are very common in western North America. A second type of volcano is the composite cone, or stratovolcano, built up of alternate layers of lava and ash (lava particles). Mount St. Helens is a composite cone. So are many other spectacular volcanic peaks, including Mount Rainier and Mount Hood in the Cascades and Japan's Mount Fuji. The summits of many composite cones tower 1800 to 2400 meters above their bases. Shield volcanoes are built almost entirely of lava flows. They are broad, gently sloping structures that resemble a warrior's convex shield laid on the ground face up. Many of the world's largest mountains are shield volcanoes; clusters of them form the Hawaiian Islands. Mauna Loa, on the island of Hawaii, is the world's largest active volcano. Its base is more than 4500 meters below the surface of the ocean, and its crest is 4170 meters above the sea level—a total height of more than 8500 meters. Lava domes, the fourth type of volcano, are built up of thick, pasty lava that tends to pile up in a rough knob rather than to flow outward as it emerges from its vent. The main factors that determine whether lava will erupt in a smooth flow or a violent explosion are its chemical composition and the characteristics of the gases that are dissolved in it. Magma is a mixture of several oxides, mainly silicon dioxide. The rock can be principally basalt, andesite, dacite, or rhyolite, which are distinguished one from the other, among other things, by the fact that each has a larger proportion of silicon dioxide than the one before. Basaltic magmas usually erupt in a highly fluid state, and their dissolved gases escape easily, so that the lava flows freely. The more viscous rhyolites and dacites usually erupt explosively. Andesites may erupt in explosions or in flows that are not as smooth as those of basaltic lava.[17]

EXERCISE 2-5   The following passages (extracted from longer reports) were submitted as is by students. As you'll discover, they require very careful reading in order to be understood. Revise them—by rearranging or rewriting sentences, or perhaps by dividing them into two or more paragraphs—so that they can be more quickly and easily understood.

A   ADVANTAGES AND DISADVANTAGES
The balance system is the simplest and cheapest system. However, it has drawn numerous complaints from users in California. The nozzles are heavy, hard to hold in place, and hard to remove due to the necessary airtight seal. The major problem has been that gasoline has been sucked out of the customer's tank back into the storage tank. This occurred when customers attempted to top off their tank and in the process overrode the automatic shutoff. The brands of the nozzles (OPW) were tested by the ARB in actual stations by inserting a small vial in the vapor return hose and checking for liquid in the vial after filling the vehicle. In the ARB tests the nozzles did not fail. However, in the MS test the nozzle failed. A $3\frac{1}{2}$-gallon container fitted with a standard auto filler pipe was used. The container had a mechanism to allow the inside pressure to be increased, and a 22.5% failure rate was reported. In one trial the container was filled to 15 gallons. The vacuum and assisted systems do not permit this overfilling, and no force is required to hold the nozzle. The prices of the three systems vary significantly. The balance system costs $\frac{1}{2}$ the amount of the assisted system, and the vacuum system is the most expensive.

B TECHNICAL DISCUSSION

The most important part of this modification program is the addition of a larger engine to the Corsair. The eighteen-cylinder Pratt & Whitney R-2800-CB-13 and three-bladed propeller that are the standard fit for an FG-1D are rated at 2500 hp, though there are cases where R-2800s have supplied up to 3300 hp (Greenamyer, 1969) in modified form. One possible alternative to the R-2800, a Curtiss-Wright R-3350-24 Cyclone 18, has been substituted for an R-2800 in an F8F-2 Bearcat. Rated at 2800 hp, this particular example has been pushed to 3800 hp (Shelton, 1971). Another alternative is the twenty-eight-cylinder Pratt & Whitney R-4360-4 Wasp Major. These were raced in the late 1940s on F2G Corsairs with four-bladed propellers. These engines were rated at 3000 hp, though in races they produced up to 4000 hp (Cleland, 1949). After some careful checking, I have chosen to recommend fitting a Pratt & Whitney R-4360-B-20 with a four-bladed propeller, to your Corsair. While rated at 3500 hp, I expect it to produce over 4000 hp after modification. One particular advantage in using the R-4360-B-20 is that it won't increase the aircraft's frontal area, being the same diameter as an R-2800. In contrast, an R-3350 would increase the frontal area by 11% because of its larger diameter. However, in fitting this larger engine, there will be a weight penalty of 1130 pounds. To keep the aircraft's center of gravity from moving too far forward, the firewall will have to be moved back at least a foot.

REFERENCES

1 Joel Gurin, "Chemical Cars," *Science 80,* September/October 1980, p. 96.

2 Julie Wei, "Membranes as Mediators in Amino Acid Transport," *The Research News* **31**(11-12):3 (1980).

3 Adapted from Blanchard Hiatt, "Coal Technology for Energy Goals," *The Research News* **24**(12):3 (1974).

4 Henry Simmons, "DNA Topology: Knots No Sailor Ever Knew," and Ben Patrusky, "Gene Segments on the Move," *Mosaic* **12**(1):9, 41–42 (1981).

5 Karl G. Pearson, *Industrialized Housing,* Institute of Science and Technology, University of Michigan, Ann Arbor, 1972, p. 39.

6 Henry Lansford, "Vulcan's Chimneys: Subduction-Zone Volcanism," *Mosaic* **12**(2):46 (1981).

7 J. Waser, K. Trueblood, and C. Knobler, *Chem One,* McGraw-Hill, New York, 1976, p. 589.

8 Edward Edelson, "The Neuropeptide Explosion," *Mosaic* **12**(3):15 (1981).

9   Henry Hooper and Peter Gwynne, *Physics and the Physical Perspective,* Harper & Row, New York, 1977, p. 207.

10 *Mosaic* **11**(5):cover 4 (1980).

11 Hooper and Gwynne, op. cit., p. 223–224.

12 Blanchard Hiatt, "Big Computing, Tiny Computers," *The Research News* **30**(4):3 (1979).

13 Alan Millner, "Flywheels for Energy Storage," *Technology Review,* November 1979, p. 33.

14 Joseph Morgan, *Introduction to University Physics,* vol. 1, Allyn and Bacon, Boston, 1963, p. 17.

15 Richard Sonntag and Gordon Van Wylen, *Introduction to Thermodynamics: Classical and Statistical,* Wiley, New York, 1971, pp. 9–10.

16 Julie Wei, "Pure Mathematics: Problems and Prospects in Number Theory," *The Research News* **30**(3):31 (1979).

17 Adapted from Lansford, op. cit., p. 50.

ADDITIONAL READING

David Kieras, "The Role of Major Referents and Sentence Topics in the Construction of Passage Macrostructure," *Discourse Processes* **4**:1–15 (1981).

David Kieras, "Initial Mention as a Signal to Thematic Content in Technical Passages," *Memory and Cognition* **8**:345–353 (1980).

Bonnie J. F. Meyer, *The Organization of Prose and its Effects on Memory,* North-Holland, Amsterdam, 1975.

R. M. Gagne and V. K. Wiegand, "Effects of a Superordinate Context on Learning and Retention of Facts," *Journal of Educational Psychology* **61**:406–409 (1970).

Patricia A. Carpenter and Marcel A. Just, "Reading Comprehension as Eyes See It," *Cognitive Processes in Comprehension,* Lawrence Erlbaum, Hillsdale, N.J., 1977, pp. 109–139.

3

PARALLELISM

All writing—whether of books, of reports, of articles, of sections of articles, of paragraphs, of sentences, etc.—consists of both form and content. Reading is made easier to the extent that the form *reflects* the content: we can then use formal features as cues to guide us as we try to absorb and interpret the content. Selective readers in particular depend on features of form to guide their reading. Such features are easy to spot and thus serve as signposts, allowing the reader to zero in on and quickly process those parts of a text that are of particular interest.

One way in which form reflects content is when we use grammatical parallelism in the writing of lists. If you write out a list, "Things To Do," that looks like this:

pick up mail

call Hazlitt

work on tailgate project

don't forget to ask RJ for $$

you're using parallelism because you've put each item on the list in the same grammatical form (verb + one or more nouns). If you write the list like this, on the other hand:

mail needs to be picked up

call Hazlitt

tailgate project

don't forget to ask RJ for $$$

you're not using parallelism.

Of course, when you're just writing to yourself, you can afford to be less careful than when you're writing to someone else. After all, *you* know what you mean, don't you? In writing to other people, however, there will be many times you'll find yourself writing out lists, and in such cases it's important that you make it optimally easy to read them. Here's where parallelism has an important role to play.

Before we describe this role, however, perhaps we should clarify what we mean by the word *list*. What exactly is a list? As we refer to it here, a list is any set of two or more independent items, i.e., any two or more items sharing a coordinate (related but nonoverlapping) relationship. This definition includes not only obvious cases such as the one described above but also others that may not be so readily apparent. For example, the following sentence contains a list since two different but related functions (*metabolite transport* and *metabolite signaling*) are mentioned:

> Evidence that the receptor sites on the membrane can serve both for metabolite transport and for metabolite signaling has raised many questions as to the mechanisms of membrane information-transfer.[1]

The following sentence also contains a two-item list:

> A pollution-free inertial-electric system has greater energy efficiency than an internal combustion engine.

Here the items forming a list are the two systems that are being compared.

3.1
TYPES OF LISTS

Lists are especially common in scientific/technical writing. We find lists of experimental apparatus, lists of instructions, lists of task objectives, and so on. Some lists are *fully formatted* as such, with alphanumeric sequencing and vertical alignment used as formal features to make the list stand out. Here is an example, from a company memo:

> REFS:(a)   Government invoice #79-1018A
>
> (b)   Bonded Stores invoice #31285
>
> (c)   DECAS letter 5/20/79
>
> (d)   Bonded Stores invoice #44590

Some lists are *partially formatted,* in the sense that they use either alphanumeric sequencing or vertical alignment but not both, as in this example:

> Factors taken into consideration include the following: (1) size of lot, (2) parking requirements, (3) need for elevator, (4) cost per gross square foot, and (5) expected annual return per gross square foot.

Some lists are completely *unformatted,* having neither alphanumeric sequencing nor vertical alignment:

> Compared to standard bipolar types, VMOS transistors offer higher input impedance, faster switching times, wider operating range, and smaller chip area.

In all cases, however, whether a list is fully formatted, partially formatted, or unformatted, *all of the items making up the list should be cast in the same grammatical form.*

Putting the items of a list in parallel grammatical form allows a busy reader to easily perceive the nature of these items and of the list as a whole. It is especially important to use this principle with unformatted lists. In the VMOS example above, notice how easy it is to glance at the sentence and see what the four criteria of comparison are: . . . *input impedance . . .switching times . . . operating range . . . chip area.* By contrast, notice how difficult it is to perceive the nature of the list, and the items making up the list, in the following badly written example:

NEGATIVE EXAMPLE
The TFC engineers and I found the selection of Hybrid Analog Transmission using FDM (Frequency Division Multiplexer) to be highly reliable, improved security of communications; complete ground isolation, freedom from crosstalk, sparking, short circuit loading; RFI (radio frequency interference), EMI (electric magnetic field), and EMP (electric magnetic polarization) immunity.

What are the major items constituting the list in this example? What role are they playing? You may be able to figure out the answers if you work at it, but many readers—especially managerial readers—are not willing to go to such effort.

As a general rule, you should always endeavor to use parallelism when you are presenting a list in written form. Often this will be a relatively straightforward matter, involving fully-formatted lists made up of simple items. In such cases, all you need do is make minor adjustments so that every item has the same grammatical form. Suppose, for example, you have written a draft of a report about wastewater processing and have included this statement:

NEGATIVE EXAMPLE
The principal processes are:

a)   coagulating and flocculation

b)   removing the solids

c)   nitrogen-removal

d)   disinfection

Surely it would be an easy matter to recast this list in a more grammatically parallel form:

The principal processes are:

a)   coagulation and flocculation

b)   removal of solids

c)   removal of nitrogen

d)   disinfection

Instead of mixing verb forms (-*ing*) and noun forms (-*al* and -*tion*) as in the first version, the second version uses all noun forms. This results in a more sharply defined list; it looks more like the work of a careful, precise technical professional.

But the proper use of parallelism can do far more than just enhance your professional image: it can also help prevent misinterpretations of your writing. To see what can go wrong when parallelism is *not* used, consider the following:

NEGATIVE EXAMPLE
This filter has two important functions: to reject impulse noise signals and passing low frequency command signals without amplitude or phase distortion.

This excerpt could easily be misread (especially by a busy nonspecialist reader) as meaning that two things are rejected: the impulse noise signals and the passing of low frequency command signals. Notice that *reject* and *passing* can be combined to produce such a misreading. Even those readers who can figure out what the writer is trying to say may have to slow down their reading just to make sure what's being rejected.

Such ambiguity is completely unnecessary. By putting the two key verbs in parallel form, the writer can create a much clearer version:

This filter has two important functions: *rejecting* impulse noise signals and *passing* low frequency command signals without amplitude or phase distortion.

Other parallel forms would work equally well: *to reject . . . to pass; the rejecting of . . . the passing of.*

EXERCISE 3-1   Each of the following contains a list that lacks parallelism. See if you can devise an improved version.

A   The building is 140 feet in length, 78 feet wide, and has a height of four stories.

B   Test results indicate that soil #1 is likely to settle, since high plasticity is equivalent to highly compressible.

C   The widest employment of DDT is in the control of insects of public importance (as a mosquito larvicide, a spray for malaria eradication, and to control typhus by dust application).

D   The reason for this is that most small businesses have a lower budget for their managers than do government or industrial managers.

E   In this particular case the most important variables are the following:
(1)   pressure and temperature of the boiler
(2)   what type of fuel is required
(3)   the amount of oxygen
(4)   fuel temperature

F   As a result of the above problem, this report addresses the following tasks:
a.   To redesign the mix for the concrete slabs
b.   An evaluation of the compressive strength with the use of test cylinders for various designs
c.   Determining both the theoretical and actual material costs per cubic yard

G   We recommend the purchase of the New Orleans heat exchanger because (i) it can be obtained for $25,000, a savings of 80% over the new cost, (ii) production could be increased by 28% if needed, 8% higher than requested, and (iii) it will help recoup losses incurred in our benzene plant.

H   Advantages of this system are:
1.   Automatically controlled
2.   Less operating cost

Disadvantages are:
1.   May cause slugging of liquid refrigerant to compressors
2.   Complex automatic controls
3.   Substantial replumbing of existing system is required

I   This buoy terminal has three components that rotate as a single unit:
(a)   The Rotating Mooring Bollard allows floating mooring lines to lie on the lee side when weather conditions become bad.
(b)   Products are carried by a Rotating Cargo Manifold from the terminal-to-tanker hoses to the multiproduct distribution unit in the center of the buoy.
(c)   A Rotating Balance Arm not only maintains the buoys on an even keel but also provides an accommodation ladder.

J   The purpose of this report is to present evidence that the contracting officer acted within the range of his authority and his final opinion was binding.

3.2

MISLEADING PARALLELISM

Given the fact that readers use parallelism as an aid in perceiving and interpreting a list, it is important that you *not* use parallelism in situations where lists are *not* involved. Otherwise, readers are apt to be misled and may misanalyze the passage *as* a list, which in turn can lead to comic interpretations or to unnoticed misinterpretations. Here is an example:

NEGATIVE EXAMPLE

Richard Clarke, senior systems programmer, asked me to develop a magnetic tape management system, to reside permanently on the computer, to give better control, and to coordinate the numerous magnetic tapes.

The parallel use of infinitives in this sentence makes it appear that a four-item list is being presented:

Richard Clarke, senior systems programmer, asked me:

- to develop a magnetic tape management system

- to reside permanently on the computer

- to give better control

- to coordinate the numerous magnetic tapes

This would mean, among other things, that it is the writer, not the system, that will "reside permanently on the computer, give better control," etc., surely a strange state of affairs! Of course, most readers would probably figure out what the writer means, but this requires extra time and effort on the reader's part—precisely what good writing avoids.

In cases like this, the writer should break up the misleading parallelism so that only those items that really are part of the list appear in parallel form. In this particular case, only the last three items belong together; the first item should be kept separate. If the last three items were written as relative clauses, say, instead of as infinitives, they would be distinguished from the first item:

Richard Clarke, senior systems programmer, asked me to develop a magnetic tape management system:

- which will reside permanently on the computer

- which will give better control

- which will coordinate the numerous magnetic tapes

This can be simplified by factoring out the relative pronoun and auxiliary verb (*which will*) and making them part of the lead-in:

Richard Clarke, senior systems programmer, asked me to develop a magnetic tape management system which will *reside permanently on the computer, give better control,* and *coordinate the numerous magnetic tapes.*

Compare this version to the original one and see what a difference there is in readability.

EXERCISE 3-2   Each of the following passages contains misleading parallelism. Correct each passage by making appropriate grammatical changes.

A   My present occupation is repairing typewriters, printing machines, and duplicating machines.

B   You learn many reasons why our product failed by reading, observing, talking and listening to our sales people.

C   This technology consists of three methods of scrambling, which are the coupling of light source and optical fiber in low-order mode, splicing, bending and tightening the fiber near its connectors, and installing the scrambler into the existing fiber optic in intervals of 1 km along the route.

3.3
PARALLELISM IN PARAGRAPHS AND LARGER UNITS
Perhaps the most important use of parallelism—one which poor writers usually neglect—is to indicate the presence of listlike structures in paragraphs and groups of paragraphs. Two or more sentences may be related coordinately within a paragraph; two or more paragraphs may be so related. Since these kinds of larger lists are customarily unformatted, parallelism is usually the only formal cue that can be used to guide the reader.

Here is an example of how the failure to use parallelism to indicate paragraph structure can make it difficult to read quickly and selectively, if not accurately:

NEGATIVE EXAMPLE
All-Savers Certificates will not benefit all investors. Investors exceeding a deposit of $7931 ($15,861 joint return) would have an after-tax yield far lower than alternative investments such as money market funds or Treasury bills. Alternative investments would also yield better after-tax yields and no penalty if the certificate was redeemed within the 1-year maturity period.

A common strategy used to skim-read paragraphs is to first read the topic sentence and then read the beginning words of the following sentences to see what kind of support these sentences provide for the topic sentence. In the case above, a person using this technique can easily be misled into perceiving the paragraph as having a general-to-particular structure. The first sentence is clearly the topic sentence. The second sentence supports it by describing a class of investors who would benefit more from alternative investments than from All-Savers Certificates. The third sentence then appears to provide additional details about alternative investments. The diagram below illustrates this structure:

All-Savers Certificates will not benefit all *investors*.

Investors exceeding a deposit of $7931 . . . would have an after-tax yield far lower than *alternative investments*. . . .

Alternative investments would also yield better after-tax yields and no penalty. . . .

Notice how the chain of repeated words links one sentence to the next, each one appearing to be subordinate to the one above it. In actuality, however, the writer means to say that there are not one but *two* classes of investors who will not benefit from All-Savers Certificates: those who exceed a deposit of $7931 and those who redeem their certificates before 1 year has elapsed. The purpose of the paragraph (as we were informed by the writer) is to describe these two classes. The second and third sentences, therefore, constitute a two-item list, with each of the sentences describing one of the two classes. If the writer had cast these two sentences in parallel form, it would be much easier for readers to see this listlike structure:

> All-Savers Certificates will not benefit all investors. Investors exceeding a deposit of $7931 ($15,861 joint return) would have an after-tax yield far lower than alternative investments such as money market funds or Treasury bills. Investors redeeming the certificate within the 1-year maturity period would have to pay a penalty and would also have a lower after-tax yield than with alternative investments.

The diagram below illustrates why this revision has a more transparent listlike structure:

All-Savers Certificates will not benefit all *investors*.

Investors exceeding a deposit of $7931 . . . would have an after-tax yield far lower than alternative investments. . . .

Investors redeeming the certificate within the 1-year maturity period would have to pay a penalty and would also have a lower after-tax yield. . . .

The use of parallelism enables the reader to see very quickly that the second and third sentences constitute a two-item list.

EXERCISE 3-3 Each of the following passages contains information that can be presented in the form of one or more lists. Reconstruct each passage accordingly, using appropriate parallelism.

A   In order to meet the job requirements it became clear that a microcomputer would be required to do this type of work. Microcomputers are very compact and portable. They are easily programmed and can be used for a wide variety of data processing. These computers are normally quite easy to operate and capable of storing large amounts of data both inside and outside of the computer itself. The most attractive part of using a microcomputer is that it is very inexpensive in comparison to other large-scale systems.

B   Reuse of treated water is most applicable where large amounts of water are used and the wastes are not too contaminated. Industrial wastes may be heavily contaminated and therefore may not be very suitable for reuse. The location of the treatment plant and the possible transport of the renovated water are also important factors. The treatment process works most efficiently and economically when dealing with a steady flow of wastewater. A very important point is whether the wastewater will be reused only once or whether it will be reused many times. Multiple recycling results in a buildup of certain dissolved materials, especially inorganic ions, that may make demineralization necessary. Most reuses do not lead to a high degree of recycling.

C   By checking the instrumentation used for this experiment, it was noticed that the collimator is not close enough to the x-ray film, and therefore a lot of neutrons have been lost in this way. But if we minimize this distance, fewer neutrons will be lost by leakage. Also, for better and sharper pictures we need a high resolution which is obtained when neutron beams are as parallel as possible; to achieve a higher resolution we have to either decrease the collimator or increase its length. In this case we cannot increase the collimator length, since it is as long as the depth of the reactor pool, which is a constant. Therefore the only alternative is to decrease the diameter of the collimator for a better resolution.

D   The new design meets all of the important criteria. The new design uses the same sulfur dioxide scrubbing process used in the present scrubbing system. Therefore, the new design gives the same sulfur dioxide removal rate as the present system. The new design includes a regeneration loop for the absorption reactant, thereby cutting the absorption reactant consumption considerably. The new design also terminates with a solid waste product. Thus, a very manageable waste product is produced.

REFERENCES

1   Julie Wei, "Membranes as Mediators in Amino Acid Transport," *The Research News* **31**(11–12):27 (1980).

ADDITIONAL READING

Francis Christensen, *Notes Toward a New Rhetoric: Nine Essays for Teachers,* Harper & Row, New York, 1967.

Part III

ELEMENTS IN PREPARING TO COMMUNICATE

4

AUDIENCES

Before we begin our discussion of audiences, we would ask you to read the partial technical reports presented in Figures 4-1 and 4-2. Each was written by an engineer. Try very hard to spend *no more than 1 minute on each report*. When you have finished reading the two, take another minute or two to decide which you feel is better and why.

Most people at the beginning of a technical writing course prefer the report in Figure 4-1; most at the end prefer the report in Figure 4-2. Why does this happen? If we look more carefully at the two reports, we notice that the one in Figure 4-1 looks very businesslike: it has a nicely indented list and lots of impressive data. Surely it provides just what the senator's public relations officer has asked for—data on the Y-Ships.

However, if we take a second look at this report and consider its audience and use, we get a somewhat different picture. The public relations officer, or someone else on the senator's staff, will be using this report to write a speech for the Y-Ship launching. Since one usually finds the most important information first, if the speech writer is not paying careful attention, he or she may produce an unusual speech:

> Good afternoon, ladies and gentlemen. We are gathered today to launch a Y-Ship of overall length 572′, beam 82′0″, depth to main deck 45′6″, designed draft 28′4″. . .

Obviously, such a speech would not fit the occasion. The information needed for an appropriate speech is at the bottom of the report segment: that this Y-Ship is the first of a new class, that it will link the east and west coasts with the Orient, that these ships are the first merchant vessels in the world to be built almost entirely of low-alloy, high-strength steel, and that they are unusually light and have greatly increased cargo space.

Given the purpose for this report, why do you suppose the writer puts the technical information first and the more general information second? Perhaps because the writer is much more interested in the technical information? This writer has just spent months designing the Y-Ships, and each of the numbers in that list represents many decisions the writer takes pride in. The problem here is that the writer is writing for the writer, not for the reader.

In contrast, if we look more carefully at the report to Lt. Cousins in Figure 4-2, we notice that it is designed for the reader. It puts the information that Lt. Cousins needs at the beginning and saves the details for later; it tells

FIGURE 4-1 Beginning of informal report by a naval architect.
(Used with the permission of J. C. Mathes.)

To: XXXX, Public Relations   December 6, 1979

From: YYYY, Naval Architect

              Y-SHIPS: VESSEL CHARACTERISTICS

I understand Senator Q's Office has requested data on the Y-Ships for use in connection with his participation in launching the PRESIDENT Y-SHIP at Pascagoula,
Mississippi.

Following are physical characteristics of these ships:

Length overall	572'-0"
Beam	82'-0"
Depth to Main Deck	45'-6"
Designed Draft	28'-4"
Maximum Draft	30'-7"

Displacement at 30'-7" Draft	21,230 Tons
Cargo Deadweight	10,000 Tons
General Cargo Capacity	770,000 Cu. Feet
Refrigerated Cargo Capacity	48,000 Cu. Feet
Liquid Cargo	40,000 Cu. Feet or 1,000 Tons
Container Capacity (8' × 8' × 20')	144

Passengers Carried	12 in Eight Staterooms
Crew	45

Propulation: Steam turbine developing 24,000 Horsepower driving a single
22'-6" diameter five-bladed propeller.

Cruising Speed: 23.0 Knots
Cruising Radius: 11,600 Nautical Miles

PRESIDENT Y-SHIP is the first of a new class of vessels known as the Y-Ships
(Design C4-S-69a). The design was developed by ABC Company, Naval Architects,
New York, to meet design and performance characteristics and service requirements established by ST Lines for vessels operating on Trade Route Number 17.
This route links both the East and West Coasts of the United States with the Orient
and involves a round trip in excess of 30,000 miles. The 23-knot continuous sea
speed capability is optimum for the long ocean legs in this service.

The Y-Ships are notable in several respects. They are the first merchant vessels in
the world to be built almost entirely of low-alloy, high-strength steel. The result is
a weight saving of approximately 330 long tons as compared with a similar ship
built of conventional shipbuilding steel. This is reflected in an equivalent increase
in cargo capacity.

FIGURE 4-2 Beginning of informal report from a sales engineer.
(Used with the permission of J. C. Mathes.)

January 29, 1969

CGC Boutwell
c/o U.S. Coast Guard Base
427 Commercial Street
Boston, Massachusetts

Attention: Lt. (j.g.) G. L. Cousins

Subject: Clutch to disengage turbine starting pump

Gentlemen:

Thank you for the courtesy extended our representative, Ed Driscoll, on his recent
visit aboard the Boutwell, at which time he discussed with you your requirement
of the clutch to be used to disengage the turbine starting pump from the main
generator engine.

We understand you wish to mount the clutch on a shaft which would be turning
720 RPM, and that the duty of the turbine pump is 117 HP. Based on this, the
torque requirement is 853 pound feet, and a clutch having 16.3 HP per 100 PRM is
indicated. Twin Disc Mode #CL-310 is rated 873 pound feet, and 16.6 HP per 100
PRM, and 135 HP normal duty. The #CL-310 therefore would seem to fill the bill
quite nicely.

We refer you to Twin Disc Bulletin #326-B, enclosed. On page 10 you will find the
description, capacities, etc., of this clutch. You will also note, it is available in $2\text{-}\frac{1}{4}$"
and $2\text{-}\frac{7}{16}$" bore. Accessories are described on page 11. Two possible spider drive arrangements are suggested, as described in Figures 2 and 3 on the back cover of the
bulletin. The spiders and their dimensions are indicated on pages 18 and 19. We
are pleased to quote as follows:

1. XA5752 Model CL-310 Twin Disc Clutch in standard bore of $2\text{-}\frac{1}{4}$"
   or $2\text{-}\frac{7}{16}$" $131.20
2. Part #3507 Throwout Yoke 2.25
3. Part #3039 Hand Lever 1.49
4. Part #1144-E Operating Shaft 2.18

where to find the details but does not clutter up the beginning of the report
with them. Notice, however, what is at the beginning: a polite thank you for
a recent visit to the ship, a statement about the report's topic, the specifications that need to be met, the acceptability of the proposed clutch (it will
meet the specifications), and the costs of the clutch. These are the kinds of
information Lt. Cousins needs to do his job: he must be able to see that the

proposed clutch will do the job, be installable, and be affordable. Thus, Figure 4-2 presents a report written for the audience. It orders information in a helpful way, pushing generalizations to the top of the report and details to the bottom. It doesn't waste the time of a busy reader.

How do we learn to write for our audience? We must first learn the characteristics of the audience and then figure out what the audience needs to know. Let us consider each of these in turn.

4.1
CHARACTERISTICS OF REAL-WORLD AUDIENCES

If we look at the audiences for whom we write in school, we notice they have the following characteristics. They usually consist of one reader (the teacher) whom the writer knows. This reader is careful, trying hard to figure out what the writer has to say (usually the teacher is trying to decide how much a student has learned in order to assign a fair grade). In addition, the teacher reads the entire paper, is usually interested in the subject, and knows more about it than the writer. If something isn't clear, the teacher can often supply missing information or see where an unclear argument is going. Finally, since the purpose of school writing is to "show what you know," the teacher often puts a great deal of emphasis on the details that show in-depth knowledge.

In contrast to this ideal situation, the writer in a real-world setting has an infinitely more difficult audience. First, this real-world audience often contains many readers having backgrounds very different from each other's and from the writer's: some readers may even come to a report years after it was written. Thus, it's no longer easy to "psych out the professor," that is, to put into a paper just what you know a single reader wants. Second, your readers may have different—even conflicting—desires; often as a writer you can't satisfy everyone. Third, your readers often have little knowledge or interest in your subject and usually read only parts of the report, frequently skipping from section to section. Sometimes, however, your readers are experts in your field and have great interest in the report; such readers might even read your report very carefully, perhaps trying to find holes in your argument. Fourth, although you have spent your entire academic life learning how to pack details into a paper, your real-world audience is usually more interested in generalizations. In fact, some readers don't want details at all, since details just confuse their grasp of the subject. Finally, most readers, especially middle- and high-level managers—the most important readers—read under the worst possible conditions: they have many pressing things on their mind, they are constantly interrupted, and they must read *quickly* in order to get through their mail each day.

Since managerial audiences are so important, let us consider them in more detail.

4.2

THE NATURE OF MANAGERIAL AUDIENCES

Managers are the most important audiences for technical communications because they make decisions that affect projects and careers. Thus, if scientists, engineers, and technicians hope to influence managers favorably, they had better understand how managers work and what will catch their attention.

The nature of managerial work has been studied by a number of researchers, and one of these, Henry Minzberg, has summarized some important characteristics of managers that emerge from the research literature:

1   *Folklore:* The manager is a reflective, systematic planner.
    Fact: Study after study has shown that managers work at an unrelenting pace; that their activities are characterized by brevity, variety, and discontinuity; and that they are strongly oriented to action and dislike reflective activities.

2   *Folklore:* The effective manager has no regular duties to perform.
    Fact: In addition to handling exceptions, managerial work involves performing a number of regular duties, including ritual and ceremony, negotiations, and processing of soft information that links the organization with its environment. [Managers have a wide variety of duties, interests, and needs.]

3   *Folklore:* The senior manager needs aggregated information, which a formal management information system best provides.
    Fact: Managers strongly favor the verbal media—namely, telephone calls and meetings.

4   *Folklore:* Management is, or at least is quickly becoming, a science and a profession.
    Fact: The managers' programs—to schedule time, process information, make decisions, and so on—remain locked deep inside their brains. Thus, to describe these programs, we rely on words like *judgment* and *intuition,* seldom stopping to realize that they are merely labels for our ignorance.[1]

Minzberg also notes that managers fill many roles at one time. By implication, these roles and the constraints noted above largely determine how a manager reacts to communication. According to Minzberg, the fundamental managerial roles are *interpersonal roles:* figurehead, leader of the unit, liaison to external units. However, since they have many contacts from their interpersonal roles and their power, managers also have important *informational roles:* they are the "nerve centers of information." As such, they monitor and disseminate information and serve as the spokesperson for their units:

The processing of information is a key part of the manager's job. In my study, the chief executives spent 40% of their contact time on activities devoted exclusively to the transmission of information; 70% of their incoming mail was purely informational (as opposed to requests for action).[2]

Finally, managers have *decisional roles:* they are initiators of change—supervising up to 50 new projects at a time—disturbance handlers, resource allocators, and negotiators.[3] In short, as Leonard R. Sayles has put it, a manager

> is like a symphony orchestra conductor, endeavouring to maintain a melodious performance in which the contributions of the various instruments are coordinated and sequenced, patterned and paced, while the orchestra members are having various personal difficulties, stage hands are moving music stands, alternating excessive heat and cold are creating audience and instrument problems, and the sponsor of the concert is insisting on irrational changes in the program.[4]

The effects of this environment on communication are severe; managers obviously have little time or attention to spare for careful reading or listening. For instance, Minzberg notes that

1   Half the activities engaged in by the five chief executives of my study lasted less than nine minutes, and only 10% exceeded one hour.[5]

2   These five chief executives treated mail processing as a burden to be dispensed with. One came in Saturday morning to process 142 pieces of mail in just over three hours, to "get rid of all the stuff." This same manager looked at the first piece of "hard" mail he had received all week, a standard cost report, and put it aside with the comment, "I never look at this."[6]

Given this almost impossible set of handicaps, what can the writer do to make a memo or report more easily read and more fully understood? The writer must certainly consider the situation and background of the audience and the uses to which the communication will be put. In addition, the writer has a number of built-in aids for making texts easier to read and understand, including the conventions that will be treated in the rest of this book.

4.3

A PROCEDURE FOR AUDIENCE ANALYSIS

Identifying the audiences, their characteristics, and their needs is one of the most important jobs communicators have; it determines what kind of information must be provided and how it should be phrased. However, identifying the audiences can be difficult since audiences vary with the situation and with the type of communication involved. For instance, technical reports and memos written within an organization usually have a wide range of audiences, perhaps including laboratory technicians, technical specialists, managers, lawyers, and specialists in marketing and finance. In contrast, technical articles written for specialized journals have a narrow range of audience, usually just the specialized readers of the journal. Thus, to help technical communicators identify their audiences effectively, the five-step procedure outlined below has been provided. If technical communicators consciously follow these five steps, they should produce a good assessment of audiences and more effective communication.

Identify the Communication's Uses and Routes

To identify effectively all of their audiences, communicators should think carefully about the uses their communication will have and the routes it may travel, since these will often be more diverse than expected. For instance, consider the situation of a computer engineer working for a company that makes small computers to control industrial machines. The engineer had been asked by his immediate supervisor to improve the program by which a user tells a computer, the Model ABX, what to do. The engineer wrote a new program for the ABX and sent it off to his supervisor with the note, "Here's the program you asked me to write last week to improve the control of the ABX." The supervisor forwarded the engineer's note and program to the head of the Programming Department, who then forwarded it for review to a second computer engineer, to the Head of the Hardware Design Department, and to the Head of the Marketing Division. The second computer engineer knew a bit about the project because she had talked to the first engineer about it briefly over lunch one day, but the two department heads knew nothing about it. Since they could not review the new program themselves, they forwarded it to people in their departments who they hoped could do the review: a junior engineer in Hardware Design and a project engineer in Marketing.

All of these reviewers were faced with the task of evaluating the first computer engineer's new program. The second computer engineer had to check the program for programming adequacy. The Hardware engineer needed to see how difficult it would be to implement the new program, since his department would have to redesign some circuitry to accommodate it if the company adopted it for the ABX. The Marketing engineer had to evaluate the sales potential of the new program. How much would it increase the cost of the ABX? Would it be something customers would want and be willing to retrain their employees to use? In such a situation, this engineer normally surveyed customers, calling them up, explaining the features and costs of a new program, and getting their response to it.

Unfortunately, the first engineer had not anticipated these reviews and so had not included an overview or adequate documentation for the new program or a discussion of its improvements on the old program. Thus, the reviewers could not review the new program as it stood. To do their jobs, they had to go to the writer and interview him to get the information that should have been provided in a short report or memo attached to the new program.

The communication situation at first seemed straightforward for the first engineer. He assumed that he had to report only to his immediate supervisor, who had handed him the assignment. In fact, he communicated with—in addition to his supervisor—his department head, the heads of two other departments, two engineers in different departments, and one in his own. In addition, if the report had been adequate, the Marketing engineer would probably have used it as the basis of a communication to a number of customers. Thus, as illustrated by this example, *technical communicators should identify the uses their communication will have and the routes it will travel.*

Identify All Possible Audiences

Technical people work in organizations, and thus their work has implications beyond any particular time or setting. For instance, they may work on a project that goes on for years and passes through many stages, as illustrated below.

$$\text{Problem Perceived} \rightarrow \text{Problem Evaluated} \rightarrow \text{Product Designed} \rightarrow \text{Product Evaluated} \rightarrow \text{Product Constructed} \rightarrow$$

$$\text{Product Tested} \rightarrow \text{Product Refined} \rightarrow \text{Product Retested} \rightarrow \text{Product Produced} \rightarrow \text{Product Improved} \rightarrow \text{etc.}$$

During each of these stages, someone will provide information about the project which may be used immediately by someone else in the communicator's organization, or by someone outside the organization, or later on by someone connected with the project. Thus, as a second step in audience analysis, *technical communicators should try to identify all of the possible audiences, current or future, for a given communication.*

Identify the Concerns, Goals, Values, and Needs of the Audience

Technical people operate in organizations composed of units with different concerns, goals, values, and needs. Engineers in industrial settings, for instance, must cope with, at least, a marketing unit, a manufacturing unit, an engineering or design unit, and a management unit. That these units have different perspectives is illustrated by the following observation:

> Sales-minded persons tend to recommend the development of machines which resemble those being marketed by successful rival firms—partly because the competitive models are known to be in good demand and partly because a

program of imitation takes a minimum of time to execute. The design engineering department, on the other hand, tends to propose the development of machines which are, at the very least, improvements over the competitor's design, or, in many instances, machines which are totally new by comparison to the offerings of the current market. Although the end result is usually more desirable, the development time required is always much greater and less ponderable because of the unknown factors which may be involved. These viewpoints may be summed up by saying that the sales department takes a short-run view while the design engineering department takes the long-run view. Before we yield to our natural bias as engineers and exponents of progress (whatever that is) and condemn the sales department for its short-sightedness, let us bear in mind that the money is made in the short run, and in the long run we are all dead.[7]

Thus, to ensure that they communicate effectively to their audiences, technical communicators should be aware of the differing perspectives of their audiences; that is, *technical communicators should identify the concerns, goals, values, and needs of their audiences.*

Make Communication Appropriate for Managers
Because technical communicators work in organizations, any project on which they work involves the integration of many perspectives and goals. This integration must be achieved at the technical level but is directed at the managerial level. Thus, to make appropriate integration possible and to gain the needed support of management for their own projects, *technical communicators should make their communication accessible and useful to busy managers.*

Identify the Audience's Preferences
for and Objections to the Arguments
The integration of perspectives and goals mentioned above will necessarily involve compromises or even occasional setbacks when conflicts between units occur. Thus, because there may be others within their organization who are competitors for scarce resources and potential critics of their work, *technical communicators should try to identify those arguments and approaches that will be most effective with their audiences and to anticipate any objections that might be raised.*

REFERENCES

1   Reprinted by permission of *Harvard Business Review.* Excerpt from Henry Minzberg, "The Manager's Job: Folklore and Fact," *Harvard Business Review,* July-August 1975, pp. 50–55. Copyright © 1975 by the President and Fellows of Harvard College: all rights reserved. (Minz-

berg has also written *The Nature of Managerial Work*, Harper & Row, New York, 1973.)

2   Minzberg, op. cit., p. 56.

3   Minzberg, op. cit., pp. 54–59.

4   Leonard R. Sayles, *Managerial Behavior*, McGraw-Hill, New York, 1964, p. 162.

5   Minzberg, op. cit., p. 50.

6   Minzberg, op. cit., p. 52.

7   Russel R. Raney, "Why Cultural Education for the Engineer?" Speech to the Iowa-Illinois Section of the American Society of Agricultural Engineers, April 29, 1953, quoted in John G. Young, "Employment Negotiations," *Placement Manual: Fall 1981,* College of Engineering, The University of Michigan, Ann Arbor, 1981, p. 49.

ADDITIONAL READING

James Adams, *Conceptual Blockbusting: A Guide to Better Ideas*, Freeman, San Francisco, 1974.

Thomas J. Allen and S. E. Cohen, "Information Flow in Research and Development Laboratories," *Administrative Science Quarterly* **14**(1):12–19 (1969).

Aristotle, *The Rhetoric of Aristotle,* translated by Lane Cooper, Prentice-Hall, Englewood Cliffs, N.J., 1932, 1960.

Susan Artendi, "Man, Information, and Society: New Patterns of Interaction," *Journal of the American Society for Information Science* **30**(1):15–18 (1979).

C. West Churchman, *The Design of Inquiring Systems: Basic Concepts of Systems and Organization,* Basic Books, New York, 1971.

Mary J. Culnan, "An Analysis of the Information Usage Patterns of Academics and Practitioners in the Computer Field," *Information Processing and Management* **14**(6):395–404 (1978).

Linda Flower, *Problem-Solving Strategies for Writing,* Harcourt Brace Jovanovich, New York, 1981.

William J. J. Gordon, *Synectics: The Development of Creative Capacity,* Harper & Row, New York, 1961; Macmillan, New York, 1968.

R. Johnson and M. Gibbons, "Characteristics of Information Usage in Technological Innovation," *IEEE Transactions in Engineering Management* **EM-22**:27–34 (1975).

J. C. Mathes and Dwight W. Stevenson, *Designing Technical Reports,* Bobbs-Merrill, Indianapolis, 1976.

Henry Minzberg, *The Nature of Managerial Work,* Harper & Row, New York, 1973; also "The Manager's Job: Folklore and Fact," *Harvard Business Review,* July-August 1975, pp. 49–61.

M. A. Overington, "The Scientific Community as Audience: Toward a Rhetorical Analysis of Science," *Philosophy and Rhetoric* **10**(1):1–29 (1977).

James W. Souther, "What Management Wants in the Technical Report," *Journal of Engineering Education* **52**(8):498–503 (1962).

Richard E. Young, Alton Becker, and Kenneth Pike, *Rhetoric: Discovery and Change,* Harcourt Brace & World, New York, 1970.

5

BASIC TYPES AND PATTERNS OF ARGUMENT

Once a communicator has become aware of a problem, she or he must investigate and construct arguments that a given solution is appropriate and preferred over all others. Since these activities often require significant effort and time, this chapter provides tools for making them more efficient.

An argument is a claim that something should be believed or done, plus proof or good reasons for believing or doing it. This definition suggests that to make effective arguments, you must be able to

1   Make appropriate claims

2   Find and recognize proof or good reasons

3   Know when you have enough proof

As first steps to gaining and refining these skills, you ought to be familiar with the basic strategies of argument, the basic types of argument, and the accepted patterns for arranging arguments.

5.1
EXPECTATIONS ABOUT CLAIMS AND PROOF

When they first begin writing on the job, many engineers and scientists fail to provide enough support, or "backup data," for the claims they make in their reports. This often makes the reports ineffective or even useless when audiences disagree with the writer's conclusion, see it as threatening, or need a solid foundation of support on which to base their own decisions.

For instance, consider the following Foreword and Summary dashed off by a junior engineer.

Subject: Establishing a computer file for our library: Cost estimate

FOREWORD
Frequent reference to the books in our library is essential for the research workers in the office. However, because of imperfections in the card cataloguing system, precious time is wasted in locating books. To solve this problem, we have decided to use the computers for book searching; this will be faster and far more convenient. I have been requested to study the feasibility of such a project under our current budget. The purpose of this report is to estimate the cost for establishing a computer file for our library.

SUMMARY

For my test run on the computer, I experimented with 50 books from the library and estimated the cost for establishing a computer file for these 50 books. Since there are approximately 5000 books in our library, that cost figure was then multiplied by 100 to give the total cost figure for the whole library. It is estimated that $2000 will be needed to establish the computer file for the library.

Assume that you are the department manager to whom this memo was sent, that your department has limited funds, and that you can spend absolutely no more than $2000 on this project, if you can spend even that. Further, you can't waste your money: if you invest in the project, you must get a completed library system in return. A partially completed project will be useless and divert money from other needed projects.

Under these very realistic conditions, what would you want to see as proof for the claims made in the summary? Wouldn't you want to know that the sample of 50 books was a representative sample? How could the writer convince you? Wouldn't you like some proof that the estimated cost for the project is accurate? How would you expect to find that out? Perhaps by seeing a breakdown of the total cost into its component parts—$1500 for labor, $400 for computer time, $100 for punch cards to enter the data? Even if the writer provided those figures, how could you be convinced that the $1500 charge for labor would be sufficient, that you wouldn't be asked to provide more money later on?

Similarly, consider the next Foreword and Summary, written by a test engineer to report the mileage for a weight-reduced T-car.

FOREWORD

Increasing the fuel economy of our automobiles has been the top priority of our company for the past 2 years. In your memo dated October 6, 1981, you requested that I perform a simulated road test for our T-car with a weight reduction of 1000 lb and compare its highway mileage rating (MPG) to our present model. I have completed the test and have placed the results in our file. The purpose of this report is to present the test results and my recommendation.

SUMMARY

Upon completion of five simulated road tests, I have determined that the weight-reduced T-car will only give an average 2.5-MPG increase in highway mileage while its lightness creates handling and safety problems. Therefore, for safety reasons, I recommend that sources other than solely reducing the weight of the car be investigated.

If you had to make decisions about fuel economy based on this summary, wouldn't you want proof that the five road tests were sufficient, that the mileage increased only 2.5 miles per gallon, and that there were handling and safety problems with the car? If you were a hostile audience—perhaps you believed in the weight-reduced car, say, or even helped design it— wouldn't you need to see the same kinds of proof before you believed the claims made in the Summary?

The following exercises provide practice in identifying claims which need to be proved, and then the rest of this chapter considers ways of proving such claims.

EXERCISE 5-1 This exercise presents Forewords and Summaries from four different reports. Predict what claims or numbers need to be proved to make a convincing argument. The fourth sample for analysis also contains a Details section, which you should evaluate for adequacy of proof.

A   FOREWORD
The wind turbine project is a new undertaking of Energy Systems. The goal of the project is to design a marketable system which makes wind energy accessible to private residents. To aid in the selection and development of a preliminary design, Mr. Zondervan asked that I research the practicality of using a flywheel for energy storage. He also requested that I identify the organizations doing development work with flywheel rotors so that further information could be obtained if a flywheel is incorporated in the design. This report addresses these requests by identifying characteristics of flywheel energy storage which affect its practicality and by listing those institutions involved in flywheel research.

SUMMARY
New flywheel designs show much promise as energy storage devices, making possible energy densities up to 40 Wh/bl and reducing costs to as low as $50 per kWh. These figures compare favorably with other storage devices, such as batteries. In addition, the new designs have solved major problems with bearings and energy conversion, making flywheel energy storage quite practical.

Flywheel development projects are being carried on by the following people at the listed institutions.

1.   David Rabenhorst, The Applied Physics Laboratory of Johns Hopkins University

2.   J. A. Rinde, Lawrence Livermore Laboratory, Berkeley, Calif.

3.   A. E. Raynard, AiResearch Division of Garret Corporation, Torrance, Calif.

4.   Frances Younger, Wm. Brobeck and Associates

5.   R. P. Nimmer, General Electric Company, Corporate Research and Development, Schenectady, N.Y.

B   FOREWORD

In the past six months, the Production Department has received complaints from workers on the main floor that their working environment has become too noisy. They complain that intense noise from machines around them is nerve-racking and that they can communicate with one another only by shouting. Furthermore, some workers have even said that their hearing is deteriorating. Consequently, I measured the sound level in various spots on the main floor to locate the source of the problem. The purpose of this report is to recommend methods for reducing the noise level on the main floor.

RESULTS

I collected data and compared my results with those obtained in December, 1979; the total noise level produced by the machinery has increased by 30 percent. I then discovered that the noise is produced by the vibration of large housings and coverplates and by friction between loose and worn-out parts of machinery. In addition, the noisy condition on the main floor is worsened by reverberations from bare concrete ceilings and metal walls, which reflect a high percentage of incident sound waves.

RECOMMENDATIONS

My recommendations for reducing the noisy condition on the main floor are as follows:

1.   Add acoustic material on the concrete ceiling and metal walls to reduce reverberation.

2.   Add vibration-damping material beneath large housings and coverplates to dissipate vibration energy in the form of heat.

3.   Replace worn-out parts of machines and fix loose components.

4.   Provide earmuffs to prevent damage of the workers' ears caused by prolonged working in noisy condition.

5.   Consider noise specifications when buying new equipment to reduce noise production in the future.

C   FOREWORD

The present economic trends have caused smaller fraternities to have problems competing financially with larger fraternities, dormitories, and co-operatives. Specifically, Alpha Epsilon Zeta Fraternity must address the possibility of expansion into nearby apartments not only to remain competitive but also to gain the many social advantages of a larger membership. The purpose of this report is to demonstrate the economic feasibility of such an annex and to recommend its optimal size and location.

SUMMARY

Rent prices vary only slightly within walking distance. Therefore, the best location is the closest one, that being Plaza Apartments at 605 E. Manroe St.

Rent per member living in the annex will cost the fraternity substantially more than for those living in the lodge. However, excess revenue generated in the board account will counter that expense. The annex must hold at least six members to be successful, and anyone in addition could even cause a slight reduction in bills.

D FOREWORD

In your letter of September 30, 1981, you requested me to obtain information regarding the competitiveness of our new MGC-200 videodisc player. Consequently, I proceeded to test a number of the comparable products and especially the Pioneer VP-1000, which is the top-selling videodisc player. This report contains the results of my studies and recommendations for improving our product.

SUMMARY

The results of my research show that even though the quality of the MGC-200 is superior to most similar products, it will not be competitive with the top-selling Pioneer VP-1000. However, some modifications will substantially improve the specifications of our product. The recommended modifications listed below will cost less than $2000 and will postpone the planned production for approximately two more weeks.

1)   Minimize the power dissipation in the microcomputer playback control system by reducing the number of integrated components in the system.

2)   Reduce the signal to noise ratio to 40 dB. This will cut the price by about 10% and will not have significant effect on the quality of the product.

3)   Provide an optional remote control system. As explained in the details, this will not increase the cost significantly.

4)   Reduce the weight by using a smaller and lighter frame.

DETAILS

The specifications of both the MGC-200 and the VP-1000 are tabulated below for comparison:

		MGC-200	VP-1000
1)	Cost	$984	$749
2)	Weight	57.3 lb	38.6 lb
3)	Power Consumption	78 W	71 W
4)	Video Bandpass	4 MHz	4 MHz
5)	Audio Bandpass	80 MHz	83 MHz
6)	S/N Ratio	50 dB	40 dB
7)	Disc Type	3M	3M, MCA

As pointed out in the summary, an optional remote control system can easily be included. The RU-1000 remote control can be used either through the IRAB sensor and amplifier as a completely wireless infrared remote control or plugged into a remote control jack and onto the photo coupler as a wired remote unit when noise is a problem.

I would appreciate being informed of your decision.

5.2

THREE BASIC STRATEGIES OF ARGUMENT

Over 2000 years ago, the Greek philosopher and rhetorician Aristotle identified three main strategies or bases for argument:

1   Logic and reason

2   The character and credentials of the communicator

3   Emotion

If you look through many types of technical writing, you will notice that most use (or try to use) arguments based overtly on *logic and reason.* For instance,

> Pierlest Inc. should adopt a proposed pollution control device because (1) it will cut pollution to an acceptable level, (2) it will be easy to install, (3) it has a reputation for being very reliable, and (4) it will be cost effective.

In contrast, relatively few arguments are based overtly on arguments of *emotion,* though some are. For instance, the argument "Pierlest Inc. is violating federal pollution standards and will be closed down unless it greatly reduces pollution" appeals to the emotion of fear, the fear of being closed down.

You will note that most arguments are combinations of two or all three strategies, with one predominating. For instance, when you write a report recommending a particular solution to a problem, you probably give a series of arguments based on logic and reason to support your recommendation. However, you also make an argument at least partly based on your character and credentials. Since you were hired because of your qualifications, your recommendation is often accepted at least partly, and sometimes fully, because your readers believe that you're competent to speak on your topic. (It is, of course, important to *look* competent and to reinforce your credentials.)

You will also note that in different situations, different strategies of argument will be more important. As indicated above, most technical communication appeals heavily to logic and reason because this strategy of argument is usually appropriate and convincing. However, sometimes technical communicators need to base arguments on emotion. For instance, when scientists and engineers confront deeply emotional issues, such as the fear a community might have over the location of a nuclear power plant, overreliance on logical arguments can be ineffective. Scientists have often argued very logically and "objectively" about a nuclear plant's backup systems for controlling dangerous situations and the very slight chance of an accident but never convinced the listeners that a plant should be located in their community. For many, the "fact" that a plant has only a one-in-a-million chance of blowing up does not outweigh the fear-producing knowledge that the reactor at Three Mile Island almost broke through several security systems to become a radioactive hazard to the surrounding communities.

EXERCISE 5-2 Read the following passages and identify the strategies being used in each.

A   **Social Control of Research**

When I promised to speak here on the topic of social control of research, I thought it would be a much easier job than it has turned out to be. I have some practical experience in the management of research programs in the federal bureaucracy, but more relevant was seven years helping the Congress deal with research and development (R&D) matters. There, I learned the intricacies of grooming and passing authorization bills for large research programs, and negotiating complex conference agreements. On a number of occasions I had translated what society—represented by its elected officials—wanted in legislation that established new programs of research and development, or institutions to control such programs and their results, such as the Office of Technology Assessment. Another important part of that congressional experience was informing and cajoling those with responsibility for the appropriations process, in order to assure funding for those authorized programs. I had even received the satisfaction which came from coordinating the override of a presidential veto, only the 89th such veto override in the history of our Republic.

My confidence that social control of research would be an easy topic was further increased by my knowledge of and involvement with a number of AAAS programs on such topics as scientific freedom and responsibility, the scientific and legal interface, regulation of recombinant DNA research, analysis of the federal research and development budget, and other policy issues.

It was with substantial optimism that I set about to brush up on the more formal knowledge extant on the subject. So I decided to back off from my original intention to provide a thoughtful and catholic review of the topic. Rather, I shall attempt to identify its more important aspects from the twin perspectives of the legislative process and ongoing AAAS activities.[1]

B   **Recombinant DNA Research**

James D. Watson of Harvard, Nobel Laureate for his discovery of the molecular structure of DNA, [said] . . . that theoretically all forms of higher animal life may be capable of clonal reproduction, and that the details of the research related to that, as well as to its implications, had not so far been communicated to the public to any substantial degree. He went on to ask,

> Does this effective silence imply a conspiracy to keep the public unaware of a potential threat to their basic ways of life? Could it be motivated by fear that the general reaction would be a further damning of all science and thereby decreasing even more the limited money available for pure research? Or does it merely tell us that most scientists do live such an

ivory tower existence that they are capable of rationally thinking only about pure science, dismissing more practical matters as subjects for the lawyers, students, clergy, and politicians to face up to in a real way.[2]

C   **Hard Times for Basic Research**

Despite the impressive growth of the National Science Foundation's budget from its initial $500,000 in 1952 to its almost $900 million in FY 1978, the fortunes of basic research, viewed in relative terms, have not been great. In the last decade funds for basic research have been of the order of 0.15 percent of the Gross National Product and have steadily declined. As Shapley, Phillips, and Roback indicate in their analysis of the FY 1978 R&D budget, funds for basic research, expressed in 1972 constant dollars, dropped 20.6 percent between 1967 and 1975 and, despite increases in both the Ford and Carter budgets, have not yet returned to their 1967 level. The growth rate in R&D funding of 11.3 percent between FY 1976 and FY 1978 has meant about 5 percent more for basic research in the academic sector, hardly anything to quicken the pulse.

Despite the general understanding that the federal government should carry the major responsibility for the funding of basic research, and, indeed, does provide about 70 percent of the monies available for this purpose, un-yielding preoccupation with quick practical payoff on the part of both the Congress and the executive leadership has meant year-to-year uncertainty for an intellectual activity that should be built on long-range continuity. At least one fundamental reason for this seems to lie in our inability to make clear to the larger public the place of serendipity in science. The layman seems not to fully appreciate the fact that one cannot provide on cue *all* the antecedent information needed for innovative technology, and therefore that knowledge must be stockpiled like scarce materials against the day when it per chance may be needed. I am confident that this lacuna in the public's understanding of science can be filled in the long run with the right kind of precollegiate science education. But, meanwhile, we must give serious attention to creating a more vigorous, more comprehensive, and more carefully calculated program of education for members of the Congress and their legislative aides than we have been willing to give time to in the past.[3]

D   **Legislative and Regulatory Control**

With regard to whether research *should* be subject to regulation, we must immediately refer to the U.S. Constitution. The First Amendment to the Constitution reads:

> Congress shall make no law respecting an establishment of religion, or prohibiting the free exercise thereof; or abridging the freedom of speech, or of the press; or the right of the people peaceably to assemble, and to petition the Government for a redress of grievances.

The intent of the authors of the Constitution was recently summarized by Professor Thomas Emerson, Emeritus Professor at Yale Law School, who wrote:

> The process is essentially the method of science. The theory of freedom of expression, indeed, developed in conjunction with, and as an integral

part of, the growth of the scientific method. Locke, following Hobbes, based his philosophical and political theories on the premises of science. And the proponents of free expression were all men who, in the broad sense at least, put their faith in progress through free and rational inquiry. Hence the process they envisaged operates upon the same principles as those that guided the men of science: the refusal to accept existing authority; the constant search for new knowledge; the insistence upon exposing their facts and opinions to opposition and criticism; the belief that rational discussion produces the better, though not necessarily the final judgment. This process did not ignore prior knowledge or opinion, but it did insist upon the responsibility of the individual to challenge such opinion and upon the obligation of all to make reasoned conclusions based upon the evidence.

Our basic principles of academic freedom are rooted in Supreme Court decisions invoking the First Amendment. In *Swezey v. New Hampshire,* Chief Justice Earl Warren, writing for the Court, stated:

> The essentiality of freedom in the community of American universities is almost self-evident. No one should underestimate the vital role in a democracy that is played by those who guide and train our youth. To impose any strait jacket upon the intellectual leaders in our colleges and universities would imperil the future of our Nation.[4]

5.3
BASIC TYPES OF ARGUMENT

There are two main types of argument: arguments of fact and arguments of policy. An argument of fact is an argument that something *is* or *exists* (or *is not* or *does not exist*); it can also be an argument that something *is* or *is not true, necessary, justified,* etc. The key word for the argument of fact is *is* or *are:* this company *is* (or *is not*) discharging pollutants into public waters; increased domestic production of oil and natural gas *is necessary* if the United States is to maintain its standard of living without being too dependent on Middle East oil; the costs of buying a new house *are* too high for the average consumer.

In contrast, an argument of policy is an argument that something *should* or *should not* be done. The key word for this argument is *should:* this company *should* be stopped from discharging pollutants into public waters; increased domestic production of oil and natural gas *should* be encouraged in the United States so that people can maintain their standard of living (or increased production of gas and oil *should not* be encouraged in the United States because it would be too harmful to the environment); the costs of buying a new house *should* be reduced for the average consumer.

Notice that your support for an argument of policy would differ from your support for the corresponding argument of fact; in some cases you would need to make an argument of fact as a subargument in an argument of policy. For instance, if you were to argue that the costs of buying a new

TABLE 5-1   OUTLINE OF THE ARGUMENT OF FACT

Argument of fact: argument that something *is* or *exists* (or *is not* or *does not exist*)

Question or subargument of existence: does the thing actually exist; has something actually happened?

Example A   Are there Soviet troops in Cuba?

Example B   Is Pierlest Inc. discharging material from its manufacturing processes into public waters?

Question or subargument of definition: if it exists or has happened, what kind of thing or event is it?

Example A   If it is granted that there are Soviet troops in Cuba, are the troops educators and advisors or attack troops or something else?

Example B   If it is granted (or proved) that Pierlest Inc. is discharging material from its manufacturing processes into public waters, are the discharged materials regulated by law, considered dangerous, considered nontoxic even in large amounts, etc.?

Question or subargument of quality: if it exists and has been defined, how is it to be judged?

Example A   If Soviet troops are in Cuba and are educators and advisors, are they justified, necessary, present in appropriate numbers, desirable, etc.?

Example B   If Pierlest Inc. is discharging material into public waters, and if those materials are regulated by law (or dangerous, etc.), are the materials present in legal (or safe) amounts, illegal (or unsafe) amounts, desirable (or undesirable) amounts, avoidable (or unavoidable) amounts, etc.?

house should be reduced for the average consumer (an argument of policy), you would need first to argue that the current costs are too high (an argument of fact).

The Argument of Fact
Arguments of fact can be derived from three sources:

1   Questions or subarguments of existence

2   Questions or subarguments of definition

3   Questions or subarguments of quality

These are outlined and illustrated in Table 5-1. It can be seen from the examples in this table that, as you move from the question or subargument of existence to the subargument of quality, you logically build an argument. For instance, if you want to argue that a certain manufacturer should be stopped from discharging pollutants into public waters, you first need to address the question of existence (is this company discharging material into public waters?). You then need to address the question of definition (are the discharged materials regulated by law? dangerous? etc.), and finally you must address the question of quality (are the discharged materials present in illegal or unsafe amounts?). Only when all of these questions have been addressed can you argue convincingly that the company should be stopped from discharging into public waters. You make this main argument by stringing together a chain of subarguments of existence, definition, and quality:

Subargument of existence:
The company is discharging material into public waters.
↓
Subargument of definition:
The materials being discharged are regulated by law and are dangerous.
↓
Subargument of quality:
The materials being discharged are present in public waters in illegal and unsafe amounts.

Note that your main argument could fail on any of these subarguments. Your argument might fail to establish the subargument of existence. It could happen, for instance, that while there are pollutants in the public waters, another company put them there and not Pierlest Inc. In such a case, placing restraints on Pierlest would be both unfair and ineffective. Similarly, your argument could fail on the subargument of definition. It could be that Pierlest is discharging material into public waters but that these materials are not regulated by law or are not harmful. In such a case, it would probably be wasteful to spend time, effort, and money trying to regulate the company. Finally, an argument could fail on the subargument of quality. Pierlest might well be discharging dangerous materials regulated by law, but only those amounts allowed under the law. In such a case, if you still wanted to stop those discharges, you would have to argue that the law should be changed; you could not effectively argue that the company was breaking the law.

You could represent the "path" of reasoning we have just seen as a flow chart, such as the one in Figure 5-1.

The Argument of Policy
The argument of policy, the second basic type of argument, can be derived from two sources:

1   Questions or subarguments of worth or goodness

2   Questions or subarguments of expediency, advantage, or use

These are outlined and illustrated in Table 5-2.

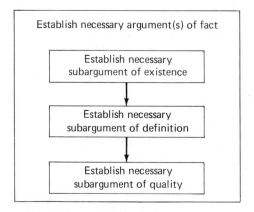

FIGURE 5-1   Partial flow chart for constructing
effective arguments: establish necessary
arguments of fact.

You should note that arguments based on expediency, advantage, or use are much more frequent in most types of technical writing—indeed, in most areas of our lives—than are arguments based on intrinsic worth, goodness, or merit. People are more likely to believe or do something if it is to their own advantage. Let us consider, for example, the arguments in Table 5-2 and the following situation. Company administrators intend to make false claims about the merits of their product in order to increase sales and profits. You want to argue instead that they should submit honest claims about the product. If you appeal only to the issue of worth (it is good and worthy to be honest), you will probably be unsuccessful. In the value scheme of these administrators, the desire for profit is probably stronger than the desire for honesty. Instead, you are more likely to succeed if you appeal to expediency and advantage: the administrators should make honest claims about the merits of their product because

1   It is advantageous in dealing with customers to have a reputation for honesty

2   Such honesty will protect the company from charges of fraud, expensive law suits, and costly penalties imposed by the government and the courts

This kind of argument better "fits" the supposed value scheme of the administrators. Can you think of a situation in which the appeal to worth or goodness would be as strong as or stronger than the appeal to advantage?

TABLE 5-2 OUTLINE FOR THE ARGUMENT OF POLICY

Argument of Policy: argument that something *should* (or *should not* be done)

Subargument of worth or goodness: is a proposed activity or course of action worthy or good in itself?

Example C The United States should not protest the presence of Russian troops in Cuba because it is right that they are there (because they have the right to be there).

Example D Pierlest Inc. should stop polluting public waters because it is right (or they have a social duty) to protect the environment.

Example E Companies should make honest claims about the merits of their products because it is good and worthy to be honest.

Subargument of expediency, advantage, or use: is the proposed activity or course of action good for the audience in that it is expedient, advantageous, or useful?

Example C The United States should agree to the presence of Russian troops in Cuba because such agreement would strengthen the presence of U.S. troops on Russian borders.

Example D Pierlest Inc. should stop polluting public waters because this would improve its public image and thus its sales.

Example E Companies should make honest claims about the merits of their products because it is advantageous in dealing with customers to have a reputation for honesty and because such honesty will protect the company from charges of fraud, expensive law suits, and costly penalties imposed by both the government and the courts (because it will be advantageous for them to do so).

The Relationship between Arguments of Fact and Arguments of Policy

You have seen that arguments of fact should be made in a logical sequence and that arguments of policy often depend upon prior arguments of fact. You can combine these insights to produce an overall plan or flow chart for ordering arguments of fact and policy, as demonstrated in Figure 5-2.

Notice that if you are only trying to make an argument of fact, you will never "get to" an argument of policy. In contrast, if you are making an argument of policy, you will need to establish all of the necessary parts of the argument of fact: the subargument of existence, the subargument of definition, and the subargument of quality. You do not, however, need to spend a lot of time establishing these subarguments if your audience already agrees to them.

For instance, suppose that you were trying in a speech to present an argument of policy about Pierlest Inc. and its pollution of public waters: "Pierlest Inc. should be required to stop polluting public waters and to follow a cleanup plan advocated by the State Department of Health." If you were talking to an audience that knew nothing about the pollution problem, you would have to carefully prove each part of the argument of fact before you

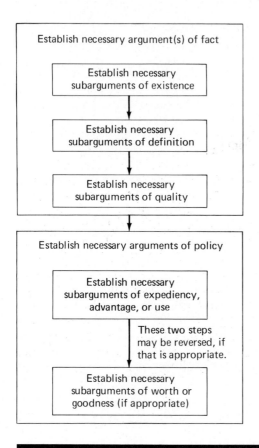

FIGURE 5-2 Flow chart for constructing effective arguments.

could argue for your policy. However, if everyone already knew that the company had been polluting the waters with a dangerous and illegal substance, you might summarize all of the subarguments of fact in a quick opening statement:

> We all know that Pierlest Inc. has been illegally dumping polyvinyl chloride into the Huron River. We are here to decide what to do about the problem. I would like to argue that Pierlest should be required to follow the cleanup plan proposed by the State Department of Health. The company has a responsibility to the people of the state to clean up the river, and this plan will allow them to do so most quickly and thoroughly and at a minimum cost.

You would then continue (1) to outline the proposed plan, (2) to demonstrate that the plan will allow quick and thorough cleanup of the river, and (3) to

FIGURE 5-3 Introduction to an argument of policy for an
audience already accepting the prior argument of fact.

We all know that Pierlest Inc. has been illegally dumping polyvinyl chloride into the Huron
River (*argument of fact, already accepted*). We are here to decide what to do about that
problem. I would like to argue that the company should be required to follow the cleanup
plan proposed by the State Department of Health (*argument of policy, to be proved*).
Pierlest has a responsibility to the people of the state to clean up the river (*subargument
of worth, perhaps needing to be proved*), and this plan will allow them to do so most quickly
and thoroughly (*subargument, advantage to the public, to be proved*) and at a minimum
cost (*subargument, advantage to company, to be proved*).

demonstrate that the cost will be minimal. The various subarguments presented are identified in Figure 5-3.

EXERCISE 5-3 This exercise consists of two parts. First, identify the arguments of fact and policy embedded in the short talk below, "The State of
Industrial R&D." Notice that much of the proof for individual arguments is
missing, since the talk merely gives a short overview, not an exhaustive case
analysis. Then, construct several short, general outlines for various arguments
of fact and policy. Be sure that you have at least one argument of fact and
at least one argument of policy.

The State of Industrial R&D

The role of research and development in industry has changed dramatically in
recent years. I can summarize what has happened in a single sentence: Industrial R&D has expanded beyond its traditional function of long-term contributor
to corporate growth and has become a real-time contributor to corporate profitability.

In times past, industry's principal technical challenges were the development of new products and the creation of new processes to manufacture
these products. As we know, times have changed. In 1977, American industry
confronts a host of technological problems that threaten its ability to remain in
business—problems such as energy shortages, environmental concerns, sky-
rocketing materials and manufacturing costs. These are problems that can only
be solved *now*.

A new set of R&D priorities has emerged for industry that has made the
present as important as the future, maybe more important. And managers of
industrial R&D face a perplexing dilemma: How to accommodate these short-
term demands without emasculating the long-term programs that are necessary
for corporate growth in the years ahead?

This, of course, is a classic predicament of resource allocation. It raises
a series of questions that are best answered through a process of research
management—a planning/programming approach that integrates industrial re-

search and development efforts with other corporate business activities. Research management is the key to apportioning finite resources across a growing list of R&D demands.

Research management is based on the simple premise that industrial R&D capability is a resource that can help a corporation achieve its business goals. Traditionally, the R&D resource is directed toward the long-term development of technology that will make possible new and/or improved products. At United's Sikorsky Aircraft Division, for example, a long-term R&D program, begun back in the late 1920s, developed the advanced articulated rotor that will lift the next generation of helicopters. This new rotor is a key to the performance, weight-lifting capability, and low vibration of the Blackhawk, tactical utility helicopters we are building for the U.S. Army. The same technology will be carried over to the S-76, a high-performance helicopter designed from the outset for civilian applications.

However, as I mentioned earlier, R&D can also make real-time contributions to corporate profitability. Here are two examples: The costs of raw materials have been climbing 2 percent faster than the overall inflation rate, and these increases represent a significant component of soaring manufacturing costs. Our Pratt & Whitney Aircraft Group is about to bring on line a hot isostatic press developed through company-funded R&D. The press will form jet engine discs by compressing superalloy powder at high temperatures and pressures. The new process needs 70 percent less material to make a turbine disc than conventional forging and machining methods.

Because research and development activities can support diverse corporate goals, it's appropriate to integrate R&D planning into overall corporate planning. This is the starting point of the Research Management process.

The result of coupling technical and business planning is a new perspective on company operations that enables both technical and corporate managers to look ahead at the anticipated benefits of industrial R&D. Specifically, they can judge whether or not the planned technical programs adequately support the company's business objectives. They can ask the kinds of questions that lead to maximum resource utilization and minimum technical risk. Are the goals of the planned R&D programs supportive of corporate strategies? Are important projects being postponed or simply overlooked? Are company activities that need technical support receiving it? If not, what additional resources are necessary to meet the needs? Is there a proper balance between R&D activities designed to develop new products, build new knowledge, and improve existing products? Are proposed R&D programs sized in keeping with their anticipated role in reaching corporate objectives? Are critical programs underfunded? Are less important projects oversupported? Are there any dead-end programs in the plan?

A comprehensive coupling analysis will help everyone involved recognize those R&D activities that are absolutely essential to achieving commercial objectives. Though these projects are on the critical path and carry maximum technical risk in case of failure, they may be small in size and low in perceived priority. At United Technologies, we've found that 10 to 20 percent of the projects in a large R&D program may fall into this category.

Identifying them in advance is the first step in managing technical risk.[5]

TABLE 5-3   OUTLINE FOR THE ARGUMENT OF FACT OR POLICY

I Problem/Introduction/Foreword
   A Direct the audience's attention toward the problem.

II Credentials*
   A If it is useful, give your credentials, i.e., explain why you can speak with authority
      on the subject, and establish common ground by pointing out shared beliefs,
      attitudes, and experiences.

III Position/Solution/Summary*
   A Briefly state your position or your proposed solution.
   B Briefly state the major reasons for advocating your position or solution.

IV Background of the Problem*
   A Point out the nature of the problem: (1) its historical background and (2) causes.
   B Explain how it concerns the audience.

V Argument for Position or Solution
   A State the criteria for judgment, i.e., the standards or characteristics any
      acceptable solution or position must meet. Include explanation where necessary.
   B State your position or solution to the problem, along with any necessary
      clarification.
   C Demonstrate the soundness of your position or solution by showing how it meets
      the criteria established in Section IIIA. This step should be accompanied by
      ample evidence: facts (illustrations, statistics, examples of successful application
      of the solution) and statements of authority. Be sure to identify the authorities if
      they aren't widely known.
   D If there are competing positions or solutions, demonstrate the superiority of yours
      by showing how the others fail to meet the criteria as completely as yours.

VI Conclusion
   A Explain briefly the benefits to be gained by accepting your position or solution or
      the dangers of rejecting it.
   B Summarize your argument: (1) restate your position or solution (IIIB); (2) restate
      the reasons your position or solution should be accepted (IIIC).

*Items II, III, and IV may be deleted depending on the situation, the needs of the
audience, and the accepted formats for a particular type of technical communication.
SOURCE: Adapted from Richard E. Young, *Rhetoric: Discovery and Change,* Harcourt, Brace & World,
New York ,1970, p. 152.

5.4

AN EFFECTIVE GENERAL PATTERN FOR ARGUMENTS

Once you have decided on the basic arguments you are going to make and
on their logical order, you need to decide *how* to present the entire speech
or report. What goes where? How should things be organized? These are
especially difficult questions if you have limited amounts of time and a large
amount of information to organize.

One psychologically satisfying and widely useful organization is presented in Table 5-3, which gives an outline for the argument of fact or policy.
This organization is so effective that it is still being used over 2500 years after
it was first developed in Greece and Rome for speaking in public forums and
courts of law. It is particularly useful for technical communicators because
almost all technical communication consists of arguments of fact or policy.
Thus, by learning this one outline, you can have at your fingertips a general
organization for essentially any speech or report.

TABLE 5-4　A SAMPLE ARGUMENT OF FACT:
"ELECTRONIC LOCK DESIGN PROPOSAL"

I　Introduction
　A　The need for effective yet flexible security systems is more critical today than ever. In many applications the traditional mechanical key or combination lock is no longer adequate. Electronic systems using electronic locks are proving to be the optimal solution, providing advantages unavailable in mechanical systems.
　B　Wiley Electronics is about to begin development of an electronic security system of its own. Mr. Silvers has asked me to design a lock to be used as part of such a system.

II

III　Summary
　An electronic lock with an 8-digit combination is the optimal lock for Wiley's electronic security system. The proposed lock is effective, simple, and versatile, and it costs only $10.35.

IV　Background of Problem
　A　Better and more efficient security systems are needed by both industry and the individual.
　　1　Most industrial complexes maintain large areas in which stringent security is essential, yet through which large numbers of employees must pass unhindered.
　　2　Many individuals also find themselves in situations in which they require a level of security beyond what is normally adequate.
　B　Strong interest in better security systems has encouraged the manufacture of many different security systems for the potential customer.
　　1　Basically, he/she must choose between ordinary key or combination locks and an electronic lock.
　　　a　A clear trend toward electronic security systems has prevailed over the last few years.
　C　Wiley Electronics is about to begin developing a security system to compete in this market.

V　Argument
　A　The criteria for choosing a system: what does a consumer expect from a security system?
　　1　The lock must keep out unauthorized persons and allow those authorized to enter freely.
　　2　The lock must be tamperproof.
　　3　The lock must have a combination that is easy to alter.
　　4　The lock must be easy to interface with a larger controlling system.
　　5　The lock must be capable of independent operation.
　　6　The lock must have an acceptable cost and market appeal.
　B　Solution: the electronic lock meets all of the criteria as well as or better than the mechanical key or combination lock.
　C　Argument
　　1　The electronic lock is as effective as the other two locks in screening authorized from unauthorized persons.
　　2　The electronic lock is more tamperproof.
　　3　Its combination is easier to alter.
　　4　It is easier to interface with a larger controlling system.
　　5　It is as capable of independent operation as the other two locks.
　　6　It has an acceptable cost of only $10.35 wholesale.
　　7　It will appeal to a wider range of the market than the other two types.
　D　Description of the Proposed General Purpose Electronic Lock
　　1　Security features of the lock
　　　a　The lock uses an 8-digit combination providing 65,536 different possible combinations for a sufficient level of security for most applications.

(continued)

TABLE 5-4   A SAMPLE ARGUMENT OF FACT:
"ELECTRONIC LOCK DESIGN PROPOSAL" (*Continued*)

      b To provide absolute security, a minor modification will allow the lock to
         operate with a 16-digit combination, providing over 4 billion unique
         combinations.
    2 Operation of the lock
      a General operation of the lock
      b Procedure for opening the lock
      c Procedure for setting a new combination
    3 Internal construction and operation of the lock
      a The push-buttons
      b Counter #1
      c The memory
      d The comparator
      e Counter #2
  E Possible alternatives: the "key card" lock
    1 Description: a "key card" lock is the type of lock that accepts a small wallet-
      size card which is inserted by the operator into a reader. The card is usually
      impregnated with a magnetic code that the card reader can detect and
      transmit as a digital signal.
    2 Evaluation: there are two possible means of processing the key card's digital
      signal, local and remote, both of which can be rejected as inadequate
      solutions.
VI Conclusion

SOURCE: Adapted from "Electronic Lock Design," *Student Report Handbook,* edited by Leslie A. Olsen,
Lisa Barton, and Peter Klaver, Professional Communications Press, Ann Arbor, Mich., forthcoming.

You should note that in any given piece of communication, Sections
II (Credentials), III (Position or Solution or Summary), and IV (Background)
of the outline may be present or absent, depending on the needs of the
audience and the accepted format for the communication. For instance, most
technical reports have a heading or title page listing the writer's name and
organizational role. These reports do not have a separate section devoted to
the credentials of the writer, since those credentials are already established
by the role. On the other hand, proposals do have a very important section
explicitly presenting the credentials of the proposers. As a second instance,
technical reports have a Summary as the second main section of text, whereas
a speaker talking to a very hostile audience may decide to hold the Summary
or the explicit Statement of Position until after all of the arguments have had
a chance to weaken the audience's resistance.

A sample argument to illustrate the general outline is presented in Table
5-4. The numbering of the sample argument corresponds to the numbering
in the outline of fact or policy in Table 5-3.

Stating Criteria

A particularly crucial section of the argument of fact or policy is the statement
of criteria (Section VA). Criteria are important because they measure the
success of *any* solution to a given problem and establish the superiority of
one. For instance, you can judge the success and superiority of the electronic

lock of Table 5-4 only if you know the criteria on which it or any such lock should be judged. By stating the criteria directly, you make it easier for the audience to see them and to agree (or disagree). If the audience agrees with your criteria, and if your position follows from them, then the audience has to agree with your solution as well. (If the audience disagrees with your criteria, it will probably also disagree with your proposed solution or position; if this is the case, you have nothing much to lose by stating the criteria and you may have something to gain.)

Stating criteria has another advantage: it makes it easier for you to create your argument in the first place. For instance, if you want to argue that your electronic lock is acceptable and if the only possible or important criteria are the six stated in Table 5-4, then it is relatively easy to (1) judge your lock by each of the six criteria and then (2) write a section on each judgment. Similarly, if you want to argue that your lock is better than any other lock, all you have to do is to see that your lock meets each criterion as well as or better than any other lock and then to write a section on each comparison. This procedure is illustrated in Section VC of Table 5-4. In some ways, once you have the criteria, the generation of the argument is easy. Note, however, that the hard part is getting the right criteria and making them clear.

Getting the right criteria is sometimes difficult, but there are several general criteria which almost always need to be considered when evaluating a proposed solution:

1   *Effectiveness:* Is the solution effective? Will it solve the problem posed? Why? How do you know?

2   *Technical Feasibility:* Can the solution be implemented? Does it require technology or resources that are unavailable? How do you know?

3   *Desirability:* Would one want to implement the proposed solution? Does it have any undesirable effects? Does it have desirable effects? Why? What are they?

4   *Affordability:* What will the solution cost to implement? To maintain? Is this cost reasonable? Is it affordable? Will it reduce costs in the future? Why?

5   *Preferability:* Is the solution better than or preferred over any other possible solution? Why?

In a particular situation, there may be other criteria that are also appropriate or necessary. As a technical expert, you must know and state these *special criteria*. It may sometimes be difficult to do this, but it should be done, especially for busy readers of important documents who rarely have time to track out arguments and to discover the criteria on which judgments are based.

FIGURE 5-4 A formal statement of criteria.

Introduction

The main conveyer line in Building 9 is used to transport three of Construction Equipment's products. However, each of the three products carried on the main conveyer line has to be switched to a secondary conveyer line which transports them to the shipping area. Currently, three workers are used to carry out this switching process. The products, each of which weighs less than 10 pounds, are lifted off the conveyer line and placed in the appropriate chutes by the workers. This is inefficient and costly, and the labor could be used elsewhere.

    The objective of this report is to introduce a successful automatic system to replace the manual system. The new automatic system must meet the following criteria:

        1. It must be compact enough to fit into a 4' × 5' area adjacent to the conveyer line.

        2. It must be economically competitive with the manual system.

        3. It should be of a pneumatic type if possible, because an air supply line is located approximately 10 feet above the conveyer line.

    In the following sections I will present the system I have designed. It will automate the current manual process and meet the stated criteria.

How does one actually state criteria? It is usually easiest for readers if criteria are stated formally as illustrated in Figure 5-4. Sometimes criteria can be stated more informally, as illustrated in Figure 5-5, the beginning of a magazine article for a general audience. See if you can identify the criteria being used to evaluate the Intermediate Capacity Transit System, as well as the other modes of mass transportation being compared with it. To give you a head start, we'll look at one sentence, "Subways, of course, are ideal—fast, unobtrusive, impervious to bad weather." The three qualities mentioned are really criteria in disguise: the sentence implies that any ideal mass transportation system should be fast, unobtrusive, and impervious to bad weather.

Sometimes criteria are so obvious that they need not be stated. For instance, look again at Figure 5-5, and especially at the fourth paragraph. Several criteria are implied in the first sentence there: "To make a quiet, unobtrusive yet affordable elevated railway was the goal of UTDC." However, not all of the criteria are stated. This article is talking about mass transit systems, and obviously a quiet, unobtrusive yet affordable elevated railway would be unacceptable if it could carry only 10 people per hour. The article might well have told the reader how many people per hour the system should be able to carry, but any reader would know that it should carry some large number. Further, given the first sentence of the second paragraph, "Yet streetcars and buses are limited to 10,000 passengers per hour," the reader might well assume that the new system should carry more or many more people than the 10,000 mentioned. Are there any other unstated criteria? Are there any criteria missing from the introduction which you think should be considered?

FIGURE 5-5   An informal statement of criteria in a magazine
article. [Charles Maurer, *Popular Science,* **217**(6), 75–76 (1980).]

Tomorrow's Train Is Running Today in Canada

We are standing by the 1.4-mile test track of Ontario's Urban Transportation Development
Corporation. Around us are empty fields that the UTDC hopes will become the foremost
center of urban mass transit in North America. The elevated train I've come to hear and
see—the ICTS (Intermediate Capacity Transit System)—is the vehicle that could bring that
plan to fruition. It's tomorrow's means of mass transportation today. Subways, of course,
are ideal—fast, unobtrusive, impervious to bad weather. But at $80 million per double-track
mile, precious few cities can afford to build them. Few can even afford to maintain the ones
they have.

  Yet streetcars and buses are limited to 10,000 passengers per hour, and to carry
that many they would have to be so close together you could almost climb out on the
vehicles' roofs and walk—and arrive at your destination faster.

  Lightweight, relatively inexpensive rail systems can fill the gap, but if they're to run
at grade level they need a right of way. And with sidewalk-to-sidewalk skyscrapers, there's
no right of way to be had—except over the streets. That's why our largest cities long ago
installed elevated street railways. Sensible monstrosities, but to judge from their noise and
looks, the same firm engineered them that handled hell.

  To make a quiet, unobtrusive yet affordable elevated railway was the goal of UTDC.
Visionary affairs like monorails were out of the question. "They're fine for fairgrounds,"
explained Dick Giles, a mechanical engineer who serves as UTDC's sales manager, "but
for a raft of reasons they're impractical." With monorails, performing maintenance, doing
switching, designing stations, and providing emergency evacuations are all far more difficult. And monorail trains require a higher, more obtrusive track.

  Magnetic levitation was also out—"too complicated, too many developmental risks."
So the train had to run on wheels. To make it unobtrusive, stations had to be short, so
trains had to be short, too—120 feet maximum (four-30-foot cars). Short trains meant fast
trains—45 mph—spaced at intervals of less than one minute apart.

  Spacing trains so closely at those speeds meant computers had to be in control. It
also meant unusually quick accelerating and braking. This, plus the ability to take steep
grades in stride—essential if the railway is to be shoehorned into existing space—ruled
out conventional motors driving steel wheels: too much slippage. And though conventional
motors can drive rubber wheels quietly and surely, the line must be kept free of ice and
snow, impossible in Canada.

EXERCISE 5-4   This exercise should develop your ability to evaluate the
overall structure of a speech or report and, particularly, to state and evaluate
criteria. In Figures 5-6 and 5-7 you will find the beginnings of two different
technical reports, "Analysis of Criteria: Sample 1" and "Analysis of Criteria:
Sample 2." For each report, write a list of criteria that could appear in the
third section of the report. (The third sections are "Why a New System Is
Needed" for Sample 1 and "Introduction" for Sample 2). Note that the
reports may have no specific criteria stated or they may have some—but not

all—of the criteria stated. You need to create a list that moves from the most important criteria to the least important ones. Be sure to consider the five general criteria discussed above—effectiveness, technical feasibility, desirability, affordability, and preferability.

Next, write an outline of an argument of fact or policy that you would like to make. As far as you can, model your outline on the one presented in Table 5-4. Be sure to state a problem at the beginning (see Chapter 6 for a suggested format), derive a list of criteria that any solution to the problem must meet, and then show that your solution meets the criteria and, if appropriate, that it meets the criteria better than other possible solutions.

FIGURE 5-6 Analysis of criteria: sample 1.

Fasano Manufacturing Co.

Interdepartmental Memo

Date:	June 24, 1980
To:	V. Voros Supervisor - Computer Operations
From:	R. Sholander Analyst - Computer Operations
Subject:	Production and Information Control System: Preliminary Report on Features and Implementation

Foreword

　　Mr. Bauer, Vice-President of Production, wants to centralize control of production and purchasing to optimize both inventory levels and assembly shops' loads. New information about purchasing and production is becoming dispersed throughout the system. This makes it difficult for managers and sales representatives to know the status of a job quickly, resulting in delays in dealing with customers. Further, inventory is often at unnecessarily high levels as stock waits to be used in production that has been delayed, and the machine shops are sometimes idle while waiting for parts because purchase orders were delayed or production is ahead of schedule. The computer operations department was given the task of researching and designing solutions to this problem. This report will present my solution and its implementation by computer.

Summary

　　The solution I propose is an integrated system combining sales forecasting, engineering, inventory control, purchasing, and job shop scheduling. The Production and Information Control System, called PICS, will provide a central data base

FIGURE 5-6   *(Continued)*

which all pertinent information will be collected. It will also provide the following:

- access only to authorized users

- a Materials Requirements Planning system

- optimal shop capacity through simulation procedures

- a status report on any job at any time in the system

- faster response to market demands

This system is well modularized and can be implemented gradually by our departments. A rough time estimate for completion would be about seven months. The cost would be minimal if PICS were operated as a manual system, but this would still require an overwhelming amount of paperwork. A gradual change to a computer system would cost approximately $150,000, but our paperwork could be handled much more efficiently if it were entered directly on a computer disk. Therefore, I suggest a rapid conversion to a computerized approach.

Why a New System Is Needed

Information concerning the inventory system is becoming increasingly more difficult to interpret and understand. Problems arise because each assembly shop keeps its own records of transactions. One problem occurs in purchasing, where orders from shops must be combined to save money in bulk ordering. This combining is often impossible if orders come in separately. A second problem is that it is hard to find the exact status of a job in production if you are not directly involved in the daily operation of that job.

The computer operations department was given the assignment of researching all possible solutions to these problems with particular emphasis on:

- a mechanized system that would facilitate a change to a computerized approach

- a central data bank where all departments contribute and receive information

- a low-cost system that could be phased in gradually

This report presents the alternative I was assigned to investigate and is divided into four sections:

- system overview

- data base requirements

- necessary changes

- cost of new system

FIGURE 5-7 Analysis of criteria: sample 2.

Babies' General Hospital

Interdepartmental Memo

Date: June 20, 1980

To: Nancy Gabrenya, Director
            Medical Information Department

From: Timothy J. Rowell, Principal Engineer
            Management Engineering Department

Subject: REQUEST PROCESSING IN THE CORRESPONDENCE UNIT:
            Recommendations to Improve Productivity and Processing Time Vari-
            ation

Foreword

The Correspondence Unit of the Medical Information Department at Babies'
General Hospital is responsible for supplying medical information to requestors.
Because requestors desire this information as soon as possible, hospital management feels that requests should be processed within two weeks. Since the system
performance is presently falling well short of this standard, a study of request
processing procedures in the Correspondence Unit was undertaken.

This paper reports the results of a three-month study (February through
April of 1980) of request processing procedures at Babies' General Hospital. The
paper examines the current operations in the Correspondence Unit, identifies the
major operational problems, and recommends system improvements which use
the currently existing data base and involve minimum implementation costs.

Summary

The request processing system in the Correspondence Unit is a manual
operation consisting of five major functions: mail processing, chart retrieving,
record screening, photocopying, and mailing. Processing a request currently takes
4.56 weeks.

The smooth flow of requests throughout this system is marred by a number
of operational problems. The average worker productivity measured in this study,
59.1%, is below the suggested 65% level. Low worker productivity has caused the
backlog of active requests to rise above 2000 requests, thus adding excess processing days. The processing times of many individual requests are increased further
by our system of assigning priorities to the requests. Finally, the processing times
for some requests are increased even further if the Correspondence Unit does not
have the record in question in its own files but must request it from other departments or clinics.

A number of system improvements are recommended to solve these operational problems:

FIGURE 5-7 *(Continued)*

To monitor worker productivity, a productivity index (P.I.) should be used:

$$P.I. = [(2.7) \text{ x } (\# \text{ mail processed})$$

$$+ (6.0) \text{ x } (\# \text{ records screened})$$

$$+ (4.5) \text{ x } (\# \text{ mailings completed})]$$

$$/ [(60) \text{ x } (\# \text{ hours worked})]$$

To insure that the backlog of active requests can be processed within two weeks, the backlog size should be maintained near 1000 requests.

To insure equitable request processing with low variance, chronological priorities should be assigned.

Finally, because of the Correspondence Unit's dependence on other departments and clinics, a two-week processing cycle is not feasible for all requests. Thus, management objectives should be to achieve a decreased processing cycle with low variance for all requests and to attain a two-week cycle for those requests with medical records located in file.

REFERENCES

1   J. Thomas Ratchford in *Proceedings of the 31st National Conference on the Advancement of Research,* Denver Research Institute, Denver, 1978, pp. 59–60.

2   Ratchford, op. cit., p. 61.

3   William Bevan in *Proceedings of the 31st National Conference on the Advancement of Research,* Denver Research Institute, Denver, 1978, p. 32.

4   Ratchford, op. cit., p. 64.

5   Wesley A. Kuhrt in *Proceedings of the 31st National Conference on the Advancement of Research,* Denver Research Institute, Denver, 1978, pp. 15, 17–18.

ADDITIONAL READING

Aristotle, *The Rhetoric of Aristotle,* translated by Lane Cooper, Prentice-Hall, Englewood Cliffs, N.J., 1932, 1960. Another good edition of this work is *The "Art" of Rhetoric,* translated by J. H. Freese, Harvard University Press, Cambridge, Mass., 1926, 1967.

Wayne Booth, *Modern Dogma and the Rhetoric of Assent,* The University of Chicago Press, Chicago, 1974.

S. Michael Halloran, "Technical Writing and the Rhetoric of Science," *Journal of Technical Writing and Communication* **8**(2):77–88 (1978).

James A. Kelso, "Science and the Rhetoric of Reality," *Central States Speech Journal* **31**:17–29 (1980).

Carolyn R. Miller, "A Humanistic Rationale for Technical Writing," *College English* **40**(6):610–617 (1979).

Steven Toulmin, *The Uses of Argument,* Cambridge University Press, London, 1958.

Richard E. Young, Alton Becker, and Kenneth Pike, *Rhetoric: Discovery and Change,* Harcourt, Brace & World, New York, 1970.

6

ORIENTING THE NONSPECIALIST
The Foreword or Introduction and Summary

In the real world, you usually communicate about things that are important to you: the proposal you are trying to get funded, the project you worked on for 6 months, a dangerous condition you have just discovered. Unfortunately, as discussed in Chapter 4, you often communicate to audiences who know little about your subject, who may perceive it as unimportant or irrelevant, and who include managers important to your career and to the success of your projects. This situation poses a problem: how do you begin a communication so that the uninformed audience will see its relevance and give it some attention? This chapter addresses that question by presenting a four-step approach for orienting an audience that can be used to begin reports, letters, papers, and oral presentations.* It results in the Foreword and Summary of a technical report or in the Introduction and Summary of a letter, paper, or oral presentation. In particular, this chapter describes

1   A general (or "full") form for posing problems

2   A compressed (or "short") form for posing problems, often found in technical reports

3   Criteria for choosing between the full and short forms

4   A form for stating solutions

6.1
A GENERAL FORM FOR STATING PROBLEMS: FOREWORDS AND INTRODUCTIONS

We all pay more attention to things that interest or seem important to us. One powerful method for developing attention in an audience has been explored by the psychologist Leon Festinger[1] and his coworkers. They argue that the mind—anyone's mind—sometimes finds itself with a sense of unease

---

*We would like to note the influence of our colleagues Dwight W. Stevenson and J. C. Mathes and our ex-colleague Richard E. Young on the evolution of this chapter and the thinking behind it.

about some conflict or inconsistency between two things it believes to be true. When such a conflict or inconsistency occurs, the mind wants it eliminated to avoid chaos, worry, and psychological discomfort. (This is the equivalent of wanting to take off your coat to reduce the physical discomfort of being too hot.)

You can use this observation to catch the audience's attention at the beginning of your communication, in the Foreword of a technical report or the Introduction of a letter or speech. You can introduce an important problem or conflict for the audience and then allow your communication to solve it. Once you've given your audiences a problem they perceive as relevant, they should be willing to hear what you have to say.

Step 1: Introducing a Problem

Richard Young, Alton Becker, and Kenneth Pike[2] have argued that a problem can be seen most easily if it is formed of two directly conflicting terms joined by a word signaling the conflict. The conflicting terms may be perceptions, facts, values, expectations, or beliefs, as indicated in Figure 6-1. Frequently, a value, expectation, or belief in our mind conflicts with an external perception or fact. This pattern is illustrated at the bottom of Figure 6-1, where a belief (the A term) conflicts with a fact (the B term):

> *Our new stereo system must appeal to the teenage market* (belief, A term). *However, it apparently does not appeal to this important market* (perception, B term).

This same pattern appears in the following Introduction from an argument that doctors "are slowly but surely extending cancer patients' lives."

FIGURE 6-1 Outline of a Foreword or Introduction by problem definition, step 1.

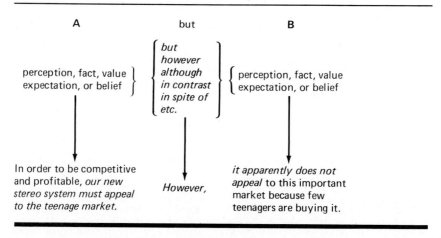

The A term of the problem statement poses the belief or assumption that an individual scientist will find the cure for cancer; the B term presents the conflicting observation that "science rarely works that way." Note that the author starts with an assumption that she believes the audience will readily share: an individual scientist will find the cure for cancer. Note also that she is dealing with a topic—finding a cure for cancer—in which the audience is probably already interested.

IS THE GREAT CANCER CURE ALREADY WITH US?
Since the National Cancer Act of 1971, there has been a continual hubbub about conquering cancer. The public has been led to believe that some morning it will wake up and newspaper headlines will glare: Cancer Cure Found! *The cure, of course, will have come about because some smart or lucky scientist anguished away days and nights until he found the key.*

But science rarely works that way. Breakthroughs come in bits and pieces. And it's exactly in this slow, undramatic way that cancer scientists have greatly increased the number of long-term or even permanent cancer remissions since cancer therapy was first tried during the 19th century. During the 1930's only one out of five cancer patients could expect a remission for five years or perhaps longer. Then one out of three could expect such a remission. Now the rate is approaching one out of two patients. And some of these patients will go on to live out their natural lifespan.[3]

This conflict between a value, expectation, or belief and a perception or fact appears again as the writer of the following Foreword sets up a problem for the managers of The Belle River Power Plant. Note that the author indicates the magnitude of the problem here—replacing coal mills is serious.

The Belle River Power Plant Unit #2 has had over forty years of on-line operation. *Although Unit #2 is still functioning, its increasing demand for both major and minor repairs indicates that its two existing coal mills will need to be replaced soon.* Thus, you requested that we solicit and evaluate bids for the design and fabrication of the two new coal mills. We understand that if you concur, our recommendation will be used as a basis for awarding a purchase order for the design and fabrication of the two new mills. Since the guidelines published by the Public Service Commission require that at least three bidders be considered for such a project, we performed a Dun and Bradstreet analysis of assets and project capabilities on every bidder and then solicited and evaluated bids from the best qualified bidders:

1) American Coal Mill Corporation

2) Babcock and Willis Incorporated

3) Western Electric Incorporated

The following report summarizes our recommendation and supplies the supporting material generated in developing our recommendations.

This author has definitely chosen to phrase the problem statement to catch the audience's attention: managers are concerned that their plants run smoothly and not break down.

Sometimes a good problem statement merely needs to state the A-but-B situation and then move on, as illustrated in the two examples above. However, sometimes a good problem statement needs to state the A-but-B situation and then go on to give details about the A or the B or both. This is illustrated in the following example:

A NEW PUZZLE IN PHYSICS: THE "COSMION"
The latest cherished principle of subatomic physics to be threatened by exper- iment is the charge independence of the strong interaction. Observations that violate it are reported in the December 30 *Physical Review Letters* as a result of work done partly at the Nuclear Research Center at Demokritos near Athens.

The strong interaction is the force that holds atomic nuclei together. It acts equally upon the electrically charged proton and the neutral neutron. *This independence of charge was crucial in convincing physicists that in the binding of the nucleus they were dealing with a new kind of force* and not a manifes- tation of electromagnetism as early thoughts on the subject had tended to suppose. *Now the charge independence doesn't look as complete as it once did* thanks to the present work. The work was done by T. E. Kalogeropoulos of Syracuse University and the Demokritos center and nine colleagues from the two institutions.

In the experiments, hydrogen and deuterium nuclei were bombarded with antiprotons to see what happened as the antiprotons met the protons in the nuclei and annihilated each other. Out of such annihilations certain debris is expected, including pi mesons and photons. The basic experimental finding is that there are too many photons, surprisingly too many, almost a whole extra photon per annihilation. The interpretation of this seemingly simple fact leads down a number of important paths in subatomic physics.[4]

At other times a good problem statement needs a line of reasoning leading up to the A or B statement. This is illustrated in the second paragraph of the example immediately above and in the following example from a student paper:

THE EMERGENCE OF GOD-FIGURES IN SCIENCE FICTION
During the twentieth century, man has enchained himself with a world of sci- ence, technology, machines and inventions. Arising with this philosophy is the feeling the universe is a machine. Man, who portrays himself as chief engineer, should have every right to probe, study, tinker with and eventually control the mechanism of the world. As scientific man learns more tricks and sheds his old "religious crutch," he is better able to rationally exercise control over nature. *Hence, one would expect science fiction, in keeping with this trend, to depict man as conqueror, leader and master of his future, free of the influence of "gods." But this often is not the case.* Much of science fiction is devoted to the

inferiority of man or his inability to execute his free will. Instead of futuristic men finding total fulfillment in expressions of individuality and personal creativity, science fiction often depicts men as passively subservient to a supreme organization as in Orwell's *1984* or Zamiatin's *We.*

The organization with its god-head is idolized and obeyed more fervently than any god ever known to religion. At other times, the controlling figure is a less tangible supernatural entity as in Lindsay's *A Voyage to Arcturus.* No matter what the form, even in science fiction man finds himself at the control of the gods.

If a communicator uses the A-but-B form to identify a problem or conflict for the audience, the communicator must be careful that the A and B terms really conflict or else the whole introduction will not make sense.

EXERCISE 6-1   The following passages were chosen from a variety of scientific and technical sources. See if you can identify the problem stated or implied in each passage, and then try to strengthen or clarify the problem statements. (Although all of these passages are understandable, many would benefit from a sharper problem statement.) Be sure that the magnitude of the problem is clear. Note that several of the passages will be outside your area of specialization; they were selected to help develop your "feel" for the nonspecialist audiences with whom you will often need to communicate.

A   **SUBJECT: ALUMINUM PITTING CORROSION:**
            A PROGRESS REPORT

Foreword: Aluminum is a desirable material for automobile cooling systems because its use provides less corrosion and up to a one-third weight savings and a one-third cost savings over conventional copper-brass cooling systems. However, aluminum exposed to chloride ion solutions undergoes localized corrosion, or pitting.

The reaction of chloride ion with aluminum metal is a well-documented phenomenon, but the interaction between the chloride ion and the passive oxide or hydroxide film which always exists on aluminum surfaces is yet unclear. This interaction is currently being studied by potentiostatic testing to determine the effect of chloride ions on the initiation of pitting corrosion. The purpose of this report is to present the results of these potentiostatic tests.

B   **CREATIVITY**

As engineers, it will be our job to find new and better ways to solve many problems. Since creativity increases one's ability to solve problems, it is logical that a more creative individual will make a better engineer. But most of us do not inherently possess a great deal of creativity. Does this mean that we are doomed to be nothing more than average engineers, hindered by our inability to think creatively? Not at all. Many researchers in this field believe that creative thinking is simply a mental process and that this process can be learned by any intelligent person. In fact, many researchers believe that most people possess a great deal of creative potential which they simply fail to develop.

So if a creative engineer is a good engineer, and if this creative ability can be learned, an important question for many engineers is: how can a person learn to increase her or his ability to work creatively? To learn how, one must first understand what creativity is.

C   SUPERGRAVITY AND THE UNIFICATION OF THE LAWS OF PHYSICS
A catalogue of the most basic constituents of the universe would have to include dozens or even hundreds of particles of matter, which interact with one another through the agency of four kinds of force: strong, electromagnetic, weak and gravitational. There is no obvious reason why nature should be so complicated, and perhaps the most ambitious goal of modern physics is to discover in the diversity of particles and forces a simpler underlying order. In particular, a more satisfying understanding of nature could be achieved if the four forces could somehow be unified. Ideally they would all be shown to have a common origin; they would be viewed as different manifestations of a single, more fundamental force.

In the past 50 years remarkable progress has been made in identifying the elementary particles of matter and in understanding the interactions between them. Of course, many problems remain to be solved; two of the most fundamental ones concern gravitation. First, it is not understood how gravitation is related to the other fundamental forces. Second, there is no workable theory of gravitation that is consistent with the principles of quantum mechanics. Recently a new theory of gravitation called supergravity has led to new ideas on both these problems. It may represent a step toward solving them.[5]

D   LIFE MAY BE HAZARDOUS TO YOUR HEALTH: I
The message we seem to be getting from the media is that our health, if not our very lives, is endangered by the very same technology that supports our existence. The series of warnings is ceaseless: DDT, PCB, polyvinyl chloride, nitrous oxides in the air, unsafe automobiles, radiation, food poisoning, sweeteners, contraceptives, fallout, fluorocarbons, noise pollution, x-rays, cosmetics, . . ., and on and on. One might begin to wonder how we ever managed to survive past the age of 25. Controversies begin and crises of this type often arise because of the uncertainties and complexities of the concept of safety.[6]

E   LIFE MAY BE HAZARDOUS TO YOUR HEALTH: II
As scientists and engineers, one of our daily concerns is with the concepts of risk and safety. It makes no difference what aspect of engineering you eventually become involved with, be it product engineering, manufacturing, quality assurance, sales and marketing, maintenance, or product supervision. In the end, something for which you have been responsible will have the potential for causing harm, big or small. It is essential that we, the bastions of technology, have a thorough understanding of risk, safety, what they involve, and how they are determined.

At one time or another we've all heard the doomsayers who preach that our technology has brought us to the brink of destruction. The solution they offer is full retreat. Before jumping to any such rash conclusions, let's step back and take an historical perspective. To make the comparison fair, we'll stick to modern times and not go back to an era when life was completely miserable.[7]

F   SCIENCE FICTION: A WARNING TO HUMANITY

From the beginning of the human race, through prehistoric times to the present, man has always sought power. He has wanted to control himself and his environment. He has wanted to be able to predict the future. In short, he has wanted to be omnipotent. The examples of this desire, this greed, to become an ideal, perfectly powerful being are many. We have, in the tale of Adam and Eve, a desire to become one's own master. More recently, we have men such as Caesar and Hitler, who wanted to control the world. Even today we see this quest for power, not only in political leaders, but in everyday, common people. People are always trying to make more money, to learn more, in order that they may improve their status.

But this quest for power is not without effect. Things go wrong, people are hurt, and the human condition deteriorates. You may ask "What can be done about this situation?" My answer is that man must be more careful about what he does; he must more closely analyze his situation, what he is planning on doing, and what its effects will be. In this paper, I will present the role that science fiction plays in showing us why we should be more careful about what we do, and what the consequences are if we are not careful.

G   THE AUTOMOBILE: FOR BETTER OR WORSE

Cadillac's chief engineer, Robert Templin, is hardly alone among his colleagues in the industry when he asks, "How much pollution is too much? How much fuel efficiency is enough? Who will pick up the tab for a zero-risk, zero-pollution car?" Never in its history has Detroit been faced with so many problems. Henry Ford himself might have opted for horse-drawn buggies instead of horseless carriages in the same situation.

On the other hand, the search for answers to perplexing problems traditionally has led to inspired and sometimes unorthodox solutions. Necessity is the mother of invention—and for the electronics industry, at least, Detroit's difficulties in meeting regulations have opened doors to electronic applications in cars that otherwise might not have opened so soon, if at all.

This special issue on the automobile reflects both the controversies about Government regulations for cars and the inroads being made by electronics. For instance, author Stork says, on page 51, "EPA is satisfied, from its analyses and studies, that the emissions standards enacted by the Congress can be met without loss in fuel economy, and that even the full statutory emissions standard for nitrogen oxides can eventually be met on that same basis." Yet, on page 47, authors Heinen and Beckman assert, "The simple truth is that there is very direct interaction between emissions and fuel economy. Probably the clearest example of that interaction is the fact that automobiles equipped to meet California's tight emissions-control regulations have consistently demonstrated about 10 percent poorer fuel economy than have comparable cars equipped to meet the less stringent Federal U.S. standards."[8]

H   PASSIVE COOLING SYSTEMS IN IRANIAN ARCHITECTURE

The internal environment of a modern building can be a comfortable one no matter how uncomfortable the external environment is. In general the reason is that energy is freely spent to heat or cool the building. In earlier times, however, when energy was not so readily available and machines such as air conditioners did not exist, the designers of buildings had to rely on other stratagems to maximize the comfort of the internal environment. For example, the traditional architecture of many cultures in climates where the temperatures are uncomfortably hot during the day and uncomfortably cool at night features buildings with thick walls of brick or stone. Such walls are both insulators and reservoirs of heat, so that during the hotter hours of the day the flow of heat from the external environment to the internal one is retarded and during the cooler hours part of the heat stored in the walls warms the internal environment and the rest is lost to the external one. The net result is a flattening of the temperature-variation curve inside the building. In a period when the energy costs of buildings are being intensively reevaluated, such stratagems clearly merit close consideration.

In Iran certain traditional building designs achieve more than a flattening of the temperature curve; they circulate cool air through the building and can even keep water cold and ice frozen from the winter until the height of the long, hot summer of the country's arid central and eastern plains. They do so without any input of energy other than that of the natural environment; hence they can be characterized as passive cooling systems.[9]

EXERCISE 6-2   Survey your experience for a hobby or some other subject you know something about. If you have worked at a job or done an experiment in a laboratory course, then pick a particular problem or experiment you could talk about. For your topic, write an Introduction by problem statement for someone who knows nothing about it. Be sure to define your audience and to keep its interests in mind in setting up the introduction.

Steps 2 and 3: Identifying Your Strategy on the Problem and the Purpose of Your Communication

Once you have established an important problem for the audience, you need to show that your communication addresses this problem. This is done by

1   Indicating the missing information you're providing and the strategy you're taking on the problem (you do this by stating the question or task you're addressing)

2   Announcing the purpose of the communication

For example, let us reconsider the example in Figure 6-1:

Our new stereo system must appeal to the teenage market.

However, it apparently does not appeal to this market.

After having stated the problem directly, we could have tried to solve it by asking questions such as the following:

What is wrong with teenagers?

What is wrong with the stereo system?

Why aren't teenagers buying the stereo system?

Where is it being sold?

Who is buying the stereo system?

If we had asked the first question, we would have assumed that something is wrong with the teenagers and have committed ourselves to this point of view. It may be that something is wrong with them, but it may be that something else is wrong, something we would never see given this perspective. The same complaint could be advanced about the second question. However, the third, fourth, and fifth questions would not have committed us to a limited point of view and thus would probably have produced better information. If, for instance, the cause of poor sales is poor advertising, the last three questions would be much more likely to guide us to this cause.

This example illustrates two important properties of such questions. First, since at the investigation stage we don't want to close off fruitful lines of inquiry, the questions recreating this stage should be open-ended, that is, they should not limit us unnecessarily and should usually begin with such words as *who, what, where, when, why, how,* or *to what degree.* Second, the questions we ask affect the solutions we reach; in fact, the questions define our *strategy* for solving a problem. For instance, all the questions above point to potential causes; they indicate that our strategy for finding a solution is to find the cause of the problem. A different set of questions would have defined a different strategy. If we had asked the following questions, for instance, our strategy would have been to improve the product:

How can we make the stereo more cheaply?

How can we improve the performance of the stereo?

How can we reduce the size and weight of the stereo to make it more portable?

Can we make the stereo look "richer" and more elegant?

Note that if the real problem had been poor advertising, the strategy of improving the product would probably have been ineffective.

How do we define our strategy and our purpose in an Introduction? Let's say that we have explored the question "Why aren't teenagers buying the stereo system?'' and decided that the reason was poor advertising, appearing at the wrong places and at the wrong times. If we were to report on this situation to a vice president and a manager at the stereo company, we might produce an Introduction similar to one of the following:

VERSION 1

In order to be competitive and profitable, our new stereo system must appeal to the teenage market. However, it apparently does not appeal to this important market since few teenagers are buying it. What caused this situation? It appears that a poor sales campaign is the only major cause.

VERSION 2

In order to be competitive and profitable, our new stereo system must appeal to the teenage market. However, it apparently does not appeal to this important market since few teenagers are buying it. I was asked to investigate this situation and to identify the cause. This report will argue that a poor sales campaign is the only major cause.

Notice that each of these has a clear problem statement followed by a statement of question or task and a statement of purpose. Notice also that in version 2 the statement of task is merely an unstated question. "I was asked to investigate this situation and to identify the cause" means "I was asked to find out what caused this situation." The process of writing an Introduction just discussed is summarized in Figure 6-2.

Another example of an Introduction by problem statement follows, written by a student to administrators at his university. He hoped that his report, from which this Introduction was taken, would help decrease or eliminate the use of the test mentioned.

In 1962, the Opinion Attitude and Interest Survey (OAIS) Test, more commonly known as the "raw carrot" test, was first introduced to the University of Michigan as part of the testing program administered during an entering student's orientation. It is a psychological test, testing for motivation, vocational aptitude, vocational interest, and certain factors relating to personality. This exam is supposed to be used as a placement device by the counselors in chemistry and math courses. However, if one surveys the counselors on campus, the test scores are not used in this manner; in fact, they are not used at all. Thus, the largest school at the University—Letters, Sciences, and Arts—stopped "coercing" its students to take this examination. Some schools have followed, but others have not—notably, Nursing, Education, and Engineering.

If the results are not used as the test examiners have projected, then why should the exam be forced on the non-LSA students in Nursing, Education, and Engineering? There is no compelling reason. In fact, the test should not even be encouraged by these colleges for large groups of students; it has internal problems and can be misused and misinterpreted, especially in com-

1 Explicit statement of problem
   • indicate magnitude of problem

2 Questions or statement of task
   • identify strategy
   • indicate missing information

3 Purpose of communication

FIGURE 6-2 Outline of Introduction or Foreword,
steps 2 and 3.

bination with other, less volatile sources of information at the counselors' disposal.

Notice that this author has considered his audience carefully in phrasing the problem statement. He has emphasized the fact that the test scores are not used and implied that the university is wasting its resources by continuing to administer the test.

The author of the above report on the OAIS also had three other complaints about the test, which he developed in his report: (1) that it was an invasion of privacy, (2) that it was poorly designed and thus gave poor results, and (3) that it wasted valuable student time during orientation week.

EXERCISE 6-3 Define some audiences for which each of these three complaints could be used in an A-but-B problem statement. You will thus have at least three audiences and three problem statements.

EXERCISE 6-4 Go back to the passages quoted in Exercise 6-1. See if you can identify the question or task and the statement of purpose in each one. Then see if you can add a question or task and purpose for the introduction you wrote in Exercise 6-2.

6.2

A SHORT FORM FOR STATING PROBLEMS

Section 6.1 presented a general form for starting letters, reports, or speeches based on an A-but-B problem statement. This format is commonly used and important. However, if you look at much communication in science and technology, you will find examples which seem to violate the A-but-B format because they seem to be missing either the A term or the B term. Consider, for instance, the following example from a report by a Research Forester to the Superintendent of a State Forest Nursery (notice how the problem statement addresses the concerns of the Superintendent and indicates the magnitude of the problem):

FOREWORD

During my visit to your nursery on September 15, 1979, you and I observed the high mortality of the two-year-old red pine seedlings in bed 19. As you stated, this would lead to a production shortage in 1980. You requested my help in finding the cause of this problem so that it could be corrected in the near future. The purposes of this report are 1) to explain that parasitic nematodes are the probable cause of the high seedling mortality and 2) to recommend a solution to the problem.

What is happening in this example and others like it is that one term in the A-but-B format is unstated, but assumed. It's assumed because it's believed by all members of the audience (the Superintendent and other employees at the nursery) and so obvious that saying it would seem ridiculous. For instance, in the example above, the author merely states the B term: "you and I observed the high mortality of the two-year-old red pine seedlings . . . this would lead to a production shortage in 1980." The author assumes any reader of the report would know the A term: one does not want a high mortality rate of red pine seedlings or a production shortage in 1980. Just stating that they might occur puts the ideas of high mortality and production shortage in clear conflict with a reader's deeply held values and assumptions that they should not occur.

The "short-form" problem statement just described is represented in Figure 6-3, where brackets have been placed around unstated but assumed material. The short form appears frequently in scientific and technical communication, since many of the problems scientists and technical people write about arise from commonly held assumptions. Almost everyone wants and is interested in

Efficiency

Low cost

Competitiveness in cost and performance

FIGURE 6-3 Outline for a short-form problem statement.

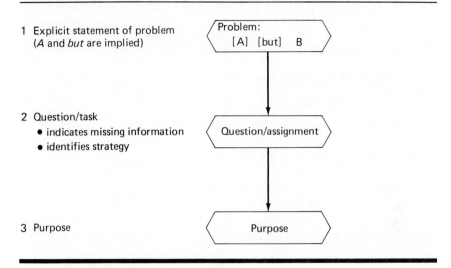

1 Explicit statement of problem
   (*A* and *but* are implied)

2 Question/task
   • indicates missing information
   • identifies strategy

3 Purpose

Products of good and even quality

Freedom from pollution violations, labor unrest, and legal action by consumers or the government

Elimination of wasted time, energy, or money

Simplicity rather than complexity in organizational structures and product design

When such values are used in a problem statement, they are usually not stated directly. However, they are there and are often the values needed to catch the interest of managers and others not directly involved with a project. Interestingly, these values also provide the context needed by someone outside a project to understand its real importance and purpose. Look back, for instance, at the example on mortality in red pine seedlings. Could you really understand why someone was worried about parasitic nematodes (the probable cause of the high seedling mortality) without calling on the values and assumptions implied in the Foreword?

We might consider another example of a short-form problem statement from a report written by a civil engineer to the Chief Engineer of a manufacturing company. The readers of this report might well include managers, architects, financial analysts, constructors, and even a vice president. The author rightly assumes that any reader would know that the manufacturing company does not want to damage its building foundations. Again, this is so obvious that an author would not state it, but it is unquestionably there in the mind of the reader.

On January 22, 1980, you requested the analysis of two samples of sand to see if they could be used as a foundation for several high-speed presses. You expressed concern that this sand, which is below the punch press foundations, might settle due to the vibration of the presses, thus damaging your building foundations. Along with the two samples, the following data was forwarded:

Void ratio of white sand (Ottawa Sand)   .65

Dry unit weight of black sand (Turkish Emery)   118 lb/ft^3

The purpose of this letter is to recommend that neither sand is sufficient for use as a foundation for high-speed punch presses. The vibratory loadings transferred from the machines will cause the foundations to settle.

EXERCISE 6-5 Examine the following Forewords and Introductions from technical articles and reports. Each is missing an overt statement of one of the conflicting terms of its problem statement. See if you can identify or easily reconstruct both the A and B terms of each problem statement.

A   SUBJECT: VACUUM FILTER SLUDGE CAKE:
ACCURACY OF PRESENT PRODUCTION DATA
Foreword: Slowdowns and necessary repairs have occurred in the incineration area of the Sewage Treatment Plant, costing two million dollars last year alone. These problems may have occurred because of inaccurate data on the amount of sludge cake being produced and delivered to individual furnaces. Thus, Dennis Moore, the Chief Project Engineer, requested that I investigate our data on sludge cake production and suggest improvements in monitoring it if those are necessary. The purpose of this report is to document the unreliability of our present monitoring technique and to recommend a better one.

B   THE SEARCH FOR LIFE ON MARS
Is there life on Mars? The question is an interesting and legitimate scientific one, quite unrelated to the fact that generations of science-fiction writers have populated Mars with creatures of their imagination. Of all the extraterrestrial bodies in the solar system, Mars is the one most like the earth, and it is by far the most plausible habitat for extraterrestrial life in the solar system. For that reason a major objective of the Viking mission to Mars was to search for evidences of life.[10]

C   SUBJECT: PROPOSED TREATMENT SYSTEM
FOR CITRUS PROCESSING WASTES
Foreword: In your letter of July 7, 1971, you asked me to suggest a treatment process for the wastewater from your new citrus processing plant. You stated that any treatment process selected should produce an effluent in compliance with federal standards and be economical to operate and maintain. Consequently, I have compared several treatment alternatives based on your data

and criteria. The purpose of this report is to recommend a process for economically and efficiently treating your citrus processing wastes.

D   DRIP IRRIGATION

Some 40 years ago Symcha Blass, an Israeli engineer, observed that a large tree near a leaking faucet exhibited a more vigorous growth than the other trees in the area, which were not reached by the water from the faucet. Blass knew that conventional methods of irrigation waste much of the water that is applied to the crop, and so the example of the leaking faucet led him to the concept of an irrigation system that would apply water in small amounts, literally drop by drop. Eventually he devised and patented a low-pressure system for delivering small amounts of water to the roots of plants at frequent intervals. The technique, as developed by Blass and subsequently refined by him and various manufacturers, consists in laying a plastic tube of small diameter on the surface of the field alongside the plants and delivering water to the plants slowly but frequently from holes or special emitters located at appropriate points along the tube. The concept, which is now called drip irrigation or trickle irrigation, has gained wide acceptance, proving to be particularly valuable in areas that are arid and have high labor costs. An unforeseen benefit is that the system works well with water that is highly saline, as water in arid regions often is.[11]

6.3

GUIDELINES FOR CHOOSING BETWEEN FULL-FORM AND SHORT-FORM PROBLEM STATEMENTS

Sections 6.1 and 6.2 describe two related forms for problem statements, the full form and the short form. The purpose of this section is to summarize guidelines for choosing one form or the other.

As already suggested, you should use the full form of the problem statement when your audience or part of your audience might not know clearly and obviously what each term is. You may choose the short form when one term is so clearly known and believed by all members of your audience that it does not need to be stated to make the problem clear. The possible choices are summarized in Figure 6-4.

For instance, in the competitive business system of the United States, the shared values include those presented in Section 6.2: the desire for low cost; efficiency; good and uniform product quality; freedom from legal prosecution, labor disputes, and pollution violations; simplicity; and elimination of wasted time, energy, or money.

In a scientific or technical field, shared knowledge, beliefs, and expectations are based on the experience common to members of the field. It would be impossible to give an inclusive list of all the shared values that could be unstated in some particular situation. Thus, a communicator must consider various audiences for a communication and produce an Introduction aimed at the least knowledgeable member of the audience.

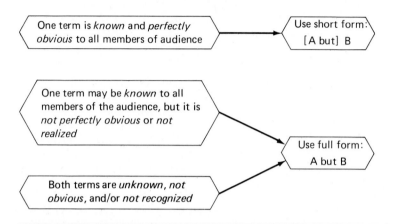

FIGURE 6-4   Guideline for choosing between full- and short-form problem statements.

6.4

INTRODUCTIONS AND FOREWORDS: A REVIEW

Communication about a project can be presented in a variety of formats and to a variety of audiences. The most common of the formats are the technical report, the letter, and the scientific or technical article.

The Foreword of a Technical Report

Technical reports may be aimed at specialists, at nonspecialists, or at both at the same time. They are usually read by a wide variety of audiences—managers, sales people, technicians, financial analysts, specialists in the field, and sometimes even the public. Thus the beginning of such a document must orient audiences of varying backgrounds, experiences, values, and assumptions. To make sure that all readers understand the problem set up at the beginning, the writer of a technical report must write to the least knowledgeable reader. This is the nonspecialist, especially the manager unfamiliar with and uninterested in the details of a particular project.

The technical report orients its variety of readers in its first two sections, the Foreword and the Summary. As with other types of introductions, the purposes of the Foreword are

1   To catch the audience's attention; to place the report in a context so that readers can see how it fits in with other communications and with overall goals; to tell *why* the project was done and *why* it is important

2   To quickly orient the audience to the subject of the report; to define *what* was done; to indicate the missing information the report provides

(these are often covered with an explanation of the writer's technical assignment)

3   To define the purpose and focus of the report

The purposes of the Summary are

4   To quickly present the main results of the project

5   To quickly present the important recommendations and implications of the project

These purposes are summarized in Table 6-1. For illustrations of Forewords, you should reread the many Forewords already presented. The Summary will be treated later in this chapter.

The Introduction of a Letter

Letters have many of the characteristics just mentioned for technical reports. They may be aimed at specialists or nonspecialists or both at the same time; they may be read by a wide variety of audiences; they should be written to the least knowledgeable audience; and, depending on their audiences, they may begin with full-form or short-form problem statements. For examples of letter Introductions, see the letters discussed in Chapter 11, "The Job Letter" and "The Business Letter."

The Introduction of a Scientific or Technical Article

Articles aimed at nonspecialists may begin with an Introduction using either the full-form or the short-form problem statement. In either event, the writer must ensure that both terms of the problem statement are clearly understood.

TABLE 6-1 OUTLINE OF THE FOREWORD AND SUMMARY
IN THE TECHNICAL REPORT

FOREWORD		SUMMARY	
PURPOSE	FORM	PURPOSE	FORM
Define context of report to catch audience's attention; tell *why* project was done and *why* it is important	[A but] B	Present main results Present recommendations and implications	Results Recommendations and implications
Indicate missing information; define subject of report; define *what* was done	Assignment or question		
Define *purpose* and focus of report	Purpose and focus		

Several examples of article Introductions aimed at nonspecialists appear in the exercises in this chapter.

6.5

THE SUMMARY

The Foreword and Summary at the beginning of a technical report are the structural features most obviously responsive to the busy managerial or non-expert reader. The Summary provides a compact statement of results, conclusions, and recommendations to help managerial readers make decisions: it tells what the writer discovered, the implications of the discovery, and the recommendations for action based on the writer's special knowledge—all in terms that the nontechnical manager will find useful.

Some of the questions that a managerial reader will have about a project or study[12] are the following:

1   What is the project's *importance* to the company? Its areas and extent of application?

2   What will it *cost*?

3   Are any *problems* projected because of it?

4   Are there any *implications* to the company from it? Any *more work* to be done? Any *resources* required or involved (people, facilities, equipment)? Any *priorities* to be added or changed?

5   Are there any important *dates* or *deadlines* associated with it?

6   Are there any important *recommendations* for future action which the writer could make based on his or her special expertise?

Writers should try to answer these questions in the Summary because they are particularly qualified to do so. They often know more about the answers than anyone else. Further, they are hired to analyze and evaluate data, not just to present it. If they make no effort to analyze their project in the larger context of the organization, they are not really doing their job. Finally, by answering such questions for managers, writers can look exceptionally competent. Relatively few technical people properly address these issues, and those who do will seem unusually useful to the organization.

What does a Summary that addresses these questions look like? For a sample, you might look at the Summary in Figure 6-5. It comes just after the Foreword, which has posed a problem: there have been frequent delays in an experiment because of the excessive time required to pump down a vacuum system. Since these delays are undesirable, the writer has been asked to reduce them. The Summary reports the writer's solution: buy a gate valve for the vacuum pump. The Summary also tells why this is a good

FIGURE 6-5 Movement from generalizations and claims to data and support in a short informal report: example 1.

Watson Telephone Laboratories, Inc.
Murray Hill, N.J. 07074

Date: February 5, 1979

To: Dr. J. C. Smith
            Manager, Low-Energy Research Group

From: Catherine Doe
            Mechanical Systems Engineer
            Positron Research Group

Subject: Positron Beam Experiment: Reducing Delays

Foreword: In your letter of January 23, 1979, to Dr. D. Newman, Director of the Positron Research Group, you expressed concern about the numerous delays in the positron beam experiment. Dr. Newman attributes most delays to the ten hours required to pump the vacuum system down from atmospheric pressure to an acceptable operating level and has asked me to investigate means of reducing these delays. This report recommends the purchase and installation of a gate valve for the Vac-Ion vacuum pump as the most appropriate method for reducing the delays.

Summary: The purchase of a gate valve for the Vac-Ion pump was found to be the optimal solution to the delay problem in the positron beam experiment. Addition of the gate valve will reduce evacuation time to three or four hours, cost no more than $600, and present no installation problems. The use of a larger fore pump was seen as the only alternative approach, but this would require extensive modification of the vacuum system, cost approximately $1100, and give no improvement in the evacuation time.

Details: The excessive time required to obtain a suitable vacuum, 10^{-7} torr or less, is primarily due to contamination of the ion pump. When the vacuum system is opened to the atmosphere, the entire inner surface of the system and the ion pump itself become contaminated. The contaminated surface then outgasses considerably upon evacuation, keeping the pressure high for more than ten hours. The two possible solutions to this problem are to avoid the contamination altogether or to decontaminate more quickly.

The proposed solution would avoid contamination by the installation of a gate valve on the Vac-Ion vacuum pump. This would allow the ion pump to remain in operation even while the rest of the system is open to the atmosphere, thus preventing contamination of the pump. This would be a major improvement, since only the relatively small surface area of the rest of the vacuum system would become contaminated, and evacuation time would be reduced to three or four hours. A suitable gate valve costs $500 to $600.

The rejected alternative, the purchase of a larger fore pump, would speed the decontamination of the vacuum system, but cost too much. Replacing our present 120 ft.^{3}/min. fore pump by an 800 ft.^{3}/min. pump would reduce the evacuation time to about 4 hours. However, an 800 ft.^{3}/min. fore pump costs approximately $1100, and its installation would require substantial modification of the vacuum system.

Thus, I recommend the purchase and installation of a gate valve as the best method for reducing delays in the positron beam experiment.

If you have any questions, I would be happy to answer them.

solution—it will reduce evacuation time to acceptable levels, cost no more than $600, and present no installation problems. Notice that these advantages address the manager's concerns. Before a manager could approve the purchase of the gate valve, she or he would need to know the cost (only $600), the benefits (reduction of delay), and the implications (no installation problems). In addition, the phrase *no installation problems* implies other useful information: that there should be no more problems with the unit and no unusual work or resources required.

A second summary, shown in Figure 6-6, illustrates many of the same points. Again the Foreword has defined a problem: Orange Grove Products needs a method for treating wastewater from its new citrus processing plant, a method which meets criteria of effectiveness and cost. The Summary provides a solution: an aerated lagoon will meet both criteria better than other potential solutions. The Summary also gives general performance and cost data, the kinds of additional information a manager needs to make a decision. (A manager would not approve a solution which was ineffective or too expensive.)

Framing Summaries for Particular Audiences

Once you have defined the audiences for a report, you need to decide how to convey your information so that the audiences can understand and use it. This can be tricky when different audiences need different kinds of information or the same information presented in different ways.

For instance, consider the following situation. You are a geologist working for a township council composed of two business people, one farmer, one school teacher, and one lawyer. The township has so attracted new industry and residents for the past 8 years that water shortages have developed and future growth will be impossible without an expanded water supply. You have been hired to locate, if possible, new wells to increase the township's water supply. After some effort, you have succeeded and must inform the council of your success. Which version of your results would be most appropriate for the township council?

SUMMARY 1

Six test wells were drilled, four of which penetrated suitable sand and gravel.

1   The relationship of geologic units encountered in Test Wells 1 through 6 is interpreted and presented in Plate 1. Two aquifers were identified; an extremely thick aquifer was encountered in Test Wells 2, 5, and 6 ranging in thickness from 78 to 125 feet, and a deep aquifer was encountered in Test Well 1 extending from 150 to 185 feet. There is theorized to be limited hydraulic connection through fine-grained units between the two aquifers. Some degree of connection between the deep aquifer and upper aquifer was established by the rapid rise in water levels in Test Well 1 observed during a heavy rain in September 1979.

2   A conservative estimate of the yield from the upper aquifer is 2.0 mgd. Aquifer data indicate this could be produced by pumping any combination of two wells among Test Wells 2, 5, and 6 (assuming manganese in Test Well 2 proves acceptable). Ultimate production from a French Creek well field could prove to be in the 3.0 to 5.0 mgd range.

SUMMARY 2

We conducted studies and six test drillings to locate an area in which a reliable, high-quality municipal water well field could be developed. We have located such a site.

Results   Four of the six wells drilled were productive and penetrated sand and gravel suitable for development of municipal wells. Each of the four productive wells showed very good water-producing capability, yielding 300 to 500 gallons per minute.

Conclusions   Production wells completed at the test well sites should more than supply the township's current and future needs. Analysis of test data indicates that such production wells could produce over 2 million gallons per day, twice the township's projected requirements. Further, if the water-bearing sand and gravel are as extensive as we project them to be, yields up to 5 million gallons per day could conceivably be developed.

Unless the reader is a geologist, version 1 will probably not make any sense. It contains unfamiliar terms such as *aquifer* and *mgd*. It also contains statements which sound as though they ought to convey information but don't. For instance, Section 2 states, "A conservative estimate of the yield from the upper aquifer is 2.0 mgd." Even if readers guess that *mgd* means *million gallons per day,* they still don't know if 2.0 mgd is a sufficient water supply.

In contrast, version 2 gives the information in a form that nonspecialists can understand. It explicitly states the generalizations the nonspecialist reader needs to know:

1   The geologist has found water.

2   The geologist has found enough water to supply the township's current and future needs.

EXERCISE 6-6   Write a Summary for the reports shown in Figures 6-7 and 6-8. When you have finished, be sure to check your Summary for the important results, conclusions, implications, and recommendations. If you have recently written any other reports, go back now and check their Summaries for appropriateness and completeness. Since Summaries are so important, you should practice writing and rewriting them whenever you have the opportunity.

FIGURE 6-6 Movement from generalizations and claims to data and support in a short informal report: example 2.

To: Ms. J. Jones, Director
           Orange Grove Products
           1135 Halifax Building
           Orlando, Florida 32105

From: Leo Ming Chen, Environmental Engineer
           Camp, Dresser and McKee Inc.
           One Center Plaza
           Boston, Massachusetts 02108

Date: October 29, 1979

Subject: Citrus Processing Waste: Treatment System Proposal

Foreword

In your letter of July 7, 1979, you asked me to suggest a treatment process for the wastewater from your new citrus processing plant. You stated that any treatment process selected should:

1. exhibit performance effectiveness under average and adverse flow conditions

2. exhibit cost superiority in terms of initial cost and yearly spending

Consequently, I have compared several treatment alternatives using the data you have supplied and your criteria as a basis for comparison. The purpose of this report is to recommend a process for economically and efficiently treating citrus processing wastes.

Summary

An aerated lagoon is recommended as the most efficient and economical method for treating citrus processing wastes. Several treatment processes were considered in the selection. These include the activated sludge process, the anaerobic lagoon, and the aerated lagoon. The advantages of the aerated lagoon over the other treatment processes are as follows:

1. The aerated lagoon is the only alternative which could meet the federal pollutions standards under adverse flow conditions. It exhibits significantly better performance under all conditions through more consistent BOD reduction and higher organic loading potential.

2. The aerated lagoon affords significantly lower initial and yearly costs due to its ease of construction, operation, and maintenance. Per lagoon, the estimated initial cost is only $114,000 and the annual operating cost $22,800, approximately half as expensive as the most economical of the other two options.

FIGURE 6-6   *(continued)*

Performance Superiority of Aerated Lagoons

      Aerated lagoons consistently produce a better quality effluent than do activated sludge processes or anaerobic lagoons. Aerated lagoons exhibit better BOD reduction and higher organic loading potential under both average and adverse flow conditions than do either of the other treatment schemes.

1. Superior BOD Reduction by Aerated Lagoons

      The standard for BOD, as published in the Federal Register of July 1, 1979, states that all discharges into receiving streams shall contain no more than 30 mg/liter of BOD. Table 1 shows aerated lagoons with 95% BOD reduction potential to be capable of producing an effluent in compliance with federal standards under both average and adverse flow conditions. Activated sludge processes and anaerobic lagoons on the other hand can only effectively treat wastewaters of average BOD values.

Table 1: BOD reduction comparison of treatment processes

Secondary Treatment System	BOD Reduction (%)	Effluent BOD of Average Strength Flow (mg/liter)	Effluent BOD of Maximum Strength Flow (mg/liter)
Aerated Lagoon	95	9.5 (OK)	28.2 (OK)
Anaerobic Lagoon	88	22.8 (OK)	79.2 (unacceptable)
Activated Sludge	85	28.5 (OK)	99.0 (unacceptable)

2. Superior Organic Loading Potential of Aerated Lagoons

      Seasonal shockloads typical of citrus processing plants are easily handled by aerated lagoons but tend to pose problems for activated sludge processes and anaerobic lagoons. Production within the plant will be a one-shift-a-day operation and may shut down completely on weekends and holidays. The volume of wastewater therefore will fluctuate through the harvesting season, which begins in October and ends in June. A brief performance comparison of how these three systems react to such periods of high loading will show the distinct advantage which aerated lagoons have over the other processes.

    1. Aerated Lagoons: Aerated lagoons, because they are not continuous flow processes, can store wastewater until it can be treated. These lagoons can therefore be designed to handle a shock load of unlimited quantity.

    2. Anaerobic Lagoons: Anaerobic lagoons also exhibit the same storage capabilities as aerated lagoons. However, because of the nature of anaerobic organisms, obnoxious odors are produced at high organic loadings. Hydrogen sulfide and mercaptans are produced at loading values exceed-

FIGURE 6-6 *(continued)*

ing 240-320 kg BOD/1000 cubic meters, a value very typical of the size of plant in question.

3. <u>Activated Sludge</u>: Activated sludge processes can only tolerate organic loading values not exceeding 260-340 kg BOD/1000 cubic meters. Higher loading rates were followed by bulking primarily caused by <u>Sphaerotilus</u> organisms.

Cost Superiority of Aerated Lagoons

Table 2 shows that aerated lagoons cost less than activated sludge systems and anaerobic lagoons in terms of initial capital as well as operational and maintenance (O&M) costs. The initial capital and O&M costs are usually reflective of the amount of equipment required for each system. Aerated lagoons require only turbine-type aerators and therefore cost infinitely less to build and maintain than activated sludge systems, which need several settling basins and numerous pumps. Anaerobic lagoons, although not needing any equipment, have a higher operating cost than aerated lagoons because of the chemicals required in stabilizing the anaerobic bacteria. The estimated cost of these various systems is based primarily on average flow and BOD loadings. More detailed cost tables appear in the Appendix.

Table 2: Cost Comparison of Treatment Processes

Secondary Treatment System	Cost (× $1000 in 1979 dollars)	
	Capital	Annual
Aerated Lagoon	114.0	22.8
Anaerobic Lagoon	212.0	44.8
Activated Sludge	725.0	39.0

Conclusions and Recommendations

My objective in this report was to recommend a process for economically and efficiently treating the citrus wastes from the new Orange Grove Products plant.

The use of an aerated lagoon is recommended if the ultimate process is to produce an effluent which consistently complies with federal standards. Such a system will also save initial capital as well as annual operations and maintenance costs.

FIGURE 6-7 Sample report 1 for analysis.

DATE: June 6, 1980

TO: Meg A. Lith, Manager
             Research Division
             Scoria Mining Company
             Menhir, Mich. 48003

FROM: Michael A. Newton, Research Assistant
             Phoenix Memorial Laboratory
             47-28 Technology Mall
             Atlanta, Ga. 30081

SUBJECT: Sediment Sample Analysis: Report of Investigation

Foreword: On 14 May you requested an analysis of two sediment samples: TG-1 and TG-2. In response to your request, we have performed a neutron activation analysis on each of the samples to determine the major constituent elements in each. This letter concerns the results of the analysis.

Details: The final analysis is based upon the collected results of two separate procedures: one for short-lived isotopes and one for long-lived isotopes. Although sample preparation and gamma ray spectra analysis for both of the procedures are identical, the process of irradiation is different. The process of analysis is detailed below.

1. Preparation: Sample preparation was performed by obtaining two 1-gram samples from each of the sediments. These samples were then sealed into individual quartz tubes for ease in handling and irradiation.

2. Irradiation: One sample tube from TG-1 and one from TG-2 underwent short half-life activation. The other sample tube from TG-1 and TG-2 underwent long half-life activation.

   2.1 Activation of Short Half-Life Isotopes: Sample activation was performed by exposing the sample tube to a neutron flux for a period of 10 seconds. The sample was allowed to sit for a period of 30 seconds to insure it had decayed to a safe handling level. The sample was then available for analysis.

   2.2 Activation of Long-Lived Isotopes: Sample activation was performed by exposing the sample tube to a neutron flux for a period of 10 hours. The sample was allowed to sit for a period of 5 minutes to insure it had decayed to a safe handling level. The sample was then available for analysis.

3. Analysis: Sample analysis was performed by individually measuring each tube for gamma ray emission. The measurement was performed with a Ge(Li) detector in conjunction with a multichannel analyzer that was

FIGURE 6-7   *(continued)*

controlled by a PDP-8 computer. The resultant gamma ray differential pulse height spectrum was analyzed to determine the following:

3.1 The elements present in the sample by referring to the photopeaks present on the spectra.

3.2 The concentration of each element present in the sample by referring to the magnitude of the gamma ray emission rate relative to a known sample's emission rate. (The concentration of each element in the known sample had been previously determined.)

Sample composition was then determined by combining the results of the short- and long-lived isotope analysis. Those results were reduced to include only the six major elements of each sediment sample.

Results: The final results of the analysis are shown below for the six major elements in each of the samples.

	TG-1	TG-2
Na-23	31600 ppm	30200 ppm
Mn-56	763 ppm	260 ppm
Fe-59	18800 ppm	46100 ppm
Co-60	4.40 ppm	15.40 ppm
Sc-46	3.63 ppm	12.30 ppm
Pa-233	25.60 ppm	6.98 ppm

Note: ppm denotes concentration in parts per million.

These measurements represent the relative concentration of the elements to within 0.01 part per million.

If any questions arise as to the analysis of the samples, please feel free to contact me.

FIGURE 6-8 Sample report 2 for analysis.

To: J. E. Mauppin
           Manager, Engineering Department

From: Christopher Perez
           Product Designer, Engineering Department

Subject: Model 704 Angle-Iron Cutters:
           Investigation of Recurring Bolt Failure

Dist: B. A. Thomas
           Manager, Manufacturing

Date: March 4, 1980

Foreword

Over the past year, Construction Equipment has been plagued by customer complaints dealing with a premature mechanical failure on the new model 704 angle-iron cutter. Apparently a bolt used in the assembly of the cutter is fracturing after minimal usage, causing the machine to be completely inoperable. I was asked by Mr. Mauppin to determine the cause of this recurring failure. The purpose of the report is to present my findings and to suggest corrective measures.

Bolt's Function

The bolt under analysis connects the lever (which raises the cutting edge) to the main body of the cutter. Once this bolt fails, the cutter is inoperable because there is no way to raise the cutting edge.

Forces on the Bolt During Operation

The bolt is subjected to three different forces during the operation of the angle-iron cutter. These forces, listed in decreasing order of importance, are

1. a torsional force, which is produced when the lever is rotated for cutting

2. a shear force, which is also produced when the lever is rotated for cutting

3. a tensile force, which is produced when the nut is tightened down

Since the torsional force is the most significant, the bolt should be designed to withstand the torque produced by the lever. Bolts which have a strong possibility of failing in torsion are usually case hardened to increase their torsional strength.

A series of fractured bolts received from one of Construction Equipment's customers was used to help determine the cause of the recurring failure. Figure 1 shows a detailed sketch of a typical bolt.

Figure 1   Sketch of Failed Bolt

Worn areas

Fracture

FIGURE 6-8 *(continued)*

As is shown in the sketch, the shank of the bolt revealed the following characteristics:

1. Fracture occurred at the first full thread. This should have been expected because of the stress-concentrating effect of the first full thread.

2. Fracture surface occurred on a plane perpendicular to the bolt.

3. Wear areas occurred where the bolt was in contact with the lever and washers.

Figures 2a and 2b [not included] are photographs (mag. 4.77x) of the mating fracture surfaces of the failed bolt. Examination of the fracture surfaces revealed:

1. a smeared rotary-type appearance in which a great amount of plastic deformation had taken place

2. a final fracture area almost exactly in the center of the bolt

Because of the deformation of the threads in the fracture region, the possibility of a fatigue fracture was ruled out.

SEM Examination of Fracture Surfaces

Further examination of the fracture surfaces at higher magnification with the scanning electron microscope revealed no imperfections in the metal. A small amount of corrosion was found toward the center of the fracture surfaces but because of its location it could not have contributed to the fracture (the corrosion must have occurred after the failure).

Hardness Test

A hardness test was conducted on the shank of the fracture bolt. A value of 74 Rb was obtained, which is quite low for a bolt used in a torsional loading situation.

Cause of Failures

The characteristics of the fractured bolt suggest that failure occurred by a single torsional overload. The source of the torsional loading on the bolt was the lever which raised the cutting edge. Therefore, it is evident that the recurring failures have been due to the inadequate torsional strength of the bolts. The first full threads on the bolts also contributed to the failures because of their stress-concentrating effect.

Recommendation

In order to increase their torsional strength, I suggest that future bolts be case hardened. This will certainly alleviate the failing bolt problem currently faced by Construction Equipment.

REFERENCES

1   Leon Festinger, *A Theory of Cognitive Dissonance,* Stanford University Press, Palo Alto, Calif., 1957.

2   Richard E. Young, Alton Becker, and Kenneth L. Pike, *Rhetoric: Discovery and Change,* Harcourt Brace and World, New York, 1970.

3   Joan Arehart-Treichel, *Science News* **107**:2 (1975), p. 26.

4   *Science News* **107**:2 (1975), p. 20.

5   Daniel Z. Freedman and Peter van Nieuwenhuizen, "Supergravity and the Unification of the Laws of Physics," *Scientific American* **238**:2 (1978), p. 126.

6   Ray Berry, "Caution: Life May Be Hazardous to Your Health," *Michigan Technic* **96**:4 (1978), p. 6.

7   Berry, ibid.

8   Ronald K. Jurgen, "The Automobile for Better or Worse," *IEEE Spectrum,* November 1977 **14**:11, p. 31.

9   Mehdi N. Bahadori, "Passive Cooling Systems in Iranian Architecture," *Scientific American* **238**:2 (1978), p. 144.

10 Norman H. Horowitz, "The Search for Life on Mars," *Scientific American* **237**:5 (1977), p. 52.

11 Kobe Shoji, "Drip Irrigation," *Scientific American* **237**:5 (1977), p. 62.

12 Richard W. Dodge has prepared an extended list of questions managers want answered for various types of projects. The list (shown below) appeared in "What To Report," *Westinghouse Engineer* **22**(4-5):108–111 and was based on information provided by James W. Souther, "What Management Wants in the Technical Report," *Journal of Engineering Education* **52**(8):498–503 (1962).

Problems	Materials and Processes
What is it?	Properties, characteristics, capabilities?
Why undertaken?	Limitations?
Magnitude and importance?	Use requirements and environment?
What is being done? By whom?	Areas and scope of application?
Approaches used?	Cost factors?
Thorough and complete?	Availability and sources?
Suggested solution? Best? Consider others?	What else will do it?
What now?	Problems in using?
Who does it?	Significance of application to Company?
Time factors?	

Field Troubles and Special Design Problems

Specific equipment involved?

What trouble developed? Any trouble history?

How much involved?

Responsibility? Others? Westinghouse?

What is needed?

Special requirements and environment?

Who does it? Time factors?

Most practical solution? Recommended action?

Suggested product design changes?

Tests and Experiments

What tested or investigated?

Why? How?

What did it show?

Better ways?

Conclusions? Recommendations?

Implications to Company?

New Projects and Products

Potential?

Risks?

Scope of application?

Commercial implications?

Competition?

Importance to Company?

More work to be done? Any problems?

Required manpower, facilities and equipment?

Relative importance to other projects or products?

Life of project or product line?

Effect on Westinghouse technical position?

Priorities required?

Proposed schedule?

Target date?

ADDITIONAL READING

Elliot Aronson, "The Rationalizing Animal," *Psychology Today,* May 1973, pp. 46–52,119.

Richard S. Burington, "The Problem of Formulating a Problem: General Considerations," *Proceedings of the American Philosophical Society* **105**(5):429–443 (1960).

Scott Consigny, "Rhetoric and Its Situations," *Philosophy and Rhetoric,* Summer 1974, pp. 175–186.

Otto Dieter, "Stasis," *Speech Monographs* **17**(4):345–369 (1950).

Richard W. Dodge, "What To Report," *Westinghouse Engineer* **22**(4–5):108–111 (1962).

J. C. Mathes and Dwight Stevenson, *Designing Technical Reports,* Bobbs-Merrill, Indianapolis, 1976.

Henry Minzberg, "The Manager's Job: Folklore and Fact," *Harvard Business Review,* July–August 1975, pp. 49–61; *The Nature of Managerial Work,* Harper & Row, New York, 1973.

Part IV

Visual Elements

7

VISUAL ELEMENTS

So far, this book has talked about the production of words, either written or spoken. There are times, however, when words alone are not the best way to transfer information or points of view, times when words need to be combined with visual aids, formatting (the use of white space and indenting), or other visual elements. For instance, as suggested later in this chapter, appropriate formatting can make a technical report much easier to read—so much easier that the formatting becomes *necessary* given the limitations on the time and attention of an audience. The same can often be said of other visual elements, such as drawings, figures, charts, or graphs, which can quickly summarize an important point or present it in a different way.*

For instance, consider the following situation. Figures 7-1 and 7-2 present the same information but in different forms, Figure 7-1 in prose text and Figure 7-2 as a visual element. The discussion in Figure 7-1 requires 134 words and forces its reader to learn its points linearly, that is, one at a time as they appear in the discussion. In contrast, the chart in Figure 7-2 presents the information quickly and holistically; that is, it presents the whole in a clearly organized pattern that can be seen at a single glance.

Psychologists have shown that you can increase the strength and memorability of a message simply by repeating it or, even better, by repeating it in a different form. Thus, when a visual presentation is added to a verbal one, the combination can produce *a much stronger and more easily remembered message* than either presentation alone. Further, a visual aid can present *a compact summary* of the main points of a verbal text. (Have you ever heard the expression "a picture is worth a thousand words"?) Finally, a visual element can often summarize in *a more memorable form* than words alone can.

Given these advantages of visual aids, a communicator ought to be able to use them effectively. This involves knowing

1 How to make a visual aid effective

2 When to use the visual aid

3 How to select the best type of visual element in a given situation (e.g., pie chart, bar graph, line graph)

4 How to integrate the visual aid into the text

---

*We would like to note the efforts and influence of our colleague Lisa Barton on the evolution of this chapter. She was a major contributor to the thinking behind Sections 7.1 and 7.3 and coauthored an early lecture on visual aids from which those sections are derived.

FIGURE 7-1 Safety records of surface and underground mining, presented in prose text form. [From Edmund A. Nephew, "The Challenge and Promise of Coal," *Technology Review* **76**:2 (December 1973), pp. 27–28.]

Safety: Room for Improvement

The safety record of surface mining—in terms of both deaths and non-fatal injuries per ton of coal mined—is spectacularly better than deep mining, which provides at best a hostile, hazardous environment for the miner.

The human costs of moving from surface to underground mining take the form of higher injury and death rates and of greater occupational hazards in general. In 1971, 86 per cent of all coal mining fatalities occurred in underground mines. In the same year, only half of the total coal production came from deep mines. Over the seven-year period 1965 to 1972, 1,412 lives were lost in the underground coal mining industry in the production of 2,335 billion tons of coal. This amounts to an average of 0.606 deaths per million tons of coal—a fatality rate more than five times greater than that of the coal surface mining industry. In recent years, falls-of-roof have accounted for 40 per cent of the deaths in underground mines and coal haulage accidents for about 20 per cent. Annual fatalities from dust and gas explosions fluctuate greatly, but over the years there has been a declining trend.

A similar safety disparity between surface and deep mines holds for non-fatal injuries as well.

During the time period from 1968 to 1971, the safety performance in deep mining of the nation's top ten coal producers ranged from 0.28 to 1.52 deaths per million man-hours. The differences are even more marked for the category of non-fatal injuries—2.72 to 72.13 injuries per million man-hours. Since the passage of the Coal Mine Health and Safety Act of 1969, most companies have greatly strengthened their safety programs, and this increased emphasis on safety may bring significant improvements in fatality and injury rates. Indeed, the wide range of safety performances cited above makes it clear that—even without a technological breakthrough—much can be done to narrow the safety gap existing between deep and surface mining.[1]

7.1

MAKING A VISUAL AID TRULY VISUAL

Take about 2 to 5 seconds to look at Table 7-1 and then cover it up. Do not look at any of the following tables or discussions. Now try to write down the main points made by the table. When you have finished this, look at the presentation of the same information in Table 7-2 and see if you can quickly add any more main points to your list. Do this before you continue.

Typically, people who read only Table 7-1 note (1) that job satisfaction declines in each of the two main groups of occupations. These readers will *sometimes* notice (2) that there is a large difference in job satisfaction between the two groups—that is, that most of the first group is relatively satisfied (93 to 75 percent satisfied) whereas most of the second group is much less satisfied (only 52 to 16 percent satisfied). Very few readers of only Table 7-1 will notice (3) that the job satisfaction of skilled printers is higher than that

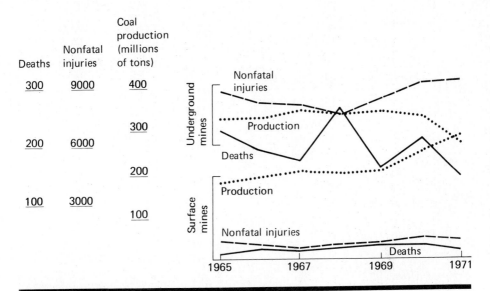

FIGURE 7-2 Safety records of surface and underground mining, presented as a visual element. [From Edmund A. Nephew, "The Challenge and Promise of Coal, *Technology Review* **76**:2 (December 1973), p. 28.]

TABLE 7-1 PROPORTIONS OF OCCUPATIONAL GROUPS WHO WOULD CHOOSE SIMILAR WORK AGAIN

PROFESSIONAL AND WHITE-COLLAR OCCUPATIONS	PERCENT	SKILLED TRADES AND BLUE-COLLAR OCCUPATIONS	PERCENT
Urban university professors	93	Skilled printers	52
Mathematicians	91	Paper workers	42
Physicists	89	Skilled auto workers	41
Biologists	89	Skilled steelworkers	41
Chemists	86	Textile workers	31
Firm lawyers	85	Blue-collar workers	24
School superintendents	85	Unskilled steelworkers	21
Lawyers	83	Unskilled auto workers	16
Journalists (Washington correspondents)	82		
Church university professors	77		
Solo lawyers	75		
White-collar workers (nonprofessional)	43		

SOURCE: Based on a study of 3000 workers in 16 industries, conducted by the Roper organization; on Wilensky's study of Detroit workers and professions; and on a study of Massachusetts school superintendents by Neal Gross, Ward Mason, and W. A. McEachern. From *Psychology Today,* February 1973, p. 39.

TABLE 7-2 ALTERNATE ARRANGEMENT FOR "PROPORTION IN OCCUPATIONAL GROUPS WHO WOULD CHOOSE SIMILAR WORK AGAIN"

PROFESSIONAL WHITE-COLLAR OCCUPATIONS	PERCENT	
Urban university professors	93	....
Mathematicians	91	....
Physicists	89	....
Biologists	89	....
Chemists	86	....
Firm lawyers	85	....
School superintendents	85	....
Lawyers	83	....
Journalists (Washington correspondents)	82	....
Church university professors	77	....
Solo lawyers	75	....
White-collar workers (nonprofessionals)	43	....
SKILLED TRADES AND BLUE-COLLAR OCCUPATIONS		
Skilled printers	52	....
Paper workers	42	....
Skilled auto workers	41	....
Skilled steelworkers	41	....
Textile workers	31	....
Blue-collar workers	24	....
Unskilled steelworkers	21	......
Unskilled auto workers	16	....

of nonprofessional white-collar workers. These last two observations (points 2 and 3) are very hard to "see" in the format used in Table 7-1.

In contrast, most readers of Table 7-2 easily and quickly note all three observations, as well as a few other, more subtle ones, simply because of the format of the table. Notice that Table 7-2 makes it *visually* quite clear that the job satisfaction ratings of the two groups overlap and that the skilled trade and factory workers as a group are less satisfied than the professionals.

This comparison illustrates an important point: to be most effective a visual aid should present information in a truly *visual* form. Table 7-1 presents information in a chart, but it does not use most of the chart's visual possibilities. It nicely arranges the information in two groups and in descending order within the two groups. However, it does not visually indicate the descending order. To really understand the chart, readers have to do a lot of mental work. For instance, they have to notice

1   That for the left column the first four categories have satisfaction levels of 93, 91, 89, and 89 percent

2   That these four levels are all quite close to each other

3   That the next levels go from 86 to 75 percent

4   That these are pretty close to each other but lower than the first group

5   That the lowest category is only 43 percent

6   That this is pretty low, too, etc.

7.2
DECIDING WHEN TO USE A VISUAL AID

Communicators often wonder *when* they should use a visual aid in a communication. Three suggested principles for deciding this are to use a visual aid

1   Where words alone would be either impossible or quite inefficient for describing a concept or an object

2   Where a visual aid is needed to underscore an important point, especially a summary

3   Where a visual element is conventionally or easily used to present data

Let us consider each of these uses in turn.

The Visual Aid for Describing or Clarifying

A communicator often has to describe an object or concept which is hard to describe efficiently in words alone. For instance, consider the following passage and then try to sketch out the image created in your mind by the passage. Do not look at Figure 7-3 before making your sketch.

SKYLAB DESCRIPTION

The Skylab cluster is an assembly of modified and specially built space hardware which together provides over 12,750 ft. of working volume and weighs in at 199,750 lbs. The Workshop is a Saturn IVB stage modified and outfitted for manned habitation; it contains living quarters, food storage and preparation and waste management facilities and the attitude control thrusters, and it carries the two ill-fated solar arrays (one shows clearly in its extended position in this drawing) whose electric input is routed to the power system in the Airlock Module (AM). The latter, serving as a passageway between the Workshop and the docking facilities, is the focus of many of Skylab's technical systems: atmospheric and thermal controls, power control and distribution, and communications and data handling. The Multiple Docking Adapter (MDA) has two docking ports—the axial port shown in use here and a contingency radial port at the bottom in this diagram; it also contains some space research equipment, including the Earth Resources Experiment Package. The Apollo Telescope Mount (ATM) is a sophisticated solar observatory—the first U.S. manned scientific telescope in space. Here as well are attitude and experiment pointing controls for the entire Skylab cluster and a solar array and associated battery system adequate to power the ATM's equipment. The drawing shows the Apollo Command and Service Module (CSM), used to transport the crew to and from orbit, docked to the MDA.[1]

As suggested earlier, try to sketch out the image created in your mind by the passage before you look at the drawing of the Skylab presented in Figure 7-3. When you are finished, compare your sketch to the drawing. If you are like most readers, your sketch captures some of the features of the Skylab: it probably shows that the *Airlock Module* comes between the *Workshop* and the *docking facilities* and that the *Apollo Command and Service Module* is attached to the *Multiple Docking Adapter*. However, it may not show that the *Multiple Docking Adapter* is the same as the *docking facilities,* and it probably will not show the appropriate sizes and shapes of the various parts of the Skylab system. Clearly, the formal drawing presents lots of information about sizes, shapes, details, and arrangements which are not presented in the words and which *could not* be presented there, given the limited

FIGURE 7-3   The Skylab cluster. [From William Schneider and William D. Green, Jr., "Saving Skylab," *Technology Review* **76**:3 (January 1974), p. 44.]

number of words in the original description and even the extra space occupied by the visual aid. In such a situation, you obviously need a visual aid.

Sometimes you will need to describe a concept or object which, even if you have an unlimited number of words, seems impossible to describe with words alone. Consider, for example, the passage below describing the pterosaur, a prehistoric flying reptile, and its near relatives.

> Pterosaurs evolved in the early part of the Mesozoic era. Their appearance preceded that of the earliest birds by about 50 million years and followed that of the earliest dinosaurs by about 20 million years. It is hypothesized that the precursors of the pterosaurs (and of the birds and of both orders of the dinosaurs, the ornithischians and the saurischians) were the reptiles called the thecodonts, which for their part evolved from the early, lizard-shaped reptiles known as eosuchians. Among the pterosaurs the rhamphorhynchoids first appeared some 50 million years before the pterodactyloids. The last pterodactyls died out with the dinosaurs 64 million years ago.[2]

If you are having trouble placing all of the prehistoric creatures just described, see if their relationships are clearer in Figure 7-4. In this case, the conceptual system described is too complex for most readers to "see" by words alone but quite easily grasped with the help of the visual aid.

The Visual Aid for Highlighting Important Points

There are many times when you need a visual aid to bring out an important point or to summarize data or a line of argument. For instance, Chapter 6 presents arguments about Introductions and Forewords: that they are based upon psychological principles and that they can be represented by a formula consisting of a problem statement in a certain format, a question or an assignment, and a thesis. These arguments included discussions and several examples and ended with a summary of the Introduction. This summary is shown in Figure 7-5, which usually recalls the information in Chapter 6 even for those readers who have not recently reviewed it.

The Visual Aid for Conventional or Easy Presentation of Data

Some types of data are almost always presented in a visual aid. These include cost summaries, budget calculations, frequency spectra, electrical circuits, architectural plans, and the relative positions of geographical areas. As a professional, you ought to be aware of the typical visual forms in your field and use them when it is appropriate.

For instance, consider the sound spectra in the following extract, a short overview of an interesting research project. In just over 500 words, a problem is introduced, the project's research model and technique are explained, the assumptions set out, and the findings reported. Note in this example that using the visual aid is not only more conventional but also more efficient than using words alone: an equally detailed verbal description of the cry spectra would take much more space. Note also *where* the visual aid is placed

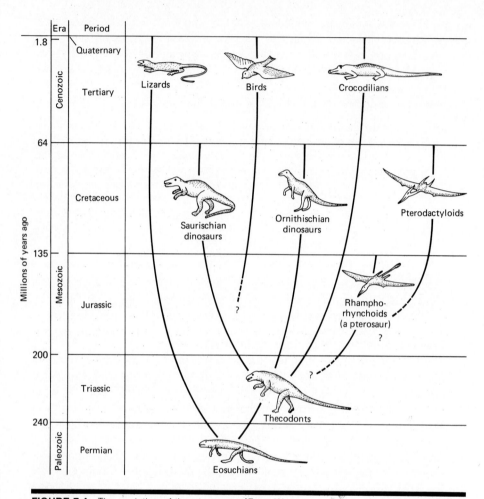

FIGURE 7-4   The evolution of the pterosaurs.[From Wann
Langston, Jr., "Pterosaurs," *Scientific American* **244**:2
(February 1981), p. 126.]

in this passage: it gives the detailed backup data needed to substantiate a
scientific claim. It would be hard to imagine another place in this brief overview where the extra proof and detail of a visual aid would be needed to
prove a central claim.

BABIES' CRIES GIVE CLUES TO DISEASES

The long, loud, crystal-shattering sounds coming from a newborn infant indicate
his good health according to current research by Howard L. Golub and Kenneth N. Stevens of M.I.T.'s Research Laboratory of Electronics. Although scientists in this country and abroad have failed for 20 years to relate cry profiles
to infant ailments, their recent tests with a computer cry model showed surprisingly accurate results.

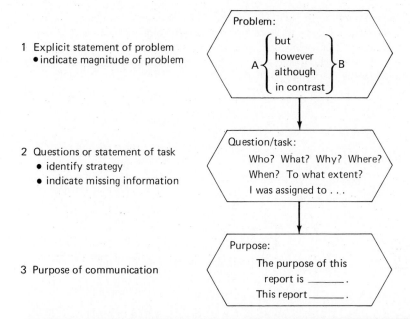

1 Explicit statement of problem
   • indicate magnitude of problem

2 Questions or statement of task
   • identify strategy
   • indicate missing information

3 Purpose of communication

FIGURE 7-5   Outline of an Introduction or Foreword by problem statement.

Mr. Golub developed the model by using the accepted acoustic theory of adult speech production and modifying it to fit babies. Since adults and children differ in anatomical proportion, ratios like pharynx length to mouth length, nasal tract length to vocal tract length, and central nervous system disparities had to be adjusted accordingly. Then, using the tape-recorded protests of 55 apparently healthy newborns during a blood test (called the P.K.U. heel stick), Mr. Golub subdivided the cries into 88 variables, such as pitch, intensity, fundamental frequencies, and resonances.

He then used a computer to compare the cries of healthy babies with those of 43 babies having known or suspected health abnormalities. Recordings of these latter cries were supplied by Dr. Michael Corwin, resident in pediatrics at the Upstate Medical Center, Syracuse, N.Y. Of these 43 cases Mr. Golub's computer analysis located 19 of 21 infants with severe jaundice and 9 of 10 babies suffering from respiratory difficulties. Fifteen healthy babies from Dr. Corwin's group were also correctly identified from their cries alone.

Mr. Golub's research team found that a specific ailment is associated with a characteristic cry pattern. Assuming that most infant pathologies alter the acoustically relevant structures, it follows that some aspects of the cry will be correspondingly changed.

For example, if the infant has respiratory distress one would expect a shorter cry due to a change in the dynamics of respiratory muscles responsible for oxygen intake, or a change in vocal tract resonances resulting from a constriction of the pharynx—a possible explanation for the tragic sudden infant death syndrome. In such a case one might also expect a higher frequency sound since the vocal tract acts much like a pipe organ acoustically—a narrow or squeezed pipe giving a higher frequency sound.

Interestingly, Golub and Stevens found precisely these effects, as illustrated by the following two charts.

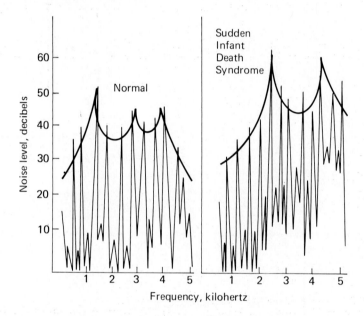

The chart at the left shows the short-time (25 milliseconds) spectrum of the mid-portion of the cry of a normal infant. The chart at the right plots the short-time (25 milliseconds) spectrum of the mid-portion of the cry of an infant that later died of sudden infant death syndrome. Notice the very high first resonance and large amount of noise at the high end of the spectrum, possibly indicating a constriction of the vocal tract near the pharynx.

As a sidelight of his research on jaundice cases, Mr. Golub discovered that some levels of biliruben—a by-product of the breakdown of hemoglobin—previously considered "safe" are likely harmful. Early treatment for such a condition could prevent brain damage. Other abnormalities likely detectable with the model include bacterial meningitis and deafness.[3]

7.3

SELECTING THE BEST TYPE OF VISUAL AID IN A GIVEN SITUATION

When you design a particular visual aid, you are consciously or unconsciously making certain decisions. You are deciding that the particular type of aid you choose (a line graph, bar chart, pie diagram, photograph) is the best *type* to

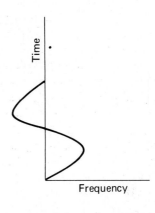

FIGURE 7-6   Preferred location of independent variable on a graph.

FIGURE 7-7   Unconventional location of independent variable on a graph.

make your point and that the arrangement and highlighting of material on the page is, again, the best to make your point. Unfortunately, there is little information available on which to base such decisions. If you are like most writers, you probably choose one type of visual aid over another simply because it is the first thing you think of using.

The purpose of this section is to sketch out some better or more conscious reasons for choosing. The section will first identify some conventions of visual perception in Western cultures and then examine several common types of visual aids to see what they do and do not show well.

Conventions of Visual Perception

Given the way Western societies view the world and read, there are a number of general statements we can make about our expectations of visual information. First, we expect written things to proceed from left to right. Note that in scientific and technical graphs, we place the independent variable on the x axis so that the more important variable moves from left to right. For instance, we plot time on the x axis and frequency on the y axis, as illustrated in Figure 7-6. This pattern is so pervasive that Figure 7-7 looks at best odd and at worst disturbing.

Second, we expect things to proceed from top to bottom, and, third, we expect things in the center to be more important than things on the periphery. Fourth, we expect things in the foreground to be more important than things in the background; fifth, large things to be more important than small things; and sixth, thick things to be more important than thin things. Note that type that is larger, thicker, or bolder than the surrounding type is usually more important: a heading, a title, or an especially important word

FIGURE 7-8 The shaping of expectation by visual arrangement:
a sample for analysis.

in a passage. Seventh, we expect areas containing a lot of activity and information to contain the most important information. Eighth, we expect that
things having the same size, shape, location, or color are somehow related
to each other. (Notice the way the curves in Figure 7-8 are grouped; all labels
have been deleted to free you to "see" this point.) Lastly, ninth, we see
things as standing out if they contrast with their surroundings because of line
thickness, type face, or color. (You should note that warm or hot colors—red,
yellow, and orange—stand out more than cool colors—blue and green.)

Notice how many of the above expectations are verified in Figure 7-9.
The dark lines tracing the bus, streetcar, and rail routes clearly stand out
from the lighter print and the white background. Conceptually, San Francisco
Bay and the Pacific Ocean are merely background for the lines of the transit
system; both the bay and the ocean are on the edge. In contrast, the most
important part of the visual aid, the transit lines, is in the center in dark, thick
lines. You might look for such features in other visual aids and try to explain
why the designer made the choices you see.

Some Types of Visual Aids and Their Uses

There are six main types of visual aids with which a scientist or engineer
should be familiar: (1) line graphs, (2) bar graphs, (3) pie charts, (4) tables,

FIGURE 7-9 The five lines of the San Francisco Municipal Transit System.

(5) photographs, and (6) line drawings. Each of these types has particular strengths and weaknesses, and to use any one appropriately, you must decide what point you are trying to make and then select the type of visual aid which makes that kind of point well.

FIGURE 7-10   River flow before (1963) and after (1977) construction of the Aswan High Dam on the Nile River. [From Julie Wei, "Aswan and After: The Taming and Transformation of the River Nile," *Research News* **31**:7 (The University of Michigan, July 1980), p. 21.]

LINE GRAPHS   Line graphs *show well* continuity and direction as opposed to individual or discrete points, direction as opposed to volume, and the importance of a nodal point, if there is one. These characteristics are illustrated in Figure 7-10. Line graphs *do not show well* the importance of one particular point which falls off a node, the relationship of many lines, or the intersection of three or more lines. If it is important to be able to trace each line on a graph, you should probably not put more than three or four on a single graph, especially if they intersect frequently, or you may produce a graph as hard to follow as the one in Figure 7-11.

BAR GRAPHS   Bar graphs *show relatively well* the discreteness or separateness of points as opposed to their continuity, volume as opposed to direction, the relationships among more than three or four items at a time, the contrast between large and small numbers, and the similarities and differences between similar numbers (you can see both that 17 percent and 18 percent are about the same and that 18 percent is a bit larger than 17 percent). These characteristics are evident in the variant of the bar graph presented in Figure 7-12 and in Figure 7-13. Bar graphs *do not show well* the absolute values of the items measured, though you can indicate absolute values with labels, as shown in Figure 7-13.

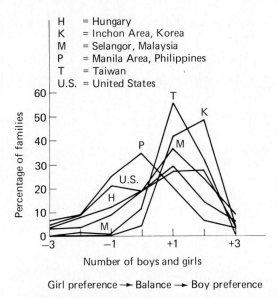

H = Hungary
K = Inchon Area, Korea
M = Selangor, Malaysia
P = Manila Area, Philippines
T = Taiwan
U.S. = United States

Percentage of families

Number of boys and girls

Girl preference → Balance → Boy preference

FIGURE 7-11 Preference of families for girls versus boys in six countries.

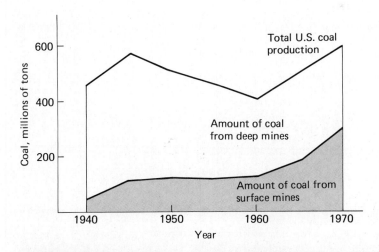

Coal, millions of tons

Total U.S. coal production

Amount of coal from deep mines

Amount of coal from surface mines

Year

FIGURE 7-12 Growth of surface mining in U.S. coal production. [From Edmund A. Nephew, "The Challenge and Promise of Coal," *Technology Review* **76**: (December 1973), p. 22.]

FIGURE 7-13 Comparison of cesium-137 fallout in the Lake Huron air-water surface, water, and sediment. [From Julie Wei, "Towards Cleaner Water," *Research News* **29** 8/9 (The University of Michigan, August-September 1978), p. 22.]

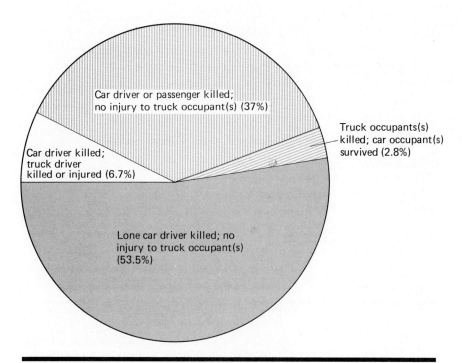

FIGURE 7-14   Distribution of fatalities in 181 fatal car-truck
crashes. [From Daniel J. Minahan and James O'Day, "Fatal
Car-into-Truck/Trailer Underride Collisions," *HSRI Research
Review* **83**: (The University of Michigan Highway Safety Research
Institute, November-December 1977), p. 7.]

PIE DIAGRAMS   Pie diagrams *show relatively well* the relationship among
three or four items which total 100 percent, the contrast between large and
small percentages, and the similarity between relatively similar percentages
(they show well that 27 percent and 29 percent are about equal). Pie diagrams *do not show well* the small difference between two similar percentages
(you can't usually see the differences between 27 percent and 29 percent).
They also *do not show well* absolute values (unless you label the parts of
the pie) or the relationships among more than five or six parts; with too many
parts, it is hard to see the relationships of part to part and part to whole.
These strengths and weaknesses are illustrated in Figure 7-14.

TABLES   Tables are convenient for presenting lots of data and for giving
absolute values where precision is very important. However, since they present
items one at a time in columns, they emphasize the discrete rather than the
continuous and make it very difficult to show trends or direction in the data.

TABLE 7-3 ANNUAL ENERGY SAVINGS FROM SOLAR ENERGY

	SOLAR SPACE AND DOMESTIC HOT WATER SYSTEMS		SOLAR TOTAL-ENERGY SYSTEMS		SOLAR BASE-LOAD ELECTRIC POWER PLANTS	
YEAR	NUMBER OF DWELLINGS (MILLIONS)	ANNUAL ENERGY SAVINGS (10^{15} B.T.U.)	FLOOR AREA OF BUILDINGS (10^4 M^2)	ANNUAL ENERGY SAVINGS (10^{15} B.T.U.)	INSTALLED CAPACITY (10^6 KW)	ANNUAL ENERGY SAVINGS (10^{15} B.T.U.)
1980	0.3	0.04	—	—	—	—
1985	3	0.4	52	0.24	—	—
1990	8.4	1.2	200	0.92	18.7	1.4
1995	14	1.9	400	1.9	69	5.3
2000	20	2.7	610	2.8	137	10.4
2005	27	3.6	850	3.8	208	15.2
2010	34	4.7	1,090	5	284	22
2015	41	5.9	1,350	6.2	363	30
2020	49	7	1,620	7.6	445	37.6

Assuming the most favorable development, annual energy savings of 52.2 × 10^{15} B.t.u.—over 1,160 million tons of oil—may be realized by the use of solar energy by the year 2020. The estimates for dwellings are based on the use of approximately 10^8 B.t.u. annually and a heating plant efficiency of 70 per cent; those for total-energy systems on consumption of 6.67 million kwh/yr in a plant of 10,000 m^2 floor area, with an efficiency of 50 per cent; those for power generation on the construction of solar plants with total capacity of 12.5 × 10^6 kwh each year beginning in 1995, the plants having 40 per cent efficiency.

SOURCE: From *Technology Review* **76**:2 (December 1973), p. 39.

Tables are not predominantly visual: the reader's mind must translate each number into a relationship with each other number, as already described in the job satisfaction example at the beginning of this chapter. Thus, for maximum *visual* impact, tables should probably be a last choice as a visual aid and used only when it is important to provide a great deal of information with precision in a very small space. As an illustration, consider Table 7-3.

PHOTOGRAPHS   Photographs *are useful* when you do not have the time, the money, or the expertise to produce a complicated line drawing; when you are trying to produce immediate visual recognition of an item; when you are emphasizing the item's external appearance (as opposed to its internal structure or a cross section); and when you are not concerned with eliminating the abundant detail a photograph provides. While photographs can be air-brushed to eliminate some undesired detail, they still are *not preferred* when you need to focus on some one aspect by eliminating a lot of detail and when you have the time and resources to produce a good line drawing.

LINE DRAWINGS The term *line drawing* includes several types of drawings which focus on external appearance, physical shape, function, or relationship. These include "simplified photos," maps, anatomical drawings, parts charts, and drawings of models (such as atomic or molecular models) or objects from any field of science or engineering. Also included are flow charts, organizational charts, schematic charts, block diagrams, architectural plans, and blueprints.

While there are many types of line drawings, all of them share certain functions. They allow you to show things which you can't normally see in a photograph because of size, location, or excessive detail. They also allow you to easily highlight a particular shape, part, or function.

For instance, consider the photograph and drawings of a eukaryotic cell and its compartments shown in Figure 7-15. Notice that the photograph, an electron micrograph, has so much detail that at first it is hard to identify the parts of the cell. Further, because of their size, it is hard even with the drawings for reference to clearly see either the cisternae and ribosomes identified in the lower right corner or the boundary of the nucleus. Obviously, the drawings included with the photograph are much clearer, allow highlighting and focusing impossible in the photograph, and can present important details missing in the photograph.

7.4

DESIGNING THE VISUAL AID

Once you have decided *where* a visual aid is needed and *what type* it should be, you must design it so that it is as relevant, clear, and truthful as possible. This will usually be at least a two-stage process: designing a rough copy and then producing the finished copy. If you work for a company which has an art or illustration department, you may be able to get a technical illustrator to produce the finished copy for you and to counsel you in the design stage. However, even if you have such help, you should be the real designer of the visual aid: you have the best knowledge of the subject and best know the purpose of the aid and the context in which it is being used.

Making a Visual Aid Relevant

Since you place a visual aid in a text to make a point, you should be sure that it makes the point you intend. For instance, suppose that you are discussing expected energy saving from the use of solar energy in the future. You have posed three possible sources of the savings—residences, total energy systems such as industrial parks and shopping centers, and solar-based electric power plants—and have broken down the specific savings as illustrated in Table 7-4.

Now that you have your data, you want to construct a visual aid to

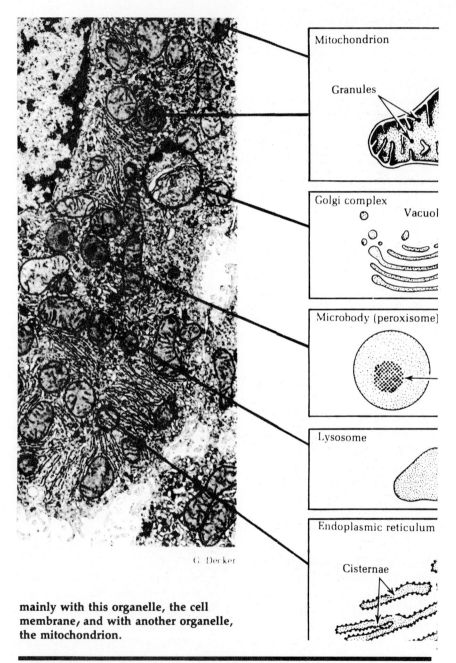

mainly with this organelle, the cell
membrane, and with another organelle,
the mitochondrion.

FIGURE 7-15   A eukaryotic cell and its compartments. [From
Albert L. Lehninger, *Biochemistry* (New York, Worth Publishers,
1975), pp. 32–33.]

TABLE 7-4   EXPECTED ANNUAL SAVINGS FROM SOLAR ENERGY

| | ANNUAL SAVINGS (10^{15} BTU) | | |
YEAR	RESIDENCES	TOTAL ENERGY SYSTEMS	SOLAR-BASED ELECTRIC POWER PLANTS
1985	0.4	0.24	—
1990	1.2	0.92	1.4
1995	1.9	1.9	5.3

show the growth in savings and the relative contributions of each source. You construct five possible versions of a visual aid, presented in Figures 7-16 through 7-20, and now have to choose the one most appropriate to your point. On what basis do you choose? What are the differences among the five visual aids?

First let us consider the bar graphs. Among the bar graphs, Figure 7-16 presents the most information in the smallest space and the clearest vision of total growth; however, in comparison to the other charts, it obscures the comparisons between items in the same year and between the same item in different years. Figure 7-17 obscures the total growth but makes the comparisons already mentioned much clearer, especially between the same item in different years. On the other hand, Figure 7-18 clarifies the comparison between items in the same year but obscures comparisons between years. The line graphs in Figures 7-19 and 7-20 have the same strengths and weaknesses as their respective bar graph counterparts, but in addition they also bring out more strongly the idea of direction and rate of change.

So how do you choose one (or two) from among the group? You pick the one which best matches the focus you wish to take in your report or talk. If you are not much concerned about total growth but want to focus on the contribution of each area for savings, then you would probably choose Figure

FIGURE 7-16   Annual energy savings from solar energy: version 1.

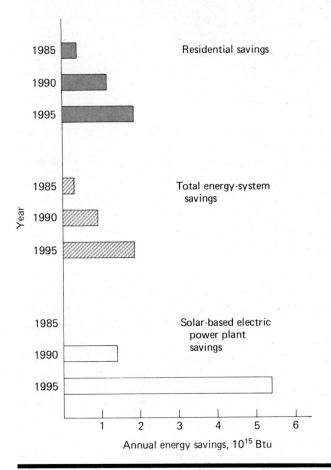

FIGURE 7-17   Annual energy savings from solar energy: version 2.

FIGURE 7-18   Annual energy savings from solar energy: version 3.

FIGURE 7-19   Annual energy savings from solar energy: version 4.

FIGURE 7-20   Annual energy savings from solar energy: version 5.

7-17. If you are interested in the growth of the contribution of each area, you would probably choose Figure 7-19. If you are primarily interested in the increase in total savings, you would probably choose Figure 7-16 or 7-20.

Since relevance and clarity are so important, let us consider one more example illustrating them, the visual aid in Figure 7-21. This visual appeared in a report outlining resources and long-term investment possibilities for a particular small city. The aid was supposed to show that significant production on a long-term basis could be achieved from shale wells and medina wells (two types of natural gas wells) and that such wells would make good long-term financial investments. Unfortunately, a reader could easily notice the

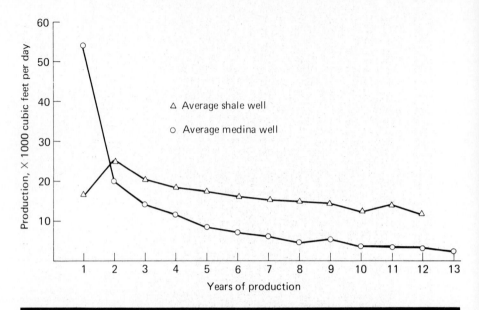

FIGURE 7-21   Production potential for average shale and medina wells: version 1.

sharp initial drop in the medina well curve and see the flat parts of both curves as low relative to the initial high production of the medina well. Such a comparison *visually* suggests that production is low after 2 years and that neither well would be a good long-term investment, although the medina well might be a promising short-term investment. However, this is the direct opposite of the intended point.

It turns out that these types of wells are financially sound investments if they produce on the average above 4000 cubic feet per day. If we add this information to the visual as a reference line, it dramatically changes the visual's impact. The modified visual, Figure 7-22, clearly (and visually) reveals the wells' adequate productivity and long-term investment potential.

Making a Visual Aid Clear

Making a visual aid clear involves two separate activities: making it conceptually clear and making it technically clear. Making it conceptually clear means having a clearly defined and relevant point and a good form for the point. Conceptual clarity is discussed above. Technical clarity is a simpler matter and will be treated here. It involves having an informative title, appropriate headings and labels, and enough white space so that an audience has the best possible chance of finding the "right" meaning for the visual aid. For

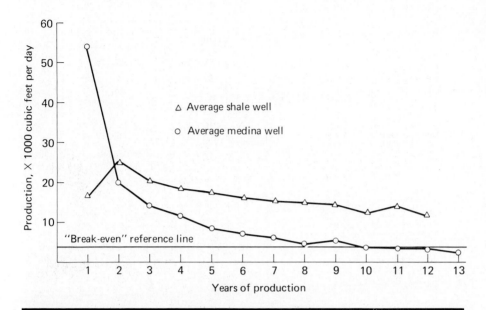

FIGURE 7-22 Production potential for average shale and medina wells: version 2. [Adapted from Moody and Associates, Inc., *Natural Gas Development Feasibility Study: Report for the School District of Erie, Pennsylvania* (Meadville, Pennsylvania, October 1980), p. 6.]

example, suppose Figure 7-21 had been entitled "Productivity Decline of Medina and Shale Wells." Given the lack of a break-even reference line, this title would have focused the reader's attention even more strongly on the downward slope of the medina production curve and on the comparatively low long-term production rates. Presumably, this would have made the visual even less effective in communicating its real point.

To really see the benefit of proper labeling and sufficient white space, look at the series of graphs presented in Figure 7-23. Graph *a* is an extremely bad example of a visual aid since it has *none* of the labeling information usually presented. Graphs *b* and *c* present more information, but still not enough to really get the message across. (Notice that graph *c* lacks enough information even though it provides everything except the title and two critical labels.) Graph *d* provides an adequate title and labels, but the grid in the background is so obtrusive that a reader can hardly see the important lines and labels. Finally, graph *e* provides adequate information and enough white space to let it be seen; from these, a careful and hardworking reader can probably figure out the message. (You should note that version *d* is typical of most student reports, which are done quickly and checked mainly for accuracy rather than readability.)

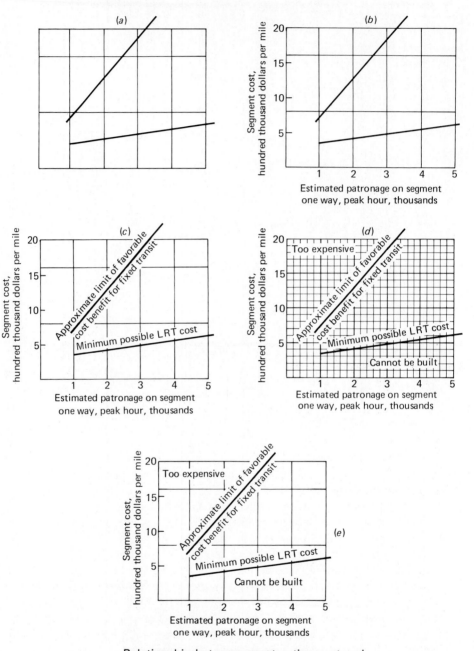

Relationship between construction cost and
patronage for light rail transit systems

FIGURE 7-23   The necessity of labels, headings, and titles in visual aids.

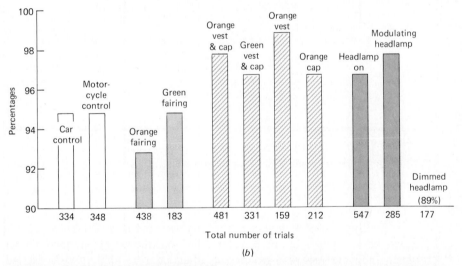

FIGURE 7-24 Elimination of unnecessary detail in visual aids.

Another use of white space to unclutter a visual aid is illustrated in Figure 7-24. Figure 7-24a has no need of three-dimensional bars; in fact, they add distracting detail. Don't you find Figure 7-24b, with its two-dimensional bars, easier to read?

You should note that color can be as misused as the graphic flourishes noted above. If you use a color other than black or white, you should be able to justify the inclusion of each additional color and of each word printed in a color other than black.

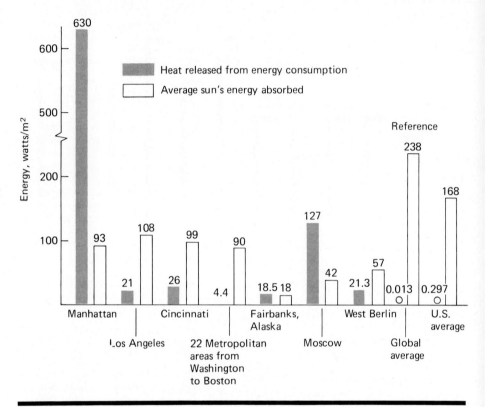

FIGURE 7-25 Use of white space to show structure in visual aid data. [From David C. White, "The Energy-Environment-Economic Triangle," *Technology Review* **76**:2 (December 1973), p. 18.]

Another use of white space showing structure in a visual aid's data is Illustrated in Figure 7-25. In this figure, the consumption-absorption units for various cities and areas are grouped together and extra white space is inserted between the city and area units. This helps to define the structure of the data for the reader. Note this same use of white space in Figure 7-27.

Making a Visual Aid Truthful

Making a visual aid truthful is important for at least two reasons. First, a visual aid which falsifies information or misleads your audience may hurt your reputation since it may lead people to see you as dishonest. (However, if you *are* honest, you probably want to be *seen* as honest.) Second, a false or misleading visual aid may hurt your argument. If members of your audience are upset or dissatisfied with your argument, they may try to discredit

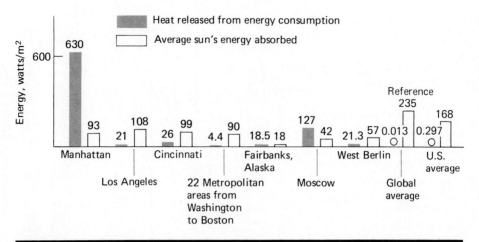

FIGURE 7-26 Distortion of significant differences by inappropriate use of scale.

it, and one of the easiest ways to do that is to point out a false or misleading spot. How can you believe the arguments of someone who is demonstrably false or misleading?

There are several ways to create a false or misleading visual aid. You can inadvertently make a poor choice in the type of visual used for presenting your information and thus obscure your point or make a point quite different from what you intended. This is probably the most frequent cause of misleading visual aids.

In a more serious falsification, you can distort the data. This is sometimes done by obscuring significant differences through inappropriate scale, as illustrated by Figure 7-26. Compare Figure 7-26 with Figure 7-25 and you will notice that Figure 7-25 preserves the zero point for reference and indicates the missing part of the scale by a cut in the y axis and in the Manhattan bar; however, it does not obscure the significant differences in energy consumption and absorption among all of the other cities and areas, as Figure 7-26 does.

Data are sometimes distorted by exaggerating insignificant differences with a suppressed zero, as illustrated by Figure 7-24. Compare the y axis of Figure 7-27 with the y axes in Figure 7-24. Notice that Figure 7-24 suppresses the zero point and greatly exaggerates the differences among the bars by the limited scale range on the y axes. It suggests that the audience responded very differently to the various safety devices. In contrast, Figure 7-27 creates a very different effect: without the suppressed zero, it suggests that the audience responded similarly to the various devices.

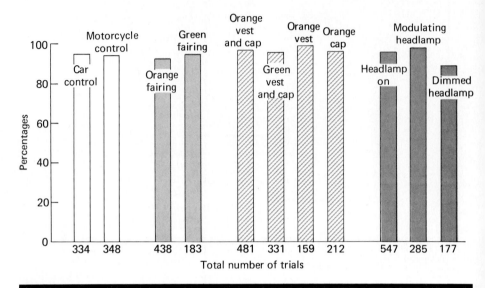

FIGURE 7-27   Preserving truthfulness by appropriate use of
scale. [Adapted from Paul L. Olson, Rich Halstead-Nussloch, and
Michael Sivak, "Means of Making Motorcycles More
Conspicuous," *HSRI Research Review* **10**:2 (The University of
Michigan Highway Safety Research Institute, September-October
1979), p 9.]

7.5
INTEGRATING THE VISUAL AID INTO THE TEXT

Once you have decided to use a visual aid in a particular spot in the text,
you must incorporate it so that it seems to belong there. This is easier said
than done.

The visual aid needs to be tied to the text and explained since it appears
in the text and must make sense to readers. In addition, if the communicator
does not explain the importance of the visual aid—its main point, limitations,
assumptions, and implications—then readers will have to provide these pieces
of information for themselves. As a general rule, when readers are put in this
position, they will—at least sometimes—see points or implications different
from those the communicator wants them to see or perhaps even completely
miss the communicator's point.

The easiest way to integrate a visual aid with the text is to explain its
main points and any special implications a reader should note. You might
refer back to the passage on cry spectra in Section 7.2 for a good example
of such an explanation. Another good example occurs in Figure 7-28, which
comes from an anthropology text describing the evolution of human beings.
Note that the explanation of the visual aid gives the thesis the aid is illus-

FIGURE 7-28 Tying a visual aid to the text with an explanation. (From Richard E. Leakey and Roger Lewin, *Origins: What New Discoveries Reveal about the Emergence of Our Species and Its Possible Future,* E. P. Dutton, New York, 1977, p. 73.)

trating: *many characteristics have evolved as a result of the positive feedback mechanism.* It also provides an analysis of the aid and its parts. Other good examples of visual aids that are well integrated and explained occur in the sample reports in the Appendix.

Making the Visual Aid Independent

In addition to being tied to the text, a visual aid also needs to be independent since it will sometimes be read by itself, without any commentary. This occurs whenever a reader leafs through a book or article just looking at the graphs and tables, or when a listener does not pay close enough attention to a speaker and suddenly finds only the visual aid left from which to find the speaker's point. It also occurs when a speaker needs to make an oral presentation, remembers a particularly fine visual from some report, and has it photocopied for the presentation. In this situation, the visual aid is totally independent of the text, and if it has not been properly titled and labeled, it may be difficult to understand.

A communicator can make a visual aid independent by following the advice in Section 7.4, "Designing the Visual Aid": make a relevant, important point and use enough headings, labels, and white space. If these can be done well, the communicator has done his or her job.

EXERCISE 7-1

A Design a visual aid to illustrate the data in each of the following four passages.[4]

1 All of the 841 fatal car-truck collisions that occurred in Michigan and Texas during the periods studied are arranged in three groups: (1) car into rear of truck or tractor-trailer, (2) car into side of truck or tractor-trailer, and (3) all other car-truck collisions (including head-on collisions and truck-into car collisions).

In the Michigan data the total of 490 collisions consists of 48 rear impacts, 46 side impacts, and 396 other types of car-truck collisions. In the Texas data the total of 351 collisions consists of 46 rear impacts, 41 side impacts, and 264 other types of car-truck collisions. The combined total for both states is 94 rear impacts, 87 side impacts, and 660 other types of car-truck collisions. Thus the total number of cases in which underride could have occurred (181) represents 21.5 percent of the total 841 cases. ["Underride" occurs when the car "rides under" the truck.]

2 The frequency of underrides is high in the 181 cases of a car's striking the rear or side of a truck, tractor, or tractor-trailer. In the Michigan data, underride occurred in 41 of the 48 rear impacts and in 30 of the 46 side impacts. In the Texas data, underride occurred in all of the 46 rear impacts and in 35 of the 41 side impacts. In the combined totals, underride occurred in 92.5 percent of the rear impacts and 74.7 percent of the side impacts. The total underrides constitute 18.1 percent of the 841 fatal car-truck accidents.

3 *What crashed:* The 181 car-truck crashes involved 126 tractor-trailer combinations and 55 straight trucks (including two tractors). The tractor-trailers were predominantly vans and flatbeds; the straight trucks were mainly stake or dump trucks. The passenger cars were categorized into three groups by size—full-size cars, intermediates, and compacts—and distributed as follows:

a. In Michigan, 22 percent of the cars involved were compacts, 20 percent were intermediates, and 58 percent were full-size cars.

b. In Texas, 36 percent were compacts, 11 percent were intermediates, and 53 percent were full-size cars.

c. For the total sample (combining the Michigan and Texas samples), 29 percent were compacts, 16 percent were intermediates, and 55 percent were full-size cars.

4 *When:* Car-truck fatal crashes are predominantly nighttime phenomena. Inclement weather is a minor contributor to these accidents, but low visibility at night is a major factor. This will be discussed under "how

TABLE 7-5 FREQUENCY OF IMPACTS AT VARIOUS POINTS OF THE TRUCKS IN THE 47 FATAL ACCIDENTS

POINT OF TRUCK	DIRECTION OF IMPACT	NUMBER OF IMPACTS
Left front wheel area	From left side	1
Left middle	From left side	1
Left rear wheel area	From left side	4
Right front wheel area	From right side	1
Right middle	From right side	5
Right rear wheel area	From right side	2
Left rear	From rear	10
Middle rear	From rear	12
Right rear	From rear	5

and why'' below, but for the moment please note that 67.4 percent of the 181 fatal crashes occurred at nighttime and 32.6 percent occurred during the day.

B   Where necessary, rewrite the passages in part A of this exercise to introduce and present the visual aids you designed.

C   Evaluate the following two passages[4] to see if any visual aids (other than the tables) are desirable. If necessary, rewrite the passage to introduce and present the aids.

1   Points of impact on the truck or tractor-trailer are of interest in considering the potential effectiveness of underride devices. While more than half of the rear impacts were relatively direct (into the rear center of the truck or tractor-trailer), 42 of the 94 occurred at the rear corner of the truck or trailer. Some side impacts occurred in wheel areas of the truck or trailer, and some occurred in places where underride was possible. Table 7-5 shows the number of impacts to various areas of the trucks involved in the collisions.

2   A variety of circumstances or combinations of circumstances precipitated the 181 fatal car-truck accidents studied here. The more common ones are tabulated in Tables 7-6 through 7-9. In rear-collision situations, the truck or tractor-trailer most often was either moving forward ahead of the car or turning left. In side-collision situations, there were three notable circumstances: the tractor/trailer was astride the road jackknifed, astride the road entering or exiting a driveway or ramp, or astride the road making a left turn or attempting a U-turn. About an equal number of times the car driver or the truck driver failed to obey a traffic signal— usually with disastrous consequences for the car occupants.

TABLE 7-6 CIRCUMSTANCES LEADING TO FATAL CAR-INTO-TRUCK UNDERRIDE ACCIDENTS IN MICHIGAN AND TEXAS

MOVEMENT OF INVOLVED TRUCK	PLACE CAR STRUCK TRUCK	NUMBER OF INCIDENTS
Moving forward, ahead of car	Rear	7
Turning left	Rear	8
Slowing or stopping for signal	Rear	3
Crossing centerline or lane line	Side	2
Stopped or parked	Side	1
Jackknifed in path of car	Side	1
Entering or exiting driveway in path of car	Rear	1
	Side	1
Making U-turn in path of car	Rear	2
	Side	4
Running stop sign or failing to yield	Rear	1
	Side	5

TABLE 7-7 CIRCUMSTANCES LEADING TO FATAL CAR-INTO-TRACTOR-TRAILER UNDERRIDE ACCIDENTS IN MICHIGAN AND TEXAS

MOVEMENT OF INVOLVED TRACTOR-TRAILER	PLACE CAR STRUCK TRACTOR-TRAILER	NUMBER OF INCIDENTS
Moving forward, ahead of car	Rear	22
Turning left	Rear	19
	Side	1
Slowing or stopping for sign	Rear	11
Stopped or parked	Rear	5
	Side	1
Jackknifed in path of car	Side	11
Running stop sign or failing to yield	Side	7
Crossing centerline or lane line	Side	1

TABLE 7-8 CAR MOVEMENT IN FATAL CAR-INTO-TRUCK ACCIDENTS

MOVEMENT OF INVOLVED CAR	PLACE CAR STRUCK TRUCK	NUMBER OF INCIDENTS
Running stop sign or failing to yield	Side	5
Crossing centerline or lane line	Side	6

TABLE 7-9   CAR MOVEMENT IN FATAL CAR-INTO-TRACTOR-TRAILER ACCIDENTS

MOVEMENT OF INVOLVED CAR	PLACE CAR STRUCK TRACTOR-TRAILER	NUMBER OF INCIDENTS
Running stop sign or failing to yield	Rear	2
	Side	9
Crossing centerline or lane line	Rear	2
	Side	13

D   Field Manufacturing Company produces three major waste products during its manufacturing process: waste A, waste B, and waste C. The company has developed a 10-year program for treating and refining these wastes and then reusing them in the manufacturing process. To aid in completing this exercise, a summary of the cost of treating and refining each waste product appears in Table 7-10 (assume that the 1982 fiscal year has just ended).

1   The supervisor of each waste treatment program—Supervisor A, Supervisor B, Supervisor C—wants to request more money. As part of the argument, each supervisor wants to compare the funding for her or his waste treatment program to the funding granted to the other two programs. You are to design the best visual aid(s) for each supervisor to use in arguing that her or his program is underfunded in relation to the others. (You will need at least one visual aid for Supervisor A, one for Supervisor B, and one for Supervisor C.)

2   The Head of the Waste Treatment Department must present a financial status report on the waste treatment programs to the company's president. The report must describe the economic history of the programs, the current economic status of each, and predictions for the future. Design the best visual aid(s) for the presentation.

E   Construct visual aids for the information in Tables 7-11 and 7-12. Define a generalization or claim you want the aid to support and make the aid as *visually* supportive as possible.

F   Read through at least one report you have written and evaluate its use of visual aids. Make sure that you have used a visual aid in each necessary or appropriate spot, that you have chosen the best type of visual aid for your purpose, that your visual aid is as *visual* as possible, that it does not distort your data, and that it is appropriately integrated into the text. Add or revise (or delete) any visual aids as necessary. When you have finished, trade papers with a friend and evaluate each other's work.

TABLE 7-10 COSTS OF REFINING WASTE PRODUCTS A, B, AND C (IN THOUSANDS OF DOLLARS)

WASTE PRODUCT	EXPENDITURES 1979	INCREASE IN EXPENDITURES 1979–80	EXPENDITURES 1980	INCREASE IN EXPENDITURES 1980–81	EXPENDITURES 1981	INCREASE IN EXPENDITURES 1981–82	EXPENDITURES 1982	TOTAL (10-YEAR) COST OF PROJECT
A	153.4	16.6 (10.8%)	170.0	19.0 (11.8%)	189.0	20.5 (10.8%)	209.5	2661.3
B	48.5	0.5 (1.03%)	49.0	2.5 (5.1%)	51.5	3.5 (6.8%)	55.0	1237.0
C	178.0	9.0 (5.05%)	187.0	6.0 (3.2%)	193.0	1.2 (0.62%)	194.2	5169.7
Total (A + B + C)							458.7	9068.0

TABLE 7-11 GROSS SALES FOR YOUGAN DAIRY PRODUCTS, 1967–71

YEAR	FRESH MILK PRODUCTS	DRY MILK PRODUCTS	DIVERSIFIED PRODUCTS	DOLLARS (THOUSANDS)
1971	56.7%	15.5%	27.8%	2074
1970	54.0%	17.0%	29.0%	1750
1969	67.0%	20.0%	13.0%	1229
1968	72.0%	23.0%	5.0%	986
1967	74.0%	24.0%	2.0%	932

TABLE 7-12 RELATIVE STEEL PRODUCTION IN THE COMMON MARKET COUNTRIES

	1956	1962	1963	1964	1965	1966
Germany	26.564	32.563	31.597	37.339	36.821	35.316
France	13.442	17.234	17.554	19.781	19.599	19.591
Italy	6.076	9.757	10.157	9.793	12.681	13.639
Belgium	6.367	7.351	7.525	8.725	9.162	8.916
Luxembourg	—	3.456	4.010	4.032	4.559	4.390
Holland	1.000	2.087	2.354	2.659	3.145	3.309

7.6

FORMATTING CONVENTIONS THAT MAKE READING EASIER

There are many features of technical writing that make it look different from most writing we see in newspapers, books, and personal letters. Look, for instance, at Figure 7-29, the beginning of a typical engineering report. (The full report appears in Appendix C.) You will notice that it has some very interesting formatting features:

1   Single-spacing

2   Short paragraphs

3   Lists

4   Headings (underlined titles)

5   Numbers to mark the various paragraphs

6   Liberal use of white space

All of these features occur frequently in scientific and technical writing because they are functional: single-spacing saves space, and the others make a text easier to read, especially for busy and inattentive readers. Headings clearly announce the contents of a section so that busy readers can skip that section if they don't need its details. Short paragraphs and white space make a report easy on the eye, even though it may be single-spaced. The numbering, indentation, and lists provide clues to the organization of the report:

FIGURE 7-29 Formatted version of the
Discussion of a technical report.

1. FAILURE DUE TO TEMPERATURE AND
MOISTURE AND PROPOSED PREVENTION
MEASURES

　　One of the main problems facing LCDs is their strong sensitivity to temperature and moisture. These factors can change the behavior of LCDs rather drastically and result in gradual degradation of the device. Two basic categories can be
recognized in this case: damages due to combination of high temperature and high
humidity and damages due to high temperature only.

　　1.1　Damages Due to a Combination of Temperature and Moisture
　　　　The most damaging environment that an LCD is likely to experience is
produced by the combination of high temperature and high humidity. Depending on how the display is constructed, three failure modes can occur
under this condition: failure of the polarizers, failure of the adhesive on the
polarizer and reflector, and, finally, failure of the liquid crystal mixture.

　　　　1.1.1. Failure of the Polarizers
　　　　　　The most common failure occurs with the polarizers. Regular
polarizing material will begin to degrade after a short period of time (4
days) when subjected to a temperature of 50°C and 95% relative humidity. Once moisture penetrates the plastic backing of the polarizing
material, it affects the polarizing properties. The digit area of the display will start to lose contrast, gradually turning from black to brown
until it finally disappears completely.

　　　　　　1.1.1.1. Preventing Failure of the Polarizers
　　　　　　　　To provide protection against the effects of high tempera-
ture and humidity, the use of a polarizing material known as
"K-sheet," introduced by Polaroid Corporation, is recommended.
This material has polarization properties which are relatively
unaffected by moisture.

　　　　1.1.2. Failure of the Adhesive
　　　　　　Moisture can also attack the adhesive on the polarizer and reflector and cause it to begin peeling away from the glass. If the alignment is lost, dark patches will appear in the viewing area like those
as already described.

　　　　　　1.1.2.1. Preventing Failure of the Adhesive
　　　　　　　　To prevent this kind of failure, the use of a truly hermetic
seal is recommended to protect the adhesive.

　　　　1.1.3. Failure of the Liquid Crystal Mixture
　　　　　　If any of the components in the liquid crystal mixture (such as
Schiff-base compounds) are attacked by moisture, the liquid crystal
molecules will start to break down, thereby lowering the clearing point
of the mixture and the effective upper operating temperature of the
display. Due to the ionic by-products of the chemical reactions, the
electrical currents of the display will also show a very significant increase.

FIGURE 7-29   *(continued)*

        1.1.3.1. <u>Preventing Failure of the Liquid Crystal Mixture</u>
            The procedure described in part 1.1.2.1 should be adopted.

1.2. <u>Failure Due to High Temperature Only</u>
    Failure can occur when an LCD is exposed to high temperatures only.
The twisted nematic display requires the glass plates to be specially treated
to impress a uniform alignment on the liquid crystal molecules next to the
glass surface. The quality of this alignment can be degraded by high temperature because of decomposition of the interface layer.

      1.2.1. <u>Preventing Failure Due to High Temperature</u>
         The procedure used in part 1.1.2.1. should be adopted.

2. <u>FAILURE DUE TO LOSS OF CONTACT</u>
    Problems can arise due to loss of contact between the connector and the
indium oxide leads on the LCD. The user will often blame the LCD for an electrical
failure such as an open segment when in fact it is the connector which is responsible.

    2.1. <u>Preventing Failure Due to Loss of Contact</u>
      This problem was solved by the introduction of the DIL displays with
substrate chips by Liquid Crystal Displays in 1975. They offer a display that
can be soldered directly in a circuit board to reduce the possibility of any
disconnections.

---

they help the reader see what sections provide proof for the generalizations, and they allow a reader to skip freely from section to section without reading everything.

To get a good idea of how helpful these simple formatting considerations can be, look at the unformatted version of the Discussion section of the report, presented in Figure 7-30. Do you agree that it is much more difficult to read? Do you agree that formatting makes the version in Figure 7-29 and Appendix C more functional, that is, easier to read and understand?

One thing that new technical writers have to learn is that the formatting conventions just discussed are widespread as well as functional. Short paragraphs, for instance, are common in much good technical writing. This may seem strange to those who were taught that a "good" paragraph should be at least 100 to 200 words long to be fully developed. Notice, however, that short paragraphs are not necessarily undeveloped paragraphs. You can spread out a 200-word proof over several short indented paragraphs and create a fully developed section, as illustrated in Figure 7-29. Such a section is much easier to read and probably more meaningful than the long-paragraph version presented in Figure 7-30.

FIGURE 7-30 Unformatted version of Figure 7-29.

One of the main problems facing LCDs is their strong sensitivity to temperature and moisture. These factors can change the behavior of LCDs rather drastically and result in gradual degradation of the device. Two basic categories can be recognized in this case: damages due to combination of high temperature and high humidity and damages due to high temperature only.

The most damaging environment that an LCD is likely to experience is produced by the combination of high temperature and high humidity. Depending on how the display is constructed, three failure modes can occur under this condition: failure of the polarizers, failure of the adhesive on the polarizer and reflector, and, finally, failure of the liquid crystal mixture. The most common failure occurs with the polarizers. Regular polarizing material will begin to degrade after a short period of time (4 days) when subjected to a temperature of 50°C and 95% relative humidity. Once moisture penetrates the plastic backing of the polarizing material, if affects the polarizing properties. The digit area of the display will start to lose contrast, gradually turning from black to brown until it finally disappears completely. To provide protection against the effects of high temperature and humidity, the use of a polarizing material known as "K-sheet," introduced by Polaroid corporation, is recommended. This material has polarization properties which are relatively unaffected by moisture. Moisture can also attack the adhesive on the polarizer and reflector and cause it to begin peeling away from the glass. If the alignment is lost, dark patches will appear in the viewing area like those as already described. To prevent this kind of failure, the use of a truly hermetic seal is recommended to protect the adhesive. If any of the components in the liquid crystal mixture (such as Schiff-base compounds) are attacked by moisture, the liquid crystal molecules will start to break down, thereby lowering the clearing point of the mixture and the effective upper operating temperature of the display. Due to the ionic by-products of the chemical reactions, the electrical currents of the display will also show a very significant increase. Liquid crystal mixture failure can be prevented in the same way as adhesive failure. Failure can occur when an LCD is exposed to high temperatures only. The twisted nematic display requires the glass plates to be specially treated to impress a uniform alignment on the liquid crystal molecules next to the glass surface. The quality of this alignment can be degraded by high temperature because of decomposition of the interface layer. Failure due to high temperature can be prevented in the same way as adhesive failure.

Problems can arise due to loss of contact between the connector and the indium oxide leads on the LCD. The user will often blame the LCD for an electrical failure such as an open segment, when in fact it is the connector which is responsible. This problem was solved by the introduction of the DIL displays with substrate chips by Liquid Crystal Displays in 1975. They offer a display that can be soldered directly in a circuit board to reduce the possibility of any disconnections.

7.7

FORMATTING CONVENTIONS THAT MAKE WRITING CLEARER

In addition to making reading easier, the two visual elements discussed in Section 7.6, white space and formatting, can also make your writing clearer. For instance, consider the memo in Figure 7-31 and try to write down the list of items the reader is asked to provide. Then compare your list to that in the version of the same memo presented in Figure 7-32. They might not match.

The paragraph structure of the first version allows many readers to miss one or more of the items. However, in the second version, the formatted list and the extra white space separating the items in the list clarify the number of items and force the reader to see them individually. They also force the writer to decide more carefully how the information is to be presented. For instance, does the writer want all of the information specified in paragraph 2 of Figure 7-31 presented together, in a "clump," or does the writer really want the information segmented into categories 1, 2, and 3, as suggested in

FIGURE 7-31   Sample memo—unformatted version.

SPELLBUCH UNIVERSITY
COLLEGE OF ENGINEERING

October 29, 1979

MEMORANDUM

To: Department and Program Heads
              College of Engineering

From: College Curriculum Committee

Subject: Developing Student's Problem-Solving
              Ability: Request for Information

The Curriculum Committee is quite concerned about the emphasis given to developing the student's capabilities in problem solving and synthesis. Virtually all the descriptions of new and revised courses submitted to the Committee for approval identify the technical material to be disseminated but do not identify techniques to be used in problem solving in particular. The Curriculum Committee feels that probably the amount of open-ended problem solving in the College should be increased, but the Committee has limited data on the nature and extent of problem solving now incorporated in College courses.

To more accurately assess the current situation, the Committee is requesting you to provide by November 29, 1979, a list of undergraduate courses (i.e., through 400 level) in your program which incorporate a significant amount of design and open-ended problem solving or which discuss problem-solving techniques for either open-ended or closed problems. An example of such a course is attached. "Significant

---

FIGURE 7-31   *(continued)*

---

amount" should be interpreted as at least one or two weeks devoted to the subject including design, analysis, synthesis, and evaluation stages rather than simple assignment of an occasional short problem. Please make a rough estimate of the fraction of effort in these courses which is devoted to open-ended problem solving and design. In addition, please identify which of the listed courses are required in your program and estimate how many are taken by the "average" student.

Please let us have the names of faculty teaching the courses and of faculty who are particularly interested in this topic. The definition of a course as problem solving is obviously not clear-cut, but please make a reasonable interpretation of this classification.

The Committee would also appreciate your giving thought, and possibly some answers, to the following (open-ended!) questions:

What resources could the Curriculum Committee provide to help promote problem-solving courses? Possible examples include organizing seminars, compiling a bibliography of literature related to problem-solving technique and pedagogy, and enlisting the help of the Center for Research on Learning and Teaching. Would involvement be greater if help and advice in developing these courses were available? Are there special computer needs for these courses?

We would appreciate it if we could have your response to the above questions by November 29, 1979. Thank you for your consideration and help on this questionnaire.

HSF:mb

Attachment

Figure 7-32? If the writer wants segmented information, he or she may need to do a lot of analysis and reclassification of the data received from readers of Figure 7-31; presumably this extra work would be avoided by sending out the version shown in Figure 7-32.

Further, a quick or careless reader is also more likely to notice that information is requested in Figure 7-32 because the second paragraph states this in a very obvious way.

In many situations, formatting information into obvious lists or breaking it up into clearly labeled sections helps a reader. When formatting, white space, and labels are used, the reader and writer are more likely to understand each other and to respond appropriately.

FIGURE 7-32 Sample memo illustrating the use of formatting, white space, and headings.

SPELLBUCH UNIVERSITY
COLLEGE OF ENGINEERING

October 29, 1979

MEMORANDUM

To: Department and Program Heads
           College of Engineering

From: College Curriculum Committee

           Developing Student's Problem-Solving
           Ability: Request for Information

The Curriculum Committee is quite concerned about the emphasis given to developing the student's capabilities in problem solving and synthesis. Virtually all the descriptions of new and revised courses submitted to the Committee for approval identify the technical material to be disseminated but do not identify techniques to be used in problem solving in particular. The Curriculum Committee feels that probably the amount of open-ended problem solving in the College should be increased, but the Committee has limited data on the nature and extent of problem solving now incorporated in College courses.

To more accurately assess the current situation, the Committee is requesting you to provide by November 29, 1979, the following information:

1. a list of undergraduate courses in your programs which incorporate a significant amount of design and open-ended problem solving or which discuss problem-solving techniques for either open-ended or closed problems (we realize that the definition of a course as having a "significant amount of problem solving" is not clear-cut, but we have appended an example of such a course and would request that "significant amount" be interpreted as at least one or two weeks devoted to problem solving and its techniques (including focus on design, analysis, synthesis, and evaluation) rather than as the simple assignment of an occasional short problem),

2. a rough estimate of the fraction of effort devoted to open-ended problem solving and design in these courses,

3. the identification of those courses listed in #1 above which are required in your program,

4. an estimate of how many of the courses listed in #1 are taken by the "average" student,

5. the names of faculty teaching these courses who could serve as resource persons within the College, and

FIGURE 7-32   *(continued)*

6. the names of faculty particularly interested in problem solving and its techniques

The Committee would also appreciate your giving thought, and possibly some answers, to the following (open-ended!) questions:

1. What resources could the College Curriculum Committee provide to help promote problem-solving courses? (Possible examples include organizing seminars, compiling a bibliography of literature related to problem-solving technique and pedagogy, and enlisting the help of the Center for Research on Learning and Teaching.)

2. How could the College or the Departments involve more faculty in expanding problem solving in our courses? Would involvement be greater if help, release time, and/or advice in developing these courses were available?

3. What, if any, are the special computer needs for problem-solving courses in your department?

We would appreciate it if we could have your response to the above questions by November 29. Thank you for your consideration and help on this questionnaire.

HSF:mb

Attachment

REFERENCES

1   William Schneider and William D. Green, Jr., "Saving Skylab," *Technology Review* **76**:3 (January 1974), p.44.

2   Wann Langston, Jr., "Pterosaurs," *Scientific American* **244**:2 (February 1981), p. 126.

3   *Technology Review* **81**:8 (August-September 1979), p. 79.

4   Adapted from Daniel J. Minahan and James O'Day, *HSRI Research Review* **8**:3 (The University of Michigan Highway Safety Research Institute, November-December 1977), pp. 4–5, 8.

ADDITIONAL READING

Rudolf Arnheim, *Art and Visual Perception: A Psychology of the Creative Eye,* University of California Press, Berkeley, 1974.

Rudolf Arnheim, *Visual Thinking,* University of California Press, Berkeley, 1969.

Gregg Berryman, *Notes on Graphic Design and Visual Communication,* William Kaufmann, Los Altos, Calif., 1979.

R. N. Haber, "How We Remember What We See," *Scientific American* **222**:5 (May 1970), pp. 104–112.

Michael Macdonald-Ross, "How Numbers Are Shown," *AV Communication Review,* **25** (Winter):4 (1977).

Michael Macdonald-Ross, "Graphics in Text," *Review of Research in Education,* vol. 5, L. Shulman (ed.), F. E. Peacock, Itasca, Ill., 1978.

A. J. MacGregor, *Graphics Simplified: How To Plan and Prepare Effective Charts, Graphs, Illustrations, and Other Visual Aids,* University of Toronto Press, Toronto, 1979.

Cheryl Olkes, "Typography/Graphics," *Document Design: A Review of the Relevant Research,* American Institutes for Research, Washington, D.C., 1980, pp. 103–110, 163–166.

Mary Eleanor Spear, *Practical Charting Techniques,* McGraw-Hill, New York, 1969.

Patricia Wright, "Presenting Technical Information: A Survey of Research Findings," *Instructional Science* **6**:93–134 (1977).

Part V

Major Genres of Technical Reporting

8

THE ORAL PRESENTATION

In many fields of science and technology, the ability to communicate technical information orally is just as important as the ability to write well. Indeed, those in management positions often find themselves making oral presentations more often than written ones; they find oral presentations faster, easier, and more suited to immediate feedback and clarification. A number of studies suggest that oral communication skills are of crucial importance, in fact, in determining who gets promoted to upper management levels.[1–5]

Managers are not alone, however, in their need to have good oral communication skills: engineers, scientists, and other technical professionals also need these skills. This can be seen, for example, in the results of a recent survey reported in *Engineering Education*.[6] In this survey, 367 engineers in chemical, civil, electrical, geological, mechanical, metallurgical, and mining engineering were asked to rate 30 specific communication tasks for job-related importance. Some of these engineers were working as managers, but most were not. Of the 24 tasks that were rated important by the respondents, 14 were writing tasks and 10 were oral tasks. Of these, five involved the giving of a formal oral presentation (project proposal presentations; oral presentations using graphs, charts, and/or other aids; project progress report presentations; project feasibility study presentations; and formal speeches to technically sophisticated audiences); the remainder were of a more informal nature, involving one-to-one talks or small-group discussions.

This chapter is devoted mainly to giving you advice on how to make an effective formal oral presentation. Preparation is the key: if you plan your presentation properly and practice it beforehand, you'll find that the actual delivering of an oral presentation will be relatively easy; if you don't prepare, it could be a real struggle. So, by all means, give yourself plenty of time to prepare an oral presentation. In particular, follow the procedure we've laid out for you step by step, taking care not to omit any of the steps. That way, you'll not only give an effective presentation but may even *enjoy* doing it!

There are times, however, when you may not be given sufficient time to prepare—when your boss, for example, says that an important group of visitors is arriving in half an hour and you are to give them a briefing. What do you do then? At the end of the chapter, we'll offer some suggestions.

8.1

GIVING A FORMAL ORAL PRESENTATION

Preparation

The basic principle to keep in mind in preparing any kind of oral presentation is that all listeners have a limited attention span and cannot be expected to follow everything you say. Their attention will probably wander from time to time, even if your presentation is only 10 minutes long. So, if you want to make sure that your listeners will come away from your talk with your main points clear in their minds, you must organize your presentation in such a way that these main points stand out. Here is how to do it:

1   *Analyze your audience, and limit your topic accordingly.* What do your listeners already know about the topic? What do they need or want to know about it? How much new information about it can they absorb? If you tell them what they already know, they'll be bored, but if you go to the opposite extreme and give them too much information too fast, they'll be overwhelmed and simply "short circuit."

2   *Determine your primary purpose.* Is there some main point you want to get across? Is there something you want your listeners to believe, or be able to do? Whatever it is, you should have it clear in your mind so that you can build your presentation around it.

3   *Select effective supporting information.* What kind of information will best support your main point? What kind of information will appeal to your audience? They will probably be able to remember no more than three or four main supporting points and no more than two or three supporting details for each of these. So choose wisely!

4   *Choose an appropriate pattern of organization.* Often, your supporting information can be presented according to a single dominant pattern of organization. For example, if you're trying to tell your listeners how to do something, you may want to organize your information into a list of instructions, chronologically ordered. On the other hand, if you're describing an experiment, you'll probably want to use the standard descriptive pattern: Introduction, Materials, Procedure, Results, Discussion. There are many possible patterns of development to choose from, some of which are described in Chapters 2 and 5. Whichever one you choose should be appropriate to the subject matter, to your primary purpose, and to your audience—and it should be followed consistently.

5   *Prepare an outline.* Keep it brief: main points, main supporting points only. Arrange these points according to the pattern of development you chose in Step 4. Do not write out a complete text unless the technical content or circumstances warrant it.

6   *Select appropriate visual aids.* Visual aids, when properly designed, have tremendous power as attention-getters. If you want to emphasize a point, by all means try to do it visually as well as orally. As a general practice, in fact, it's a good idea to emphasize all major points and major supporting points with visual aids. Properly sequenced, the aids will then also serve as cue cards for you. This does not mean simply putting the outline of your speech on flip charts. Words and phrases alone do not really have visual impact, at least not in the way that graphs, sketches, three-dimensional objects, and other pictorial displays do. Visual aids should be truly *visual!* The more visual they are, the more effective they will be in capturing your audience's attention. (Review Chapter 7 on the principles of designing effective visual aids.)

If you have any reason to believe, however, that your audience may have trouble understanding your pronunciation—for example, if you are a nonnative speaker of English or if you are addressing a group of nonnative speakers—then it would be a good idea to write out your main points and some keywords for your audience.

7   *Prepare a suitable introduction.* It's essential that your listeners have enough background information to understand and appreciate your presentation. Are you addressing some problem? Make sure you define it so that your listeners know exactly what it is and can appreciate your proposed solution. Are you taking sides over an issue and arguing for your point of view? If so, make sure your listeners know exactly what the issue is. In short, if you have any doubts about the audience's background knowledge, be sure to provide a basic orientation and to define important terms.

8   *Prepare a closing summary.* Listeners are typically very attentive at the beginning of a presentation, less attentive as it wears on, and then suddenly more attentive again as it comes to an end. In other words, they perk up at the end, hoping to catch a final summarizing comment or recommendation. This is a well-proven phenomenon, common to all of us: just think about the times you've listened to someone else's presentation and have come away remembering his or her final words best of all.

In preparing your oral presentation, therefore, you should plan to take advantage of this fact of human psychology—you should prepare a good, solid closing summary. Take this opportunity to repeat and thus reemphasize your most important "bottom-line" conclusions and recommendations, along with the major reasons for them. Make these closing comments crisply and emphatically.

EXAMPLE You are a systems analyst employed by the Coronado Sugar Company (CSC). For a number of years the company has been using outside help in recording the amount of sugar beets received at its four processing plants. Since this service is quite costly, your supervisor has asked you to investigate the feasibility of using the company's own computer for this purpose. You have done so and are now ready to prepare for a brief oral presentation of your findings. Accordingly, you go through the following eight steps:

1   *Audience analysis.* You expect the audience to include, in addition to your supervisor, the managers of the four processing plants and the company comptroller. All of these people are familiar with the company's system of recording individual deliveries on weight tickets; they are also familiar with the daily and weekly summaries that are produced from these weight tickets. They do not, however, know how these summaries are put together (that's what the outside firm has been doing for $15,000 a year), and they do not know much about computers. Your supervisor and the company comptroller will be concerned mainly with how much money can be saved by adopting a new system. The four plant managers, on the other hand, will want to know mainly whether or not a new system will disrupt plant operations in any way.

2   *Purpose.* Your investigation has revealed that there are three in-house methods that would be substantially more cost-effective than using an outside firm and that any of these methods could be implemented without disrupting company operations. One of these methods would actually save the company more than $100,000 over 8 years, and you would like to see it adopted. However, if you push this one method to the exclusion of the others, your presentation may sound too much like a sales pitch. So it might be a good idea to broaden your scope a bit and promote the idea that any of the three in-house methods would be better than the present system. This will give your listeners a range of options from which to choose. Surely at least one of the options should appeal to them.

    Your primary purpose, then, will be simply to convince your listeners that the present system of compiling beet receiving data should be replaced by an in-house method. Once you've done that, it should be relatively easy to sell your own particular preference from among the three competing options.

3   *Supporting information.* Your supervisor (the decision maker in this case) and the comptroller will be most impressed, of course, with the cost figures, which favor the three in-house options over the present system; be sure to show both the short- and long-term cost advantages. The four plant managers will want to know how the three alternative methods work; in particular, they will want to be reassured that an in-house system will not disrupt their plant operations. All of your listeners

will be pleased to hear that at least one of these in-house methods (the one you prefer) will provide a fringe benefit in the form of making extra equipment available for other uses during the off-season.

4   *Pattern of organization.* Your audience's main concern will clearly be how the four methods compare in terms of cost benefits and impact on plant operations. So the best way to present your supporting information would be according to a comparison-and-contrast pattern of organization.

5   *Outline*   I   Problem statement
                        Need for compiling summaries
                        Present method

                  II   Alternative methods
                        Method 1
                          Costs
                          Advantages and disadvantages

                        Method 2
                          Costs
                          Advantages and disadvantages

                        Method 3
                          Costs
                          Advantages and disadvantages

                  III Summary
                        All three alternative methods are more cost-effective
                          than the present system.
                        Method 2 is the most cost-effective in the long run.

6   *Visual aids.* Your audience will want to see a projected cost breakdown for each of the three alternative methods, as well as a summary graph comparing all four (Figure 8-1). The summary graph is particularly important, as it will allow you not only to emphasize your main point (that any of the three alternative methods would be cheaper than the present system) but also to show how Method 2 is ultimately the most cost-effective.

7   *Introduction.* You'll want to engage your audience's interest from the outset by explaining the full nature of the problem: CSC needs to compile regular summaries of beet receiving data for various purposes, but the present system of doing so is far too expensive. Don't get bogged down in too much background information: your listeners already know a lot about the present system, and much of what they don't know (for example, how the outside firm puts together the data) they don't really need to know. Also, this is supposed to be a *brief*

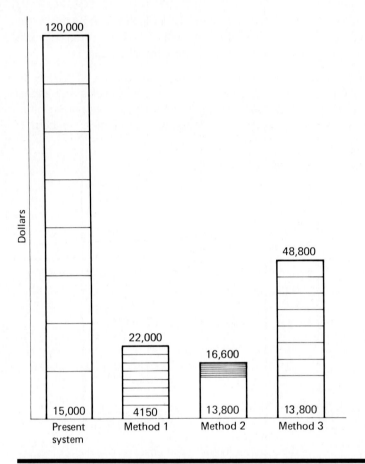

FIGURE 8-1   Comparison of processing methods: total cost for
8 years.

presentation, so don't waste your audience's time telling long-winded
jokes or anecdotes. Instead, get right to the point. After describing the
problem, tell your listeners that you've investigated various possible
solutions and have found three that deserve their consideration.

  Your introduction might go something like this:

> As you all know, every fall we collect an enormous number of weight
> tickets from our four processing plants. In fact, we've been averaging
> about 100,000 of them over the past five years. From all of these tickets
> we have to compile regular summaries—daily summaries and weekly
> summaries of beets received for each plant, and then final summaries of
> delivered amounts for each grower. The daily summaries are particularly
> important because we need these to maintain smooth plant operations,

keep to our shipping schedule, and allow for any sudden breakdowns. This means, of course, that the data have to be processed fast and accurately, and in the past we've always felt we needed the MFC people to do it. After all, they're the pros. And indeed, they've done a fine job for us. The problem is that their service is very expensive: $15,000, to be exact, for just the two months we use them.

So last month Glenn asked me to look into the possibility of using our own company computer to do the job. I've now completed my investigation and am happy to report that there are indeed three different ways we could use our own computer to get the job done without spending nearly as much money or sacrificing any efficiency. One of these methods, in fact, could save us as much as $100,000 over the next eight years, without any adverse impact on plant operations. There are even some fringe benefits to be gained. . .

8   *Summary.* You'll want to reiterate the fact that all three in-house methods are more cost-effective than the present system, and you'll want to use your summary chart (Figure 8-1) to emphasize this point. More specifically, you'll want to repeat your claim that Method 2 is the most cost-effective of all, noting that not only would it save CSC the most money in the long run but it would also allow the company to invest in equipment which could be used for other valuable purposes in the off-season. Be sure to make it clear to your listeners that these are not only your conclusions but also your strong recommendations:

. . .and so I recommend strongly that we stop using MFC's services and create instead our own in-house program, using our own equipment and our own people. It would take us a little time to get it up and running, but we have the time and will start saving money the minute it's going. It will also increase our computing capability in all sorts of other ways. If you have any questions, I'd be glad to try to answer them for you.

Practice

Nothing is more helpful to the ultimate success of an oral presentation than practice. Not even the best of speakers can give a totally effective presentation without practicing it first. Practice allows you to spot the flaws in a presentation and eradicate them. It enables you to work on making smooth transitions from section to section, instead of awkward stops and starts. And practice gives you an idea of how long your presentation will take; if it's too long, you still have time to make changes so that you can ultimately deliver it at a tempo that's comfortable for you. All of these benefits promote greater self-confidence, which in turn leads to a more convincing, emphatic, effective style of delivery.

The best way to practice a talk is by rounding up a few friends and trying it out on them. Ask them to hear you all the way through, taking as many notes as possible but not raising any questions until you've finished. Then ask them for a complete "postmortem"; take note of spots where they had trouble following you, and immediately try out some other approach to see if it makes things clearer. Note also what the strong parts of your talk are: maybe you can use the techniques used in the strong parts in other places.

If you'd prefer not to force your friends to listen to you stumble through a practice session, a good alternative would be to use a videotape recorder and then critique your own performance. This is particularly effective in allowing you to spot nervous mannerisms that you might not be conscious of while actually performing. An audiotape recorder, while not as useful as a videotape recorder, can also be used, especially for the purpose of listening to yourself read from a manuscript (see point 6 below).

Here are some specific things to work on while practicing your oral presentation:

1   *Devise ways of repeating your important points without being too repetitive.* Since your important points should all contribute to a single cumulative effect, it's a good idea to reiterate these points occasionally as you go along—especially in summary form at the end of your talk. However, since exact repetition of wording can become annoyingly monotonous the third or fourth time around, try to vary your wording.

2   *Create smooth transitions between sections.* Take note of places where the flow of your presentation seems to break down, and see if you can't insert a phrase or two to act as a bridge. (If you can't, there may be a fundamental flaw in the overall structure of your presentation; in that case, try to reorganize it.)

3   *Familiarize yourself with the equipment you'll be using.* It's embarrassing—and annoying to the audience—to waste precious time by fumbling around with slide projectors, television monitors, microphones, and other equipment when you're supposed to be giving your presentation. It's even worse if your presentation depends crucially on some piece of equipment and you can't get it to operate at all. Therefore, *if you plan on using any equipment, check it out ahead of time and become familiar with it!* Learn how to use it so that it will help your presentation, not hinder it, and be prepared with some backup system just in case.

4   *Prepare yourself for questions.* Listeners may raise questions at any point in your presentation, and it is vitally important that you answer them satisfactorily. If you don't, your most precious asset as a speaker— your credibility—may be endangered. So be sure you know your topic *well.* To test your knowledge, have some friends listen to you while you practice and have them deliberately throw tough questions at you;

if they succeed in stumping you, go do some research. It's not necessary that you have an answer for *every* conceivable question the audience might raise; there are questions for which *no one* has an answer (and you should certainly never try to fake an answer!). But if you have researched a topic well and have state-of-the-art knowledge about it, you'll be able to answer most questions and will be able to say confidently about the others, "We don't have an answer to that question yet."

5   *Develop your own speaking style.* Practice telling stories or jokes to a few friends at a time in an informal setting. Stand tall and face them as in an oral presentation; engage your listeners' interest. Use natural, animated gestures, and vary your intonation and rate of speech. In short, *be expressive! Let your enthusiasm show!* At the same time, take note of any distracting habits you might have: leaning against something, biting your nails, playing with a pencil, and so on. If you are aware of such habits, you can often take measures to keep them under control while you're "on stage."

6   *If you are going to be reading from a manuscript, work on giving it a lively intonation.* There is a strong tendency when reading aloud to adopt a monotonous style of delivery that is very boring for an audience to listen to. So practice varying your intonation as you read aloud. Raise your voice pitch for the more important words and allow it to drop for the less important ones. Pause occasionally at appropriate places. Put special emphasis on contrasting words, as in the following example:

> System 2986 is an experimental system with two processors which can work *independently* or work *together as coprocessors. One* processor is a commercially available single-board computer (Intel's ISBC 86/12A), which is also referred to as System *86. The other* processor is a custombuilt microprogrammable special-purpose processor which is built around AMD 2900 series components and referred to as System *29.* The two processors can talk to each other over the Multibus using the Multibus protocols.

Delivery

As the time draws near for actually delivering your oral presentation, you will experience what all speakers experience: nervousness! If you have adequately prepared and practiced your presentation, of course, you will probably be less nervous than if you have not. No matter how well you prepare yourself, however, you will still be at least a little bit on edge.

The key to an effective delivery is to convert your nervousness into the kind of energy that injects liveliness, enthusiasm, and animation into your speech. If you try to either suppress your nervousness or let it dominate you, the result will be either a listless, boring, "laid back" style or a jittery, key-jingling, pacing-back-and-forth one—neither of which will appeal to your audience.

How can you control your nervousness? Basically, by not being self-conscious. The best way to do this is by establishing a "conversation" with your listeners. If you concentrate on your listeners and on what you're trying to tell them—just as you do in small-group conversation—you'll forget about yourself and lose your self-consciousness. And, in the process, you'll lose your nervousness.

Here are some specific suggestions:

1   *Establish some personal contact with at least some of your listeners before you begin your talk.* Mingle with them. See what they're interested in, what kind of mood they're in. That way, you'll have some friendly faces to look at during your talk.

2   *As you give your presentation, concentrate your full attention on what you want to say to your listeners.* Stick to your outline, make sure you cover all your main points and main supporting points. Plan ahead; know where you're going, what you'll say next. Remember, your listeners are interested more in your topic than in you. If you provide a steady stream of well-organized information, they'll pay attention to you and give you support. If you don't, they'll lose interest.

EXERCISE 8-1

A   Convert one of your written reports into a 5- or 10-minute oral presentation suitable for an audience of nonspecialists.

B   Convert the same report into a 5- or 10-minute oral presentation suitable for an audience of specialists.

C   Take a technical discussion section (300 to 500 words) from one of your reports and prepare it for oral reading to a large audience.

8.2

GIVING AN INFORMAL ORAL PRESENTATION

Occasionally you may be asked to give an oral presentation on short notice and may not have time to practice or prepare any visual aids. For example, as a member of a project team you may be asked at any time to give a spur-of-the-moment progress report. Or perhaps an orientation tour is taking place and you are asked to describe your laboratory. Perhaps your group leader has been asked to present a feasibility study but has been taken ill, and you are a last-minute replacement for her.

In such cases, the best thing to do is to concentrate on the first five steps in preparing any oral presentation, formal or informal: (1) audience analysis, (2) determination of primary purpose, (3) selection of supporting information, (4) selection of a pattern of development, and (5) outlining. Each of these steps is vitally important and should be performed as carefully and thoroughly as time permits. But *keep it simple!* Don't let yourself get bogged down in complicating details. Time does not allow it, and it will only result in disorganization if not outright confusion. Instead, keep your mind on your main points and how you can best and most clearly support them.

EXERCISE 8-2   Take no more than 30 minutes to prepare a 10-minute informal oral presentation on a familiar topic. Then give it. (If you're a student, you might arrange to give the presentation to one of your teachers during office hours. If you're working for a company, try your supervisor or some colleagues or perhaps a technical writer or editor.)

REFERENCES

1   William H. Middendorf, "Academic Programs and Industrial Needs," *Engineering Education,* May 1980, pp. 835–837.

2   Richard M. Davis, "Technical Writing: Who Needs It?" *Engineering Education,* November 1977, pp. 209–211.

3   Ed Gilbert, speech to the Conference on Teaching Technical and Professional Communication, co-chaired by Dwight W. Stevenson and J. C. Mathes, Department of Humanities, College of Engineering, The University of Michigan, 1975.

4   Nicholas D. Sylvester, "Engineering Education Must Improve the Communication Skills of Its Graduates," *Engineering Education,* April 1980, p. 739.

5   Peter M. Schiff, "Speech: Another Facet of Technical Communication," *Engineering Education,* November 1980, p. 181.

6   William R. Kimel and Melford E. Monsees, "Engineering Graduates: How Good Are They?" *Engineering Education,* November 1979, pp. 210–212.

ADDITIONAL READING

Aristotle, *The Rhetoric of Aristotle,* translated by Lane Cooper, Prentice-Hall, Englewood Cliffs, N.J., 1932, 1960. Another good edition of this work is *The "Art" of Rhetoric,* translated by J. H. Freese, Harvard University Press, Cambridge, Mass., 1926, 1967.

Kevin R. Daley, "Presenting Your Viewpoint," *Chemical Engineering,* November 17, 1980, p. 287.

William E. McCarron, "Oral Briefing versus Technical Report: Two Approaches to Communication Problems," *Courses, Components, and Exercises in Technical Communication,* D. W. Stevenson, (ed.), National Council of Teachers of English, Urbana, Ill., 1981, pp. 144–156.

Thomas M. Sawyer, "Preparing and Delivering an Oral Presentation," *Technical Communication,* First Quarter 1979, pp. 4–7.

Peter M. Schiff, "Speech: Another Facet of Technical Communication," *Engineering Education,* November 1980, pp. 180–181.

9

TECHNICAL REPORTS

As noted in Chapter 1, experienced scientists and engineers rate many writing and speaking skills as being very important or even critically important to job performance. Written communications cited for their importance and frequency include project proposals, progress reports, technical descriptions, memos and short reports, feasibility studies, and, depending upon the organization, business letters, long formal reports, and technical articles. Important oral communications cited include one-on-one talks with nontechnical personnel, project proposal presentations, progress reports, feasibility study presentations, other oral presentations using visual aids, and committee work. Since these communication forms are so important, these next four chapters will describe some organizing and formatting conventions for important written communications. Chapter 8 focuses on important oral ones.

Technical reports, proposals, and business letters are read by a variety of audiences: by technical experts in the field as well as by managers and other nonexperts—the busy, inattentive, and perhaps uninterested reader described in Chapter 4. Thus, the report, letter, or proposal should be accessible to all these readers and responsive to their needs and reading habits. In part, such responsiveness appears in two important structural features: (1) the Foreword and Summary (or their analogues) in the technical report and Abstract and Statement of Problem in the proposal and (2) the placement of generalizations or claims before their support in the part of the report, proposal, or letter written for expert readers.

The technical article, which is read almost exclusively by expert readers, uses the second feature in addition to others described in Chapter 12.

9.1

FOREWORD AND SUMMARY: ORGANIZING MAIN POINTS FOR NONSPECIALIST READERS

The Foreword and Summary, which are discussed in Chapter 6, are placed at the beginning of a technical report to be accessible to the busy managerial or nonexpert reader. The Foreword orients the reader and shows that the report addresses a problem or issue important to the reader. The Summary provides a compact statement of results, conclusions, and recommendations to help managerial readers make decisions: it tells what the writer discovered, the implications of the discovery, and the recommendations for action based on the writer's special knowledge—all in terms that the nonspecialist manager will find useful. By placing the Foreword and Summary first in a technical report, the writer allows the managerial reader to stop after the first page or so and still have all of the necessary information.

9.2

CLAIMS BEFORE PROOF: ORGANIZING DETAILS
FOR SPECIALIST READERS

The Abstract

The Abstract is a very important document for the specialist reader. In combination with the title, the Abstract is the vehicle through which readers

1   Identify and locate a report

2   Decide if they need to read it or not

In the second case, the Abstract is a sorting device; it should sort out all readers who don't need to read a report from those who do. To fulfill the identifying and locating function, the Abstract must contain the bibliographic information needed by libraries and information retrieval systems for cataloguing. To fulfill the sorting function, the Abstract must in addition contain a condensation of the report.

The bibliographic information needed in an Abstract varies depending on the report's purpose, audience, and expected circulation. For instance, a report which will circulate only inside the organization which produced it (an in-house document) might need only

1   The name of the author

2   The title of the project

3   The date it was written

4   The project under which the document was produced

5   Perhaps a special organizational number

On the other hand, a report available to people outside the organization which produced it might require all of that information plus

6   The name of the organization which produced it

7   The name of the organization which commissioned it (if any)

8   The contract number under which the document was produced (if any)

9   A security classification

10   Key words and special numbers for cataloguing in an information retrieval system

Items 7 through 10, for instance, are necessary in reports on government-sponsored research.

As a report writer, you will have to find out what types of information are required for your report. You can do this by looking at similar reports which have already been published to see what kinds of information they include, but the best methods are (1) to consult the "information or instructions to authors" published by the organization for which you are writing or (2) to consult with your librarian.

The condensation of the report is probably what we consider to be the heart of the Abstract. It reviews the investigation, main conclusions, and recommendations in the report. All of this needs to be done in 50 to 250 words, depending upon the writing situation and the word limit set on the Abstract. The Abstract needs to be totally self-contained because it will often be listed in abstracting or retrieval sources totally separate from the report.

Since so much needs to be done in so few words, there is a great tendency to write very general condensations, such as the following:

> G. E. Lambert and D. F. Bryan. 1978.
> The Influence of Fleet Variability on Crack Growth Tracking
> Procedures for Transport/Bomber Aircraft
> Air Force Flight Dynamics Lab. Rep. AFFDL-TR-78-158
> 198 pp., illus., Boeing Witchita Co., Witchita, Kan. Unclass.

> The purpose of this program is to provide generalized crack growth tracking procedures for transport/bomber aircraft. A study and tests were conducted on crack growth rate characteristics, analysis schemes were developed, and tracking procedures were evaluated.

> KEY WORDS: Crack Growth Tracking, Individual Aircraft Tracking, Force Management, Damage Tolerance, Durability, Spectrum Development, Crack Growth Analysis, Crack Growth Retardation, Retardation Models, Data Acquisition Systems.

This type of Abstract is called a descriptive abstract, because it merely describes what the report is about. It does not give any specific information about the problem, method, results, or conclusions. Such an Abstract is not very useful to many readers since it lacks the specific information which would tell them (1) what they need to know about the report and its results or (2) if they really need to get the report to more closely inspect its contents. A more useful type of Abstract is the informative abstract, which gives as much of the important particular information as can be crammed into the word limit. Consider, for instance, an informative version of the descriptive abstract presented above:

> The purpose of this program is to provide generalized crack growth tracking procedures for transport/bomber aircraft. The study was composed of three major tasks: (1) an evaluation of the effects of usage parameters on crack

TABLE 9-1 THE RELATIONSHIP BETWEEN SUMMARY AND ABSTRACT*

	SUMMARY	ABSTRACT
Bibliographic data		XX
Overview of problem		XX
Review of method	x	XX
Review of results	XX	XX
Review of conclusions	XX	XX
Review of recommendations	XX	x

*x indicates optional material; XX indicates necessary material.

growth, (2) the development of generalized tracking procedures and (3) an evaluation of the techniques for implementing the individual aircraft tracking program. The KC-135A tanker was selected as the baseline aircraft for this study. Approximately thirty tests were conducted to experimentally verify predictions made from analyses and to determine the crack growth rate characteristics of the specimen material. The results of the parametric and variability studies were used to develop analysis schemes for predicting the effects of usage variations on crack growth. Four tracking procedures were evaluated: (1) pilot logs with the use of parametric crack growth rate data, (2) pilot logs with the use of parametric stress exceedance data, (3) the Mechanical Strain Recorder (MSR) and (4) the crack growth gauge. The implementation of each tracking procedure was evaluated by developing a cost model to study relative life cycle costs.

Notice that this informative version of the Abstract tells the reader much more about the report than did the descriptive version. It tells the reader specifically what the major tasks were in the investigation, what particular aircraft was used in the study, how many tests were conducted and why they were conducted, what uses were made of the results of the studies, which tracking procedures were evaluated, and how the implementation of the tracking procedures were evaluated. It did all of this in 172 words, but more important, it did this by giving some of the principal details of the report.

One last point that should be considered is the relationship between the Summary and the Abstract. In a technical report, the Summary is written for the nonspecialist managerial reader, whereas the Abstract is written for the specialist technical reader. In addition, the Abstract covers more ground than does the Summary, since the Abstract briefly presents bibliographic data and an overview of the problem, which are missing in the Summary. In comparison to the Summary, the Abstract also stresses the methodology and slights the recommendations. The relationship between the two is outlined in Table 9-1.

EXERCISE 9-1

A   Three Abstracts follow, two of them different versions for the same report. See if you can decide which ones are informative and which are descriptive.

B   Rate the Abstracts from best to worst from a user's point of view. Be prepared to justify your choices.

C   Write an informative Abstract for a report you are writing or one which you have read.

ABSTRACTS 1 and 2
Rinalducci, Edward J., and Arthur N. Beare, "Visibility Losses Caused by Transient Adaptation at Low Luminance Levels," *Driver Visual Needs in Night Driving,* Transportation Research Board Special Report 156, 1975, pp. 11–22.

Transient adaptation refers to the rapid fluctuations in the sensitivity of the eye that result from sudden changes in luminance level. The research reported here examines the effects of transient adaptation and resultant losses in visibility by using luminance levels comparable to nighttime highway lighting conditions. Research has been initiated on the problem of nonuniformities in roadway luminances in the motorist's visual environment. Experiments to examine multiple nonuniformities and the effect of nonuniformities at various distances from the line of sight on transient adaptation are planned.

Transient adaptation refers to the rapid fluctuations in the sensitivity of the eye that result from sudden changes in luminance level. The research reported here examines the effects of transient adaptation and resultant losses in visibility by using luminance levels comparable to nighttime highway lighting conditions. At low luminance levels, sudden increases produce losses in visibility equivalent to those previously found at higher levels. However, at low luminance levels, decreases produce smaller losses than those observed at higher luminance levels. The results also suggest that there is a preadapting level or range of levels below which there is little or no difference between visibility losses for 10- and 100-fold decreases and above which there is a difference. The transition appears to be a gradual one and is complete at about 8 ft-L. The findings of these investigations suggest that visibility loss depends more on the ratio of steady state thresholds, particularly at low luminance levels, than on the ratio of luminance change as previously supposed. Research has been initiated on the problem of nonuniformities in roadway luminances in the motorist's visual environment. Results indicate that the size of a nonuniformity may have little effect on transient adaptation. However, experiments to examine multiple nonuniformities and the effect of nonuniformities at various distances from the line of sight on transient adaptation are planned.

ABSTRACT 3
McKenzie, Robert Lawrence, "*Vibration-Translation Energy Transfer in Vibrationally Excited Diatomic Molecules,*" NASA/Technical Report NASA TR- R- 466, October 1976, unclass.

A semiclassical collision model is applied to the study of energy transfer rates between a vibrationally excited diatomic molecule and a structureless atom. The molecule is modeled as an anharmonic oscillator with a multitude of dynamically coupled vibrational states. Three main aspects in the prediction of vibrational energy transfer rates are considered. The applicability of the semi-classical model to an anharmonic oscillator is first evaluated for collinear encounters. Second, the collinear semiclassical model is applied to obtain numerical predictions of the vibrational energy transfer rate dependence on the initial vibrational state quantum number. Thermally averaged vibration-translation rate coefficients are predicted and compared with CO-He experimental values for both ground and excited initial states. The numerical model is also used as a basis for evaluating several less complete but analytic models. Third, the role of rotational motion in the dynamics of vibrational energy transfer is examined. A three-dimensional semiclassical collision model is constructed with coupled rotational motion included. Energy transfer within the molecule is shown to be dominated by vibration-rotation transitions with small changes in angular momentum. The rates of vibrational energy transfer in molecules with rotational frequencies that are very small in comparison to their vibrational frequency are shown to be adequately treated by the preceding collinear models.

Structuring Proof and Technical Discussions

As just discussed, a report has a short Foreword and Summary which set up a problem and give a proposed solution and cost in general terms for a manager or nonexpert reader. These are followed by details to support the proposed solution, details which the manager could read or skip depending upon a need to know. Thus, the overall report structure moves from generalizations or claims (the Foreword and Summary) to support (the details):

Generalizations (claims)

↓

Details (support)

The section of the report giving details is written for experts, readers with the technical background and interest to follow the details and fine points of the technical argument. This section is usually called the *Discussion* in a long report and the *Details* or *Discussion* in a short report. Such a section is illustrated in the *Details* section of Figure 6-5, which appears on page 105.

As does the structure of the overall report, the structure of the Discussion and its individual paragraphs moves from generalizations and claims to support and details. For example, in the *Details* section of Figure 6-5, the first paragraph poses a generalization or claim and then its support. The first sentence of the *Details* section claims that "*the excessive time required to obtain a suitable vacuum. . . is primarily due to contamination of the ion pump.*" The rest of the paragraph supports this claim; it tells *why* the time required is due to contamination. Given this structure, a busy reader could skim the *Details* section for the kind of information needed without having to read *every* word.

Another example of this movement from generalizations or claims to

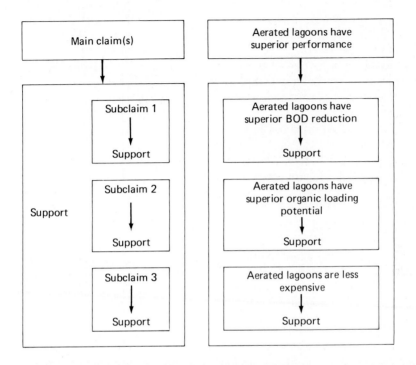

FIGURE 9-1 Organization of information for expert readers.

data and support appears in Figure 6-6, pages 108–110. Note especially the first main section after the Summary, *Performance Superiority of Aerated Lagoons*. This section begins with a heading and then a first paragraph that gives the claim being made: Aerated lagoons have superior performance to activated sludge processes or anaerobic lagoons because aerated lagoons show better BOD reduction and higher organic loading potential. Notice that the main claim (*aerated lagoons have superior performance*) is visible from the heading alone.

The main claim is then supported by the subsections which follow, with each subsection proceeding from claims (or subclaims) to support. In each case, the subclaim appears in the subsection heading—*Superior BOD Reduction by Aerated Lagoon, Superior Organic Loading Potential of Aerated Lagoons, and Cost Superiority of Aerated Lagoons*—and again in the first sentence of the section. This structure is summarized in Figure 9-1.

EXERCISE 9-2 Rewrite the reports in Exercise 6-6, page 107, to strengthen their movement in the Details section from generalizations or claims to data and support. Don't forget to consider the headings.

9.3

TECHNICAL REPORTS

The technical report is one of the most frequently used types of technical communication. However, it is also one of the most difficult for students to master since they have usually had much less experience with it than with any of the other types.

There are three subtypes of technical reports which will be discussed in this chapter:

1   The short informal report

2   The long informal report

3   The formal report

All of these reports share the structural features just mentioned: the Foreword and Summary written for managerial and nonexpert readers and the ordering of generalizations or claims before data and support in the sections written for experts.

The Short Informal Report

The short informal report is a one- to four-page, single-spaced report, usually written to someone within the writer's own company or organization. It may note the existence of a problem, propose some course of action, describe a procedure, or report the results of a test or investigation.

The term *informal* here does not mean sloppy or casual or carelessly done; an informal report should be as carefully prepared as possible—thoughtfully written, neatly typed, and thoroughly proofread for errors. The term *informal* refers to the format in which the report is presented.

The Heading of the Short Informal Report   The short informal technical report begins with a *heading section.* Typically, the heading gives the following information, where *Distribution, Enclosures,* and *References* are optional:

TO:	Name, job title
	Department
	Name of organization
	Address of organization (sometimes included)
FROM:	Name, job title
	Department
	Name of organization
SUBJECT:	Title of report (*Subject* may be replaced by *Re* or *RE,* for *Regarding*)
DATE:	Date

DIST: Distribution list of other people receiving the report; omitted if there is no distribution list

ENCL: Enclosures; other documents which are included with the report; omitted if there are no enclosures

REF: References; list of particularly important background documents; omitted if there are no such documents

Note that the heading is easy to read because of its format: liberal use of white space and aligned columns. It is also functional because the first information it gives is the information readers need first.

It is obvious why writers need to put the subject and date on the report and why they need to give the name of the reader and writer. It may be less clear why they need to give the job title, department, and organization of the reader and writer. These are provided because, at one level, documents are written not to people but to and from the occupants of particular jobs. Thus, it needs to be clear what job or position a person holds. Note that none of the information is presented more than once at the beginning of the report. If the name of the organization and/or the department are included in the letterhead, then these do not reappear under the *To* and *From* headings. If they do not appear in the letterhead, then they must appear under the headings.

Figures 9-2 and 9-3 present two different heading situations with ap-

FIGURE 9-2   Heading segment for short informal report on organizational letterhead (no department or division specified).

American
Electronics
Incorporated

To: Mr. John Nicol
            Chief Optoelectronics Engineer
            Optoelectronics Department

From: Khalil Najafi
            Assistant Optoelectronics Engineer
            Optoelectronics Department

Date: October 30, 1979

Subject: Liquid Crystal Displays: Analysis of Failure and Recommendations for Solution

Dist: Mr. Edward Jones, President
            American Electronics Incorporated

FIGURE 9-3 Heading segment for short informal report on departmental letterhead when reader and writer are in the same department.

American
Electronics
Incorporated

Optoelectronics Department

To: Mr. John Nicol
             Chief Optoelectronics Engineer

From: Khalil Najafi
             Assistant Optoelectronics Engineer

Date: October 30, 1979

Re: Liquid Crystal Displays: Analysis of Failure and Proposed Solutions

Dist: Mr. Edward Jones, President
             American Electronics Incorporated

propriate organization of information. Note that Figure 9-2 is written on organizational letterhead—no department or division is specified in the letterhead—and so *Optoelectronics Department* must appear under the names of the reader and writer. In contrast, Figure 9-3 is written on departmental letterhead, and so *Optoelectronics Department* does not need to be repeated. If the reader had been in a different department—say, in the Finance Department—then the appropriate heading section would be that of Figure 9-4. Notice that the writer's department appears in the letterhead but not under his or her name, whereas the reader's department appears under his or her name and not in the letterhead. The rest of the information remains the same.

Formatting the Short Informal Report   Each main section of the report—the Foreword, the Summary, and the Discussion (or Details)—follows the previous section with only a double or triple space between sections. A clearly visible heading indicates the beginning of each section and the contents of the section. The report is organized so that a busy managerial reader (a reader who wants to read no more words than are absolutely necessary) can stop after the Foreword (the statement of the problem) and the Summary (the main results and other important information a manager needs to know). The Discussion or Details section gives the extra information needed by technically involved readers: support for the claims in the Summary or extra

FIGURE 9-4 Heading segment for short informal report on departmental letterhead when reader and writer are in different departments.

American
Electronics
Incorporated

Optoelectronics Department

To: Ms. Karen Smithfield
           Chief Financial Analyst
           Finance Department

From: Khalil Najafi
           Assistant Optoelectronics Engineer

Date: October 30, 1979

RE: Liquid Crystal Displays: Analysis of Failure and Proposed Solution

Dist: Mr. Edward Jones, President
           American Electronics Incorporated

FIGURE 9-5 Structure of the short informal report.

| Foreword | For all readers, especially managerial and nonexpert readers |
| Summary | |

| Details or Discussion | Additional material for technical or expert readers |

details needed to implement or fully understand the solution proposed. This structure is summarized in Figure 9-5. Sample short informal reports appear in Figures 6-5 and 6-6.

The Long Informal Report
The long informal report is quite similar to the short informal report. Both are written for readers within the writer's organization; both use the *To, From,*

FIGURE 9-6 Structure of the long informal report.

```
┌─────────────────────┐
│ Foreword            │
└─────────────────────┘

                              For all readers, especially managerial
                              and nonexpert readers

┌─────────────────────┐
│ Summary             │
└─────────────────────┘

┌──────────────────────────────┐
│ Discussion                   │
│   ┌────────────────────┐     │
│   │  Introduction      │     │
│   │     to the         │     │
│   │  Discussion        │     │
│   └────────────────────┘     │
│                              │
│   ┌────────────────────┐     │   For expert or technical readers
│   │  Proof and         │     │
│   │  development       │     │
│   │  of claims         │     │
│   └────────────────────┘     │
│                              │
│   ┌────────────────────┐     │
│   │  Conclusions       │     │
│   │     and            │     │
│   │  Recommendations   │     │
│   └────────────────────┘     │
└──────────────────────────────┘
```

Subject, Date heading; and both are formatted so that each section follows the previous section with only a heading and a double or triple space between sections. However, the long informal report is different from the short report in two important ways: the long report is longer, obviously, and because of its length, its Discussion section is differently organized to make accessible the much greater amount of technical material it contains.

If you examine the long informal report in the Appendix, the similarities and differences between it and the short report are obvious. Let us focus on the differences. As illustrated in Figure 9-6, the Discussion section has three main subsections: the Introduction to the Discussion, the main Proof or Development section, and the Conclusions and Recommendations section. As a whole, the Discussion restates the material in the Foreword and Summary, but for a technically competent and interested reader. The Discussion gives the technical details, technical reasoning, and data that the technical reader needs—in contrast to the general orientation to the subject that the managerial reader needs. How does the Discussion do this?

The Introduction to the Discussion   The Introduction to the Discussion is the first part of the Discussion; it restates the Foreword in terms meaningful to the technical reader. Remember that the Foreword focuses on the implications of the problem for the managerial reader; it presents the problem as one of safety, cost, efficiency, quality control, labor relations, or compliance with legal regulations. In contrast, the Introduction to the Discussion focuses on the implications of the problem for the technical reader; it defines the technical nature of the problem or outlines technical assumptions, values, subproblems, background, or resources relevant to the problem.

For example, consider the following two examples, the first written by an electrical engineer at a company making heaters, fans, and portable lights and the second written by a harried inventory controller. In the first, the Introduction focuses on the technical nature and details of the problem; in the second, it focuses on the technical background.

EXAMPLE 1
FOREWORD
Because of increasing competition in the home appliance market, Electronic Design Division is evaluating the operating convenience and marketability of its less successful products. As part of this effort, I was asked to evaluate a proposed control circuit for our combined heater, light, and fan; to document any problems with this circuit; and if necessary to design a new control circuit. This report presents my evaluation and new design.

INTRODUCTION TO THE DISCUSSION
Ms. Durkis of Marketing has requested a simpler control switch for our combined heater, light, and fan. I have evaluated a preliminary design for a circuit which could turn on all three units by flipping a single switch and then remove power from one or more of the units by depressing another switch. The design for the control circuit was built around three relays and claimed to operate by picking these three relays in different combinations to produce the desired switching capabilities. The following sections detail the preliminary design of the circuit and its problems, a revised design, and a cost analysis of the revised design.

EXAMPLE 2
FOREWORD
During Photon's recent companywide expansion program, the inadequacy of our inventory numbering system became apparent. Thus, I was asked to design a more efficient and systematic numbering system based on our departmental needs and expected growth. This report provides an overview of the system and its response to our needs.

TABLE 9-2 DISTRIBUTION OF CONTENTS IN FOREWORD AND IN INTRODUCTION
TO THE DISCUSSION

	FOREWORD	INTRODUCTION TO THE DISCUSSION
Managerial problem	35–70%	0–20%
Technical problem or assignment	5–35%	10–80%
Forecast of contents	10–30%	0–25%

INTRODUCTION TO THE DISCUSSION

The present inventory (parts) numbering system has been in use for almost
five years. At the time of its conception, Photon did not have the computer
capabilities it has now, nor did it envision the growth it has recently experienced.
Because of these factors, the current parts numbering system is quite basic and
limited. As the quantity of parts increased over the years, the numbering system
could not efficiently keep track of them nor could it assign new part numbers
on a systematic basis. (In fact, new part numbers were assigned on a purely
random basis.)

Last month, Photon underwent one of its most comprehensive expansion programs to date and suffered critically from the present numbering system. Often
customers had to wait for us to reorder unexpectedly depleted supplies, and
our inventory department logged hundreds of hours of overtime. Thus, management decided to do away with the present system and develop a new one.
This report documents the various departmental needs identified, the com-
puter-assisted inventory control system I have designed, and features of the
system directly responding to the needs.

The different emphases of the Foreword and the Introduction to the
Discussion are reflected by the relative percentages of each section devoted
to the managerial problem, the technical problem or assignment, and the
forecast of contents. A rough comparative distribution for the two sections
appears in Table 9-2. Notice that the Foreword devotes proportionally more
space to the managerial problem, whereas the Introduction devotes more
space to the technical problem.

EXERCISE 9-3 A student complained that she was having trouble with
the following Foreword, that it was more like an Introduction to the Discussion than a Foreword (and she was right). Turn it into a Foreword by eliminating details unnecessary for a managerial reader. Feel free to rewrite sentences
or to move information around.

FOREWORD 1

Flutter is a phenomenon in which structural instabilities arise from aerodynamic
loads. In aircraft, these instabilities are often characterized by divergent oscil-

lations of the wings, resulting in structural failure and possible loss of the aircraft. Flight flutter testing involves the tracking of damping estimates of the excited vibration modes at different flight conditions.

The violent nature of flutter makes safety an important concern during flight flutter testing. As with any flight test program, cost is also a major concern. Our objective at Dryden is to provide safe testing at the lowest cost. Costs can be lowered by decreasing flight time. This requires utilizing the fastest analysis technique that will meet the accuracy requirements that safety demands.

Presently, we have available two software packages for data analysis during flight flutter testing. They are

1)   the Power Spectral Density (PSD) package, and

2)   the Random Decrement (RDEC) package

The main objective of each package is to determine damping estimates of the structural modes.

I was assigned to do a comparison study of the two packages looking at two questions:

1)   How accurate are the damping estimates?

2)   How much time is required for the analysis?

These questions are concerned with safety and cost, respectively.

The purpose of this report is to present the findings of my study and to make some recommendations concerning future testing.

Conclusions and Recommendations   The Conclusions and Recommendations section bears the same relationship to the Summary as the Introduction to the Discussion bears to the Foreword. The Summary is written for the managerial reader; it focuses on the conclusions and recommendations relevant to that reader and gives the information a manager must know to act. The Conclusions and Recommendations section, which appears at the end of the Discussion, gives information a technical reader might want in addition to that needed by the manager. It might include the most important technical results which led to the recommendations and statements technically too complex or difficult for the managerial reader. For instance, consider the *Summary* and the *Conclusions and Recommendations* presented below from a report written by a geologist to the Milton Township Council. (The *Foreword* is given first to orient you.) Unless you are a geologist, you probably won't understand all the information presented in the *Conclusions and Recommendations,* though you should easily find your needed information in the *Summary.* Remember that where possible, the *Conclusions and Recommendations* should move from generalizations or claims to data and support.

FOREWORD

Within Milton Township, the greatest industrial and residential development has occurred in areas with limited groundwater resources. This development has already created water shortages and a need to deepen existing wells; more seriously, it has greatly handicapped or prevented the development of new industry. Thus, the township council authorized Moody and Associates to locate, if possible, several high-yield water wells to provide a sufficient municipal water supply. The purpose of this report is to describe the successful location of sand and gravel materials with excellent groundwater yielding capability.

SUMMARY

We conducted studies and six test drillings to locate an area in which a reliable, high-quality, municipal water well field could be developed. We have located such a site.

Results: Four of the six wells drilled were productive and penetrated sand and gravel suitable for development of municipal wells. Each of the four productive wells showed very good water-producing capability, yielding 300 to 500 gallons per minute. Further, the wells produced water of generally excellent quality, though with a potentially high but treatable concentration of manganese. One test well showed manganese concentration now high enough to require treatment, and there is concern that manganese levels in all the wells might increase to such a level. However, this would not prevent development of the well field, and additional study will be required to determine if manganese removal is necessary at all.

Conclusions: Production wells completed at the test well sites should more than supply the township's current and future needs. Analysis of test data indicates that such production wells could produce over 2 million gallons per day, twice the township's projected requirements. Further, if the water-bearing sand and gravel are as extensive as we project them to be, yields up to 5 million gallons per day could conceivably be developed.

Recommendation: We recommend that the property tested be acquired for development as a municipal well field. The property is unique in its supply capability within Milton Township, and a well field on this property should meet any present and foreseeable water requirements within the township.

CONCLUSIONS AND RECOMMENDATIONS

Conclusions: Six test wells were drilled, four of which penetrated sand and gravel suitable for developing adequate municipal wells.

1. The relationship of geologic units encountered in Test Wells 1 through 6 is interpreted and presented in Plate 1. Two aquifers were identified; an extremely thick aquifer was encountered in Test Wells 2, 5, and 6 ranging in thickness from 78 to 125 feet, and a deep aquifer was en-

countered in Test Well 1 extending from 150 to 185 feet. There is theorized to be limited hydraulic connection through fine-grained units between
the two aquifers. Some degree of connection between the deep aquifer
and upper aquifer was established by the rapid rise in water levels in
Test Well 1 observed during a heavy rain in September 1979.

2.   A conservative estimate of the yield from the upper aquifer is 2.0 mgd.
     Aquifer data indicate this could be produced by pumping any combi-
     nation of two wells among Test Wells 2, 5, and 6 (assuming manganese
     in Test Well 2 proves acceptable). Ultimate production from a French
     Creek well field could prove to be in the 3.0 to 5.0 mgd range.

3.   It is difficult to estimate the maximum production available from the lower
     aquifer encountered in Test Well 1 because this unit was not identified
     in the other five test wells. Based on the data from the pump test of Test
     Well 1, the yield of the lower aquifer is estimated to be 0.5 to 1.0 mgd.

4.   The water quality from all wells is good, with the exception of the anom-
     alously elevated manganese found in Test Well 2. Iron and manganese
     were present in all other wells but in acceptable concentrations. If there
     is a source of manganese (none has been identified to date) in the im-
     mediate vicinity of Test Well 2, there is concern that due to the extremely
     high permeability of the aquifers, the high manganese groundwater could
     migrate toward the other wells with long-term pumping.

     Typically, the water is moderately hard to hard, with hardness ranging
     from approximately 7 grains in Test Well 1 to 10 grains in Test Well 5
     to 14 grains in Test Well 2.

Recommendations: Based on results from drilling and sample analysis, the
following tests and land purchases are recommended.

1.   A short pump test (8 hours) should be conducted on Test Well 2 to
     recheck the iron and manganese concentrations in this well.

2.   In order to stress the aquifer as heavily as possible and simulate municipal
     well field operation, a pump test should be conducted on Test Well 6
     with an attempted pumping rate of 700 to 1,000 gpm for 48 hours.

3.   In order to secure the existing test wells, provide a buffer around these
     wells, and permit some aquifer test drilling, it is recommended the Town-
     ship purchase property roughly rectangular in form extending 800 feet
     north and 800 feet south of Test Well 1 and west to the abandoned
     railroad tracks. This would be approximately 40 acres. If obtaining this
     acreage is not feasible, then it is recommended that at a minimum the
     Township obtain property extending 800 feet north and 800 feet south
     of Test Well 1 and 600 feet to the west of Test Well 1. This would be
     approximately 22 acres. It is also recommended that a long-term option
     be obtained on the Barco property south of Test Well 2 and southwest
     of Test Well 3 for potential long-range future well field development.

EXERCISE 9-4 The following Foreword and Summary were written by an engineer at Jur-Kay Carriers, a moving company specializing in long-distance hauling. Unfortunately, the engineer provided far too much detail for the busy managerial readers of this report. Your job is to write a new Foreword and Summary which give a proper (and short) overview of the problem and proposed solution. Try to make each section no longer than 100 to 150 words. Then write an Introduction to the Discussion and a Conclusions and Recommendations section. A Table of Contents for the report follows to outline the report's structure.

THREADED FASTENERS: FAILURE ANALYSIS AND PROPOSED IMPROVEMENT

TABLE OF CONTENTS

FOREWORD

On April 3, 1974, a Jur-Kay Carriers semitractor-trailer overturned on I-96 approximately 30 miles east of Grand Rapids, Michigan, and the driver was seriously injured. There were no other vehicles involved in the accident. The exact cause of the accident is unknown. However, it is believed that one of three bolts in the steering mechanism failed, causing the driver to lose control. The truck, trailer, and contents of the trailer were a total loss.

On July 27, 1974, a Jur-Kay Carriers semitractor-trailer lost control on U.S. 101, 52 miles south of Grand Rapids, involving three other vehicles in a crash resulting in two deaths. The driver of our vehicle incurred minor injuries. The accident investigation revealed that a bolt in the power steering linkage of our tractor failed, thus causing the driver to lose control. Maintenance Record #1825, dated July 27, 1974, revealed that five hours before the accident a fleet mechanic replaced the original SAE-grade-8 linkage bolt with an ungraded bolt. It can be assumed that if the bolt had been replaced with an SAE-grade-8 bolt, the accident would not have occurred.

Jur-Kay Carriers is now in litigation with a suit brought against them by the owners of the three vehicles involved in the crash.

As a direct result of these two accidents, and because for the last six months our maintenance records have shown an extremely large number of threaded fastener replacements, Mr. Paul Milani, Senior Engineer, Jur-Kay Carriers, suspected that a problem existed with threaded fastener usage in our maintenance department. On July 29, 1974, Mr. Milani requested that I evaluate this usage of threaded fasteners. The investigation has been completed.

The purpose of this report is to show that our mechanics have a dangerous lack of knowledge about fasteners.

The purpose of this report is also to present the needed material for the recommended instruction class, emphasizing:

1.   Bolt strength and preload specifications

2.   Torque requirements

3.   Proper selection of a nut

4.   Why fasteners fail

5.   What different failures indicate about the inadequacies of the specific nut-and-bolt assembly in question

SUMMARY

A preliminary examination of maintenance records dated since January 1, 1974, was conducted by Mr. Milani. The examination has shown that a large number of threaded fastener replacements and repairs are being made in the maintenance department of Jur-Kay Carriers. In addition to this, there have been two serious vehicle accidents in the last four months caused by Jur-Kay Carriers. They were both the result of threaded fastener failures. Because of these facts, it is believed that a definite problem exists and an investigation into threaded fastener usage in our maintenance department has been made.

The problem had to lie in one of four areas (or a combination thereof): either (1) our fasteners were of inferior quality, (2) there was such a lack of standardization in the fasteners we used that many types of gradings were present, resulting in utter confusion on the part of the mechanics whose job it was to select the replacement part, (3) there was simply a high degree of carelessness in the mechanics' performance of their jobs, or (4) the mechanics were simply not knowledgeable enough to make the proper grade choice when replacing a bolt.

Research revealed that the problem lay in none of the first three areas:

1.   Our fasteners are not of inferior quality.

2.   There is no lack of standardization in the grading of threaded fasteners we use.

3.   There is no degree of carelessness in our mechanics' performance of their jobs. In fact, the opposite was found to be the case.

The problem was found in the last area:

4.   Our mechanics simply do not know the field of fasteners. Indeed, sub-
        sequent observations of activities in the garage and conversations with
        the mechanics verified this conclusion.

Our mechanics lack necessary knowledge in the threaded fastener field. Thus, they are unable to make the correct choices in replacing a nut-and-bolt assembly that has failed. Low-tensile-strength bolts are being substituted for high-tensile-strength bolts, and bolts are not being used for their designed purposes. This has resulted in an unnecessarily large number of failures and replacements, not to mention the two latest vehicle accidents.

I recommend that an instruction class be initiated to inform and educate our mechanics in basic areas of threaded fastener technology. I estimate that five sessions, each an hour in duration, is sufficient to present the necessary material. Such a class would eliminate the problems involving threaded fasteners present in our maintenance department. That is, the serious problem of substituting low-tensile-strength bolts would be avoided. Our mechanics would begin using bolts for their designed purposes. We would also solve the recurring bracketing and "bolt loosening" problems caused by reusing bolts that have stretched way beyond their optimum strength.

I volunteer my services as instructor if the need is to be met. Such a class will emphasize

1.   Bolt strength and preload specifications

2.   Torque requirements

3.   Proper selection of a nut

4.   Why fasteners fail

5.   What different failures indicate about the inadequacies of the specific
        nut-and-bolt assembly in question

I also recommend that after our mechanics are taught the basic threaded fastener technology, a number of bolt-grade-marking charts be posted throughout the garage.

The Formal Report

Formal reports are usually relatively long reports written to readers outside the writer's organization. Since such reports represent the organization and spread its reputation, they are sent out with a "dressier" look—often a fancy cover and binding with the organization's seal prominently featured. Instead of the *To, From, Subject, Date* heading of the informal report, a formal report begins with a Title Page, an Abstract, a Table of Contents, and a List of Illustrations. Formal reports also begin the Foreword, Summary, and Discussion on new pages. In all of these senses, the formal report closely resembles a book.

The Title Page of the formal report includes all of the information you have already seen in the heading of the informal report: information about *To, From, Subject,* and *Date.* The major difference is in the arrangement of this information on the page. The Title Page spreads the information over the page in a visually pleasing arrangement with lots of white space between items. Sample Title Pages appear in Figures 9-7 and 9-8, which contain the typical heading information:

1   Name of receiver and/or receiving organization

2   Name of writer and writer's organization

3   Subject of report

4   Date of report

Note that these correspond to the *To, From, Subject,* and *Date* of the informal report. In addition, they also include the following funding information:

5   Funding agency (often this is the same as the receiver or receiving organization)

6   Funding contract number(s)

7   Beginning and ending dates of the project

8   File number of the report or document authorizing the project

The specific information on a given Title Page and its format may vary with the requirements of different organizations. You must find out what the specific requirements are for your situation and provide the required information in the required format.

In other ways, the formal report duplicates the long informal report. It is long and has a Discussion section which includes an Introduction to the Discussion and a Conclusions and Recommendations section. These sections restate the Foreword and the Summary, as described above, and the main part of the Discussion provides support for the generalizations in the Summary and the Conclusions and Recommendations. A sample formal report appears in the Appendix.

9.4

AN EXAMPLE OF ADAPTING FORM TO PURPOSE: THE FEASIBILITY REPORT

In most of the reports discussed in this chapter, someone has noticed a problem or undesirable situation and then proposed or asked someone else to propose a solution. Sometimes, however, it may not be clear that there

FIGURE 9-7 Title Page from a formal report.

A DESCRIPTOR VARIABLE APPROACH TO MODELING AND OPTIMIZATION OF LARGE-SCALE SYSTEMS

{ TITLE

FINAL REPORT

MARCH 1975 - FEBRUARY 1979

{ TYPE OF REPORT
  PROJECT DATES

DONALD N. STENGEL
DAVID G. LUENBERGER
ROBERT E. LARSON
TERRY S. CLINE

{ AUTHORS

SYSTEMS CONTROL, INC.
1801 PAGE MILL ROAD
PALO ALTO, CA. 94004

{ AUTHORS'
  ORGANIZATION

DATE PREPARED - FEBRUARY 1979

{ REPORT DATE

PREPARED FOR THE
DEPARTMENT OF ENERGY
OFFICE OF THE ASSISTANT ADMINISTRATOR FOR ENERGY
TECHNOLOGY
DIVISION OF ELECTRIC ENERGY SYSTEMS

{ FUNDING AGENCY

WORK PERFORMED UNDER CONTRACT NCS EX-75-0-0-2090
ET-73-0-0-2353

{ FUNDING CONTRACT
  NUMBERS

FIGURE 9-8 Sample formal report Title Page specifying funding agency.

NATIONAL COOPERATIVE HIGHWAY RESEARCH
PROGRAM REPORT **207**

UPGRADING OF LOW-QUALITY AGGREGATES FOR PCC AND BITUMINOUS PAVEMENTS

P.D. CADY, P.R. BLANKENHORN, D.E. KLINE,
AND D.A. ANDERSON
THE PENNSYLVANIA STATE UNIVERSITY
UNIVERSITY PARK, PENNSYLVANIA

RESEARCH SPONSORED BY THE
AMERICAN ASSOCIATION OF STATE
HIGHWAY AND TRANSPORTATION
OFFICIALS IN COOPERATION WITH
THE FEDERAL HIGHWAY
ADMINISTRATION

AREAS OF INTEREST:
PAVEMENT DESIGN AND PERFORMANCE
BITUMINOUS MATERIALS AND MIXES
GENERAL MATERIALS
MINERAL AGGREGATES
SOIL EXPLORATION AND CLASSIFICATION
(HIGHWAY TRANSPORTATION)

TRANSPORTATION RESEARCH BOARD
NATIONAL RESEARCH COUNCIL
WASHINGTON, D.C.   JULY 1979

is a solution or that any given solution would be worth what it would cost in time, money, and effort. At such times, it is useful to conduct a feasibility study, that is, to analyze the problem and its implications, to evaluate alternative solutions, and finally to decide on one alternative or course of action. The results of such a feasibility study are written up in a feasibility report.

The term *feasibility* has two somewhat different meanings. In its more limited sense it means technically capable of being done, executed, or effected. We shall call this *technical feasibility*. If we were considering building a tunnel through a mountain, this more limited sense of feasibility would refer to such questions as the following:

1   Can you build the tunnel? Do you have the technology to keep the soil from caving in?

2   Do you have enough people available for the job?

3   Do they have the required expertise?

4   Can you get the required building supplies and building permits?

In addition to this sense of technically capable of being done, *feasibility* also has the wider, more inclusive sense of "suitable" or "reasonable." To return to the tunnel example, this wider sense of feasibility encompasses such additional questions as the following:

1   Assuming that we can build this tunnel (that it is technically feasible), should we? Will it produce enough benefits to be worth the cost?

2   Is the cost reasonable?

3   Can we afford it, even if it is reasonable?

4   Will building or having the tunnel produce any undesirable side effects?

5   Will it produce any desirable side effects, such as increased employment in the area?

In this larger sense, the notion of feasibility covers the five general criteria discussed in Chapter 5 that are used to evaluate any proposed solution to a problem:

1   *Effectiveness:* Is the solution effective? Will it solve the problem posed? Why? How do you know?

2   *Technical Feasibility:* Can the solution be implemented? Does it require technology or resources that are unavailable? How do you know?

3 *Desirability:* Would one want to implement the proposed solution? Does it have any undesirable effects? Does it have desirable effects? Why? What are they?

4 *Affordability:* What will the solution cost to implement? To maintain? Is this cost reasonable? Is it affordable? Will it reduce costs in the future? Why?

5 *Preferability:* Is the solution better than or preferred over any other possible solution? Why?

This larger notion of feasibility also covers any special criteria necessary or appropriate for the given situation, for instance, important issues of safety, environmental pollution, conformity to particular laws or traditions, precedent setting, or justice. As a technical expert, you must be able to deal with all of the relevant issues and criteria in a feasibility study.

When all of these criteria enter into the notion of feasibility, they provide the basis for a full-fledged argument of fact or policy. For instance, when you judge a tunnel as being effective, technically feasibile, desirable, affordable, preferable, and safe, and then argue on the basis of these judgments that the tunnel is feasible, you have constructed an argument of fact supported by several subarguments of fact. If you go one step further and argue that the tunnel should be built, you have constructed an argument of policy supported by several subarguments of fact, one of which is the argument of feasibility.

Let us examine the arguments being developed in a sample feasibility report to further illustrate this point. The report concerns flight tests on airplanes—in particular, flight flutter tests. A flight flutter test studies the vibratory motion and resulting instability of aircraft wings during flight. When a plane is in flight, air passing over and around the wings creates some wing motion. If this motion is too great, the wings will vibrate catastrophically, or "flutter." If this flutter becomes too severe, the wings can actually break off. The Foreword to the report sets up the problem being addressed in a feasibility study.

FOREWORD

Flight flutter tests are inherently dangerous. To maintain a margin of safety, we must have fast and accurate in-flight analysis of the vibratory motion in the wings created by the test flight conditions. However, the computer program currently used for in-flight analysis is outdated, and its inadequacies present safety problems. Thus, a new analysis technique must be implemented to improve safety and to maintain our reputation in this area of flight testing.

$10,000 was allotted for this purpose, and a survey of the field was made to determine possible choices for the new analysis technique. I was assigned to review the choices for feasibility, select the best one and, if it seemed feasible, identify initial development and implementation concerns. This report presents my solution to our problem and provides initial information on the development and implementation of that solution.

While doing the analysis and evaluation necessary for this report, the writer realized that any proposed solution to the problem—that is, any computer program which would produce an adequate in-flight analysis—would have to meet the following criteria:

1   *Effectiveness:* The solution must be capable of providing accurate damping estimates of up to four superimposed vibration modes from the accelerometer data available.

2   *Technical feasibility:* The solution must be capable of working with the present computer (HP-5451C Fourier Analyzer), with minimal modifications. The solution should be adaptable by our staff, given their expertise and available time.

3   *Desirability:* The solution should have some flexibility in its software structure to allow modifications to the existing package should a particular test call for them.

4   *Affordability:* The solution purchase, development, and implementation costs should be less than the allotted $10,000.

5   *Preferability:* A solution that provides conservative damping estimates and requires the minimum analysis time is preferred.

The writer then evaluated three possible alternatives—the Random Decrement (Randomdec) Technique, the Time Domain Package, and the Flight Test Data Reduction System—against these criteria, finding that two of the alternatives were ultimately infeasible:

1   The Time Domain Package was rejected for failing to meet the desirability criterion. The package is proprietary, and the producer refuses to disclose the details of the software structure. Therefore, the flexibility in modifying the package is absent.

2   The Flight Test Data Reduction System was rejected for failing to meet the effectiveness criterion under certain conditions. Accurate damping estimates were difficult to obtain when two or more modes fell within a frequency range of $\Delta 5$ Hz. Since this condition is quite possible during testing, this package was rejected.

One remaining alternative, however, was feasible; it met all of the criteria and was ultimately recommended as the solution to the problem of inadequate in-flight testing:

I recommend the Randomdec Technique as the solution to our problem because it meets or exceeds all of the given criteria.

1   *Effectiveness:* This technique provides accurate damping estimates of up
    to four superimposed modes.

2   *Technical feasibility:* This technique can utilize a software structure which
    is well suited for use on the available computer. We have the personnel
    and expertise in-house to develop and implement the necessary software.

3   *Desirability:* This technique is well documented and available for use by
    the general public. A proprietary software package need not be pur-
    chased. The necessary software can be developed in-house and tailored
    to our specific needs with the flexibility desired.

4   *Affordability:* This technique is estimated to cost under the allotted figure.
    Our survey and in-house discussions gave an initial estimate for devel-
    opment and implementation costs of \$8,000 (400 hours of labor @ \$20/
    hour).

5   *Preferability:* This technique is mathematically derived to provide con-
    servative damping estimates, thus allowing some safety margin (Figure
    1). Initial estimates of analysis time were also found to be acceptable.

After conducting the analysis and evaluation, the writer had to present
the results in a feasibility report. The feasibility report follows the structure
already defined in this chapter for technical reports. It has a Foreword and
Summary for managerial and nonspecialist audiences and then a more detailed section outlining the argument for specialists. The argument in its final
form is presented in the Summary:

SUMMARY
In an attempt to find a replacement for our inadequate flutter analysis package,
a survey of the field turned up three alternatives: the Random Decrement
(Randomdec) Technique, the Time Domain Package, and the Flight Test Data
Reduction System. Each alternative was studied for effectiveness, technical
feasibility, desirability, affordability, and preferability. As a result of this study,
I recommend the Randomdec Technique as our solution: it meets all of the
criteria whereas the others fail to do so. An initial cost estimate of developing
and implementing this technique is \$8,000 (\$2,000 under budget). Implementing this technique will increase the safety of our flight flutter tests and
provide a desired margin of safety; it will also allow us to maintain a prominent
position in this area of flight testing.

A Details section follows the Summary and provides backup data and
references to support the argument stated in the Summary. The Details section first outlines the problem, then presents the argument for the Randomdec Technique, then addresses the issue of development and implementation

(a part of the writer's assignment), and finally presents the conclusions. The outline for this structure is given below and the full report appears as Figure 9-9.

Foreword
Summary
   I The Problem and Its Background
  II Argument for the Randomdec Technique
     A Criteria for Judgment
     B Selection of the Randomdec as the Recommended Solution
     C Rejection of Alternatives
  III Development and Implementation of the Randomdec Technique
     A Brief Explanation of the Technique
     B Cost and Time Required
  IV Conclusions
   V References

From this perspective, the feasibility report is much like the reports discussed earlier in this chapter. It is based on a problem; it presents an analysis of a possible solution, of some possible solutions, or of all possible solutions to a given problem; and, if possible, it recommends the best solution given the circumstances involved. Such a recommendation may constitute an argument of policy (you should adopt this solution), but at the least it makes an argument of fact (this is or seems to be feasible). If a feasibility report does not build all the way to recommendations and an argument of policy, it should at least make, as clearly as possible, the various subarguments of fact for the people who will ultimately have to decide whether to fund or approve the project being studied. If these decision makers are better qualified than the report writer to make recommendations, then leaving the recommending to them may be appropriate. However, if they are not better qualified, then the report writer should attempt to make the recommendations which must ultimately be made. Writers of feasibility reports are hired for their interpretative ability as well as for their fact-gathering ability, and in many instances the writer of the feasibility report is the one person (or one of the few persons) best qualified to make the needed recommendations.

9.5

A CONCLUDING NOTE

This chapter has presented some of the conventional ways of organizing reports. It has stressed the persuasive nature of reports—that they are basically arguments of fact or policy which the writer is presenting to the reader. A feasibility report, for instance, makes a claim—that something is or is not feasible, some course of action should or should not be pursued—and then presents the analysis and data to support that claim.

FIGURE 9-9 A sample feasibility report.

DATE: December 5, 1981

TO: Regina J. Petty
Branch Chief
Dynamics Branch

FROM: Louis J. Bogart
Aerospace Engineer
Dynamics Branch

SUBJECT: DEVELOPMENT AND IMPLEMENTATION OF A FLUTTER ANALYSIS
PACKAGE: A FEASIBILITY STUDY

ATTACHMENT: Functional diagram of the Random Decrement Analysis Procedure

cc: R. J. Jones, Chief, Safety Office
T. R. Wilson, Chief, Flight Operations

Foreword

Flight flutter tests are inherently dangerous. To maintain even a margin of
safety, we must have fast and accurate in-flight analysis of the wings' vibratory
motion created by the flight test conditions. However, the computer program currently used for in-flight analysis is outdated, and its inadequacies present safety
problems. Thus, a new analysis technique must be implemented to improve safety
and to maintain our reputation in this area of flight testing.

$10,000 was allotted for this purpose, and a survey of the field was made to
determine possible choices for the new analysis technique. I was assigned to review the choices, select the best one, submit it to a feasibility test, and, if it seemed
feasible, identify initial development and implementation concerns of the selected
technique. This report presents my solution to our problem and provides initial
information on development and implementation of that solution.

Summary

In an attempt to find a replacement for our inadequate flutter analysis package, a survey of the field turned up three alternatives: the Random Decrement
(Randomdec) Technique, the Time Domain Package, and the Flight Test Data Reduction System. Each alternative was studied for effectiveness, technical feasibility,
desirability, affordability, and preferability. As a result of this study, I recommend
the Randomdec Technique as our solution: it meets all of the criteria whereas the
others fail to do so. An initial cost estimate of developing and implementing this
technique is $8,000 ($2,000 under budget). Implementing this technique will increase the safety of our flight flutter tests and provide a desired margin of safety;
it will also allow us to maintain a prominent position in this area of flight testing.

I. The Problem and Its Background

The problem arises out of the fact that our software package for flutter analysis is outdated. The Power Spectral Density (PSD) package presently in use was
developed over ten years ago, and the level of accuracy it provides is no longer

FIGURE 9-9 *(continued)*

acceptable. The accuracy of the damping estimates obtained by the PSD method is erratic, and several superior analysis techniques have been developed in recent years (reference 1).

The importance of having damping estimates that are as accurate as possible can readily be seen from the potential danger involved in flight flutter testing. An incorrect damping estimate obtained during in-flight analysis could result in the loss of an aircraft due to a wing fluttering off.

An updating of our software is needed to reduce the risk involved and to maintain a prominent position in this area of flight testing. Funds ($10,000) have been allotted to improve our capabilities in this area, and a survey of the analysis techniques presently being used in the field has been made. The survey resulted in three analysis techniques being chosen as the most promising alternatives to our present situation. They are

1. The Randomdec Technique - broadly used by a number of aircraft manufacturers.
2. The Time Domain Package - specific package developed by the Aerotest Corporation and used by the Air Force at Edwards Test Center.
3. The Flight Test Data Reduction System - specific package developed and used by the Northrop Corporation.

Since our present PSD package has already been modified to its limits with unsatisfactory results (reference 1), the solution to our problem lies in choosing one of the above alternatives and implementing it into our flight testing system.

II. Argument for the Randomdec Technique

    A. Criteria for Judgment

        1. Effectiveness: The solution must be capable of providing accurate damping estimates of up to four superimposed vibration modes from the accelerometer data available.

        2. Feasibility: The solution must be capable of working with the present computer (HP-5451C Fourier Analyzer), with minimal modifications.

        3. Desirability: The solution should have some flexibility in its software structure to allow modifications to the existing package should a particular test call for them.

        4. Affordability: The solution purchase, development, and implementation costs should be less than the allotted $10,000.

FIGURE 9-9 *(continued)*

      5. Preferability: A solution that provides conservative damping estimates (Figure 1) and requires the minimum analysis time is preferred.

B.   <u>Selection of the Randomdec Technique as the Recommended Solution</u>

I recommend the Randomdec Technique as the solution to our problem because it meets or exceeds all of the given criteria.

      1. Effectiveness: This technique provides accurate damping estimates of up to four superimposed modes.

      2. Feasibility: This technique can utilize a software structure which is well suited for use on the available computer. We have the personnel and expertise in-house to develop and implement the necessary software.

      3. Desirability: This technique is well documented and available for use by the general public. A proprietary software package need not be purchased. The necessary software can be developed in-house and tailored to our specific needs with the flexibility desired.

      4. Affordability: This technique is estimated to cost under the allotted figure. Our survey and in-house discussions gave an initial estimate for development and implementation costs of $8,000 (400 hours of labor @ $20/hour).

      5. Preferability: This technique is mathematically derived to provide conservative damping estimates, thus allowing some safety margin (Figure 1). Initial estimates of analysis time were also found to be acceptable.

C.   <u>Rejection of Alternatives</u>

    1. The Time Domain Package was rejected for failing to meet the desirability criterion. The package is proprietary, and the producer refuses to disclose the details of the software structure. Therefore the flexibility in modifying the package is absent.

    2. The Flight Test Data Reduction System was rejected for failing to meet the effectiveness criterion under certain conditions. Accurate damping estimates were difficult to obtain when two or more modes fell within a frequency range of $\Delta 5$ Hz. Since this condition is quite possible during testing, this package was rejected.

FIGURE 9-9 *(continued)*

III. Development and Implementation of the Randomdec Technique

    A.   Brief Explanation of the Technique

                The Randomdec Technique was developed by Cole (reference 2) and has been used successfully by a number of aircraft companies (references 3, 4). The proposed functional diagram of the analysis procedure can be seen in Attachment 1. To start the procedure, the Randomdec signature is extracted from the initial vibration data (time history) using the methods detailed in reference 2. The signature is then truncated since only the beginning is needed for further analysis. A curve fitting of the signature is then performed to obtain the frequency and damping characteristics of the individual modes. However, first a power spectral density of the signature is taken to supply the curve fitting program with initial frequency estimates of the modes. After the curve fitting, the individual modes are separated and output with their respective frequency and damping estimates.

    B.   Cost and Time Required

                The familiarity of the engineers in the Dynamics Branch with both the computational procedures involved in the Randomdec Technique and the capabilities of our computer system should facilitate a quick and efficient development and implementation of the new package. Having worked extensively in programming the computer in the past, I would estimate that a maximum of 400 hours would be necessary to complete the development, implementation and check-out phases. This represents $8,000 (@ $20/hour), which is below our allowable budget of $10,000.

IV. Conclusions

                I recommend the development and implementation of the Randomdec Technique as the solution to our problem of an inaccurate and out-dated flutter analysis package. Use of the Randomdec Technique will reduce the risk involved in the potentially dangerous area of flight flutter testing and help Dryden to maintain a prominent position in this flight test field. The Randomdec Technique is being recommended because it meets the following necessary criteria:

        1. Effectiveness
        2. Feasibility
        3. Desirability
        4. Affordability
        5. Preferability

The other alternatives were rejected for failing to meet all of the above criteria.

FIGURE 9-9   *(continued)*

V. <u>References</u>

1. DFRC Memorandum 81-0611-2: Inadequacies of the Power Spectral Density Flutter Analysis Package. June 11, 1981.
2. Henry A. Cole Jr.: On-Line Failure Detection and Damping Measurement of Aerospace Structures by Random Decrement Signatures. NASA CR-2205, 1973.
3. W.J. Brignac, H.B. Ness, and L.M. Smith: The Random Decrement Technique Applied to the YF-16 Flight Flutter Tests. AIAA Paper No. 75-776, May 1975.
4. R.V. Doggett Jr. and C.E. Hammond: Determination of Subcritical Damping By Moving-Block/Randomdec Applications. In "Flutter Testing Techniques," NASA SP-415, 1975, pp. 622-46.

Similarly, other types of reports do the same thing. A progress report claims that the writer (or the writer's group) has made certain progress and then supports that claim with evidence and documentation of the progress. A final report presents the results of an investigation (the writer's claim about some state of affairs) along with the data to prove that the investigation was done carefully and appropriately, that is, with the proof that the writer's vision of "reality" is true or defensible. Thus, in all of these types of reports, a writer is making arguments of fact or policy and providing support.

The writer is doing more than providing support for current readers, however. He or she is also providing crucial documentation for future readers, for those who may need to continue, extend, or test the writer's work. This function is crucial in all reports but is especially pressing in the progress report. Progress reports are written before the entire project is completed. This documentation of what has been done is important since usually the only other sources of information about current work on the project are cryptic notes. If something happens to the writer—an accident, untimely death, job change—then those who must continue the project need to have the earlier work available in a useable form. Since most people keep notes and workbooks in a form that only they can understand, the progress report becomes a convenient vehicle for spelling out conclusions and for assembling data at regular intervals. If the writer has finished some library research and discovered something important for the project, she or he should spell out the discovery and provide a relevant bibliography. If the writer has designed a part or developed a process flow chart or derived an important mathematical relationship, that work should be documented, by including the relevant material either in the body of the report or in an appendix. That work shouldn't have to be done again if something should happen to the writer.

10

THE PROPOSAL

A proposal requests support—usually money—for work that a proposer wants to do. It is written to a funding agency, which may be inside or outside the proposer's organization. This funding agency has its own interests and goals, which may or may not coincide with those of the proposer. The proposal must convince the funding agency that the proposed activity will be a good "investment," that is, that the activity is worthy of support and will advance the agency's goals, produce quality results, and do all this better than other activities competing for the same funds. To make things even more difficult, the proposal must make these arguments to busy readers trying to divide too few resources among too many applicants.

To be successful in this environment, a proposal must quickly and clearly answer the questions a potential funder will bring to any request for support:

What does the proposer want to do?

How much will it cost?

Is the problem important and relevant to the funding agency's interests?

Will the proposed activity solve or reduce the problem?

Can the proposed activity be done? Will it duplicate other work?

Is the method or approach appropriate, clearly defined, and well thought out?

Can the results be adequately evaluated?

Is the proposer qualified to do the research? Better qualified than other proposers?

Will the results of the activity be available to others?

Are the proposed schedule of activity and budget reasonable?

The way these questions are answered will vary as particular projects are adapted to the guidelines of particular funding agencies. However, all of these questions need to be addressed at some level, usually in a format similar to the one outlined below. To be sure that a given proposal meets the expectations of a given funding agency, the proposer should carefully consult any proposal writing guidelines put out by that agency and then interview people familiar with the agency's funding interests. A proposer might even

FIGURE 10-1   Parts of a long proposal.

Title page

Abstract

Table of contents

Introduction

   Problem addressed
   Purpose or objectives of research
   Significance of proposed research

Description of proposed research

   Plan for accomplishing objectives
   Plan for evaluating results
   Schedule for project completion

Institutional resources and commitments

List of references (for six or more references)

Personnel

   Explanation of proposed staffing
   Relevant experience of major personnel

Proposed budget

   Budget in tabular form
   Justification of budget items

Appendices

   Letters of endorsement
   Promises of participation, subcontractor's proposals
   Biographical data sheets (vita sheets)
   Reprints of relevant articles, reports, background documents

call or visit the funding agency to get the most recent information on its interests and requirements.

10.1

THE ORGANIZATION OF A FORMAL PROPOSAL

Most proposals follow the general outline presented in Figure 10-1, an outline similar to that given in Chapter 5 for the argument of fact and policy. Indeed, you could argue that a proposal is an argument of policy (the funding agency should fund this proposal) based on several embedded arguments of fact (the problem is important, the activity proposed will alleviate the problem, the proposers are qualified to do the work proposed, etc.).

The general outline will be adapted and modified according to the needs of the readers and the demands of the topic proposed. For instance, long or complicated proposals might well contain all of the sections shown in Figure 10-1 in fully developed form with their own headings corresponding to the sections and subsections of the outline. In contrast, shorter or simpler proposals might contain only the sections noted in Figure 10-2. Short proposals may treat a given subsection very briefly or combine several subsections. A longer proposal to a funding agency outside the writer's organization appears in the Appendix. A shorter proposal, but still a formal one, written to someone in a different unit in the writer's organization is shown in Figure 10-6, and a very short proposal written to someone in the writer's own unit is shown in Figure 10-4.

FIGURE 10-2 Parts of a short proposal.

Title page (ideally, with abstract)

Introduction

    Problem addressed
    Purpose or objectives of proposed work
    Significance of proposed work

Plan for accomplishing objectives

Plan for evaluating results

Schedule for project completion

Institutional resources and commitments

Personnel

    Explanation of proposed staffing
    Relevant experience of major personnel

Budget

    Budget in tabular form
    Justification of budget items, where necessary

Title Page

The Title Page provides the basic *To, From, Subject, Date* information found in headings and Title Pages for other types of technical communications; it also includes financial information relevant to proposals alone. Specific formats for Title Pages vary from one funding agency to another, but most require the following:

1   The title of the proposal (as short and informative as possible)

2   A reference number for the proposal

3   The name of the potential funding agency

4   The name and address of the proposer(s): principal investigator and any co-principal investigators or project director

5   The proposed starting date and duration of the project

6   The total funds requested

7   The proposal's date of submission

8   The signatures of the project director and responsible administrator(s) in the proposer's institution

A sample Title Page is shown in Figure 10-3. It lacks items 7 and 8 because these were included on the institution's cover form which accompanied the proposal.

Abstract

Like other Abstracts, the Abstract of a proposal is short, often 200 words or less. In a short proposal addressed to someone within the writer's institution, the Abstract may be located on the Title Page; in a long proposal or one addressed to a funding agency outside the writer's institution, the Abstract will usually occupy a page by itself following the Title Page.

The Abstract is a critical part of the proposal because it provides a short overview and summary of the entire proposal; it is the only text in the proposal seen by some readers. The Abstract should briefly define the problem and its importance, the objectives of the project, the method of evaluation, and the potential impact of the project. It normally does not define the cost. A sample Abstract appears in Figure 10-6; a longer treatment of Abstracts appears in Chapter 9.

Introduction

The Introduction of the proposal, like the Foreword of the technical report, orients a nonspecialist to the subject and purpose of the document. It should define the problem being addressed (perhaps defining what it isn't as well

FIGURE 10-3   Sample Title Page for a formal proposal.

IN REPLY REFER TO:
DRDA 81-2096-P1

PROPOSAL FOR EXTENSION OF

NASA GRANT NSG 1306

MODELS AND TECHNIQUES FOR EVALUATING THE

EFFECTIVENESS OF AIRCRAFT COMPUTING SYSTEMS

Submitted to the

NATIONAL AERONAUTICS AND SPACE ADMINISTRATION
LANGLEY RESEARCH CENTER
HAMPTON, VIRGINIA 23365

Submitted by the

SYSTEMS ENGINEERING LABORATORY
DEPARTMENT OF ELECTRICAL AND COMPUTER ENGINEERING
THE UNIVERSITY OF MICHIGAN
ANN ARBOR, MICHIGAN 48109

PRINCIPAL INVESTIGATOR: John F. Meyer

PROPOSED STARTING DATE: 1 July 1981

PROPOSED DURATION: 1 year

AMOUNT REQUESTED: $39,932

as what it is) and explain the purpose and significance of the proposed project in terms appropriate for a managerial audience. If the project is simple, the Introduction may also include the few relevant details which would belong in the Background section of a more complicated proposal.

Background

As a separate section, the Background allows you to fill in important technical details inappropriate for the nonspecialist readers of the Abstract and Introduction. The Background provides a place to discuss the history of the problem, to survey previous work on your topic (a survey leading up, of course, to some problem or gap in the previous work), and to place this project in a particular context of previous work you may have done on the problem. If this proposal extends earlier work you have done, be sure to show why your previous work needs to be continued and how the proposed work differs from it. Do not spend a great deal of time justifying your earlier efforts and budgets; concentrate on the new work proposed.

If there has been previous work done on your project, you need to demonstrate that you are aware of it and understand its importance and limitations. (Sponsors don't want to pay you to reinvent the wheel.) You usually do this in a section called "Literature Survey" or "Previous Work," and you will demonstrate your competence by carefully selecting and evaluating the works you cite. If you can't select the most crucial items for your project and briefly show why they're crucial, you're probably not expert enough to know what you need to do and how best to do it. Also, you won't be able to show how your work fits into the larger scheme of things, how it builds on previous work and goes beyond it, how it is original and contributes to knowledge in the field.

Description of Proposed Research

The Description of Proposed Research is the most important section of the whole proposal. It describes what you want to do and how you intend to do it:

> The objectives to be achieved
>
> The plan for reaching those objectives
>
> The plan for evaluating the results
>
> The schedule for completing the work

This section will be evaluated carefully by the proposal's reviewers, who will be knowledgeable in the field. Their job is to eliminate all proposals whose objectives or plans are inappropriate or unclear or not well thought out. Thus, your job as a writer is to convince them that you are doing what needs to be done and are doing it in the most careful and thorough way.

When preparing the Description section you should assume that you are writing to a critical, hostile audience, for you are. You should provide all of the details a knowledgeable critic would need to assess your argument—the assumptions on which your research is based, the hypothesis you are following, the specific question(s) you are trying to answer, the particular research and evaluation methods you are using, the appropriateness of your methods for the problem proposed. In addition to providing these items, you may need to justify them, especially if there might be any question or controversy about them. You will probably want to overtly demonstrate the appropriateness of your method for solving the problem posed. If this isn't clear, probably nothing else will matter.

You also need to convince the critical reader that your proposed schedule is appropriate and realistic. You don't want to propose to do too much, given your time and resources, or it will seem that you have a poor assessment of the project and don't really know what you are doing. One way to demonstrate that your schedule is realistic is to spell it out very specifically so that the critical reader can easily see its merits and the care and thoroughness you put into determining it. If you demonstrate that you've really thought of everything, carefully, you're halfway to success.

Description of Relevant Institutional Resources

If you are proposing a project that requires special equipment, one important factor in your ability to do the proposed work is having access to that equipment. Having this equipment already available at your institution is a big plus for your proposal since a funding agency could pay you much less to do the work than it would have to pay someone who had to buy the equipment. Thus, it is to your advantage to list relevant institutional resources. Further, funding agencies often feel that proposers work harder (and institutions monitor them more carefully) if the proposer's institution has resources invested in the project.

List of References

If you have enough references to interrupt the text if you insert them as you go along, you may want to set them up in a separate section. You may also want to do this if the previous work is especially important and you want your reviewers to see that you have cited all the "right" items. As a rule of thumb, if you have more than six references, you might consider a List of References, placed before the sections on Personnel and Budget. The references are listed consecutively as they appear in the text and are listed with the author's name in normal order (first name or initials first).

Personnel

The purpose of this section is to explain who will be doing what and to demonstrate that the people listed for a proposed activity are competent to do it. This is normally accomplished in two subsections, one outlining the responsibilities of the individual participants and the structure to coordinate them and one providing short biographical sheets for the main participants. Both sections should focus on only the *relevant* qualifications of the participants.

Budget

Like the Personnel section, the Budget section has two purposes: to explain what things will cost and to justify and explain individual expenditures, especially where these are not obvious. The budget is usually summarized in a table, such as Table 10-1. (A simple proposal may have a much simpler budget.) The typical headings in a budget are Personnel, Equipment, Supplies, Travel, Computer Time (if relevant), and Indirect Costs. Items typically included under these headings are listed in Table 10-2. Note that not all proposals will need to include all items, but where appropriate the items listed in Table 10-2 should be treated in a proposal's budget.

Appendices

Appendices are reserved for necessary supporting documents which, because of their length or type, would disrupt the "flow" of the proposal. The most common Appendix items are biography sheets for more than six to eight people, letters of endorsement for the proposal, and promises of participation from important participants. Other materials may be pertinent to a given proposal, but the proposal writer should consider carefully any item included in the Appendix and eliminate anything not really needed to support the importance of the topic, the credentials of the proposers, or the ability of the proposers to carry out their work.

10.2

THE ORGANIZATION OF A SHORT INFORMAL PROPOSAL

The outlines for proposals presented in Figures 10-1 and 10-2 are relatively full outlines; they allow writers to argue convincingly to someone who doesn't know them that their problem is important, that their proposed activity will alleviate the problem, and that they are qualified to do the proposed work. Sometimes, however, proposers need to write very short, almost routine proposals meant for people in their own unit. Such proposals might concern some small thing to be done, some small problem to be noticed, some small piece of equipment to be purchased. In such a situation, a proposal using a compressed form of the outline shown in Figure 10-2 might be very appropriate.

Such a proposal appears in Figure 10-4. This proposal omits the sections on Personnel and Budget. These are, however, implied in Sections 5 and 6: one person, the writer, will work for 4 weeks, and perhaps someone

TABLE 10-1   SAMPLE 12-MONTH BUDGET

	CONTRIBUTED BY SPONSOR	CONTRIBUTED BY OUR COMPANY	TOTAL
PERSONNEL			
Project director, half-time	$ 8,000	$ –0–	$ 8,000
Project associate, ¼-time	–0–	3,000	3,000
Research assistant, full-time	6,000	–0–	6,000
Clerk-typist, full-time	5,000	–0–	5,000
Subtotal	$19,000	$3,000	$22,000
Staff benefits (15% of salaries and wages)	2,850	450	3,300
Subtotal	$21,850	$3,450	$25,300
CONSULTANTS			
Warren Duval, $100/day, 2 days	$ 200	$ –0–	$ 200
EQUIPMENT			
Methometer	$ 2,000	$ –0–	$ 2,000
MATERIALS AND SUPPLIES			
Miscellaneous office supplies	100	–0–	100
Glassware	100	–0–	100
Chemicals	100	–0–	100
Subtotal	$ 300	$ –0–	$ 300
TRAVEL			
Project director consultation with sponsor, Chicago to Washington, D.C., and return; 1 person, 2 days			
Air fare	$ 94	$ –0–	$ 94
Per diem © $33/day	66	–0–	66
Local transportation	20	–0–	20
Subtotal	$ 180	$ –0–	$ 180
TOTAL DIRECT COSTS	$24,530	$3,450	$27,980
INDIRECT COSTS (69% of salaries and wages, including staff benefits)	$15,077	$2,381	$17,457
GRAND TOTAL	$39,607	$5,831	$45,437

SOURCE: *Proposal Writer's Guide,* Division of Research Development and Administration, University of Michigan, Ann Arbor, September 1975, p. 9.

from the Hardware Department will work for a short time to provide the necessary hardware. However, the proposal does have the other sections in relatively obvious terms. It has an Introduction which states

1   The problem (we need a catchy demonstration program)

2   The objectives of the proposed work (to provide a demonstration program by simulating the control panel of a nuclear reactor)

TABLE 10-2 CHECKLIST FOR PROPOSAL BUDGET ITEMS

SALARIES AND WAGES
1 Academic personnel
2 Research assistants
3 Stipends (training grants only)
4 Consultants
5 Interviewers
6 Computer programmers
7 Tabulators
8 Secretaries
9 Clerk-typists
10 Editorial assistants
11 Technicians
12 Subjects
13 Hourly personnel
14 Staff benefits
15 Salary increases in proposals that extend into a new year
16 Vacation accrual and/or use

EQUIPMENT
1 Fixed equipment
2 Movable equipment
3 Office equipment
4 Equipment rental
5 Equipment installation

MATERIALS AND SUPPLIES
1 Office supplies
2 Communications
3 Test materials
4 Questionnaire forms
5 Duplicating materials
6 Animals
7 Animal food
8 Laboratory supplies
9 Glassware
10 Chemicals
11 Electronic supplies
12 Report materials and supplies

TRAVEL
1 Administrative
2 Field work
3 Professional meetings
4 Travel for consultation
5 Consultants' travel
6 Subsistence
7 Automobile rental
8 Aircraft rental
9 Ship rental

SERVICES
1 Computer use
2 Duplication services (reports, etc.)
3 Publication costs
4 Photographic services
5 Service contracts
6 ISR services (surveys, etc.)

OTHER
1 Space rental
2 Alterations and renovations
3 Purchase of periodicals and books
4 Patient reimbursement
5 Tuition and fees (training grants)
6 Hospitalization
7 Page charges
8 Subcontracts

INDIRECT COSTS

SOURCE: *Proposal Writer's Guide,* Division of Research Development and Administration, University of Michigan, Ann Arbor, September 1975, p. 10.

3 The significance of the work (to prove the ability of the computer programming language, Industrial Pascal, to communicate with various input-output devices and to test its capabilities to monitor and control several simultaneous tasks)

The proposal also has a plan for accomplishing the objectives, and this is outlined in Sections 2 to 5 of the proposal. It has an implied plan for evaluating the results; an observer would simply see if the demonstration program simulates the nuclear control panel by producing the actions described in Section 4. Finally, the proposal has a schedule (Section 6), and it lists resources already available in the unit (Section 2) as well as needed resources (Section 5).

In what it includes, then, this proposal provides the context and information necessary for those reviewing it while still being an appropriately short proposal for a small project.

FIGURE 10-4   Sample of a short informal proposal.

Chen Computer Systems—Interoffice Memorandum *

To: John Van Roekel
           Software Manager

From: Larry N. Engelhardt
           Senior Programmer

Date: 9 August 1979

Subject: Proposed Demonstration Project for Industrial Pascal: Specifications
           and Description

1. Introduction

The Programming Division has just finished developing an Industrial Pascal
language system for our new 3935 RacPac minicomputer system. We now need a
computer program that will allow our sales people to demonstrate the capabilities
of the RacPac system running Industrial Pascal in a "real" environment.

I propose to develop a demonstration program that will control a simple but
realistic representation of a control panel of a nuclear reactor. This demonstration
program will be written entirely in Pascal and will run on a RacPac system in an
Industrial Pascal environment.

The RacPac will be hooked into a 3910A-3920A simulator, and together the
RacPac and simulator will represent the control panel of a nuclear reactor. The
program will read and write digital and analog inputs and outputs to and from the
controls and indicators on the simulator front panel and respond in an interesting
and amusing manner. This demonstration will both prove the ability of Industrial
Pascal to communicate with various input/output devices and test its capabilities
to monitor and control several simultaneous tasks.

2. Hardware Capability

The hardware configuration on which the demonstration program will run
consists of a simulator built by Bill Lepior and a 3935-A RacPac system. The simulator is a cabinet which features two potentiometers and a meter connected to a
3910A analog I/O controller and eight lamps and eight switches connected to a
3920A digital I/O controller. Both controllers are plugged into the chassis of the
RacPac. Furthermore, the RacPac has an integral plastic keyboard and 5-inch CRT
screen controlled by an 1812 interface for operator I/O. This mix of digital and
analog I/O devices offers a good sample of control devices.

3. The Demonstration

One assembler language program has already been written to demonstrate
the simulator. Of course, a Pascal program could be written to duplicate the functions of the assembler program exactly. However, to be interesting for the customers, the Pascal demo program could be made more realistic. In addition, a
dynamic and robust demonstration program is needed to make use of all the I/O
devices of the simulator and also to exercise the multitasking features in IP in a
real-time control situation. This document proposes that the IP demo program be
designed to be a nuclear reactor control panel.

FIGURE 10-4   *(continued)*

4. Operator Controls

The IP demo program would turn the simulator, keyboard, and screen into a simplistic representation of a control panel of a nuclear reactor. For example, the eight panel switches could move damper rods into and out of the make-believe reactor. The corresponding panel lights would go on when a rod is removed. As damping rods are removed, the core temperature would rise over time. The core temperature can be displayed by the large analog meter. As more rods are withdrawn the temperature would rise at a faster rate. The "operator" can reduce the rate of temperature rise by increasing the cooling water feed rate with one potentiometer. The second potentiometer could set an alarm point on a temperature which will cause the reactor to "scram" automatically. The goal of the "operator" is to keep the reactor running in a range between letting the core cool off and scramming, or going critical.

The screen of the RacPac can display values for all of the activity of the simulation. The screen can also display helpful but ominous warning messages such as "Warning: Evacuate the state of Pennsylvania!" While the screen duplicates the analog and digital inputs and outputs, the keyboard will accept some simple commands to control execution of the program.

5. Hardware Required

The development of the demonstration program will require a 3800B for editing and compiling. Intermediate debugging could be done by plugging the simulator into a 3800B if it is equipped with an 1850 controller board in addition to its standard 1824. Final checkout will require either down-line loading of the program from a 3800B to the simulator's RacPac or burning of the program into PROMs for installation into the RacPac. The Hardware Department has both a 3800B and an 1850 controller board, and it could burn the program into PROMs and install the PROMs.

Description	Weeks Required	Total Time
Research drivers	1	1
Design program	1	2
Code	1	3
Debug	1	4

*Note to readers: <u>Pascal</u> is a computer language, and <u>Industrial Pascal</u> (<u>IP</u>) is a version of Pascal used on small computers in industrial settings to control machinery or to monitor various processes. <u>Software</u> is a computer program or programs. <u>I/O</u> is input/output, that is, the text or information put into a computer (input) or that given out by it (output). A <u>CRT screen</u> is a cathode ray tube screen, a screen similar to a television, which prints out the letters and numbers the operator types and which prints anything the operator commands the computer to calculate and print. <u>Hardware</u> is equipment, computers or the "hard" parts of computers, and a <u>PROM</u> (A programmable, read-only memory) is a small part in a computer.

10.3

EDITING THE PROPOSAL

Once you have written a proposal, it is wise to analyze it for weak spots, areas needing more proof or detail. As a guide, you might refer to the analysis below of the problems detected in 605 proposals rejected by the National Institutes of Health.[1] More than one item may have been cited in rejecting a proposal.

A   PROBLEM (58 percent)
1   The problem is not of sufficient importance or is unlikely to produce any new or useful information. 33.1
2   The proposed research is based on a hypothesis that rests on insufficient evidence, is doubtful, or is unsound. 8.9
3   The problem is more complex than the investigator appears to realize. 8.1
4   The problem has only local significance, or is one of production or control, or otherwise fails to fall sufficiently clearly within the general field of health-related research. 4.8
5   The problem is scientifically premature and warrants, at most, only a pilot study. 3.1
6   The research as proposed is overly involved, with too many elements under simultaneous investigation. 3.0
7   The description of the nature of the research and of its significance leaves the proposal nebulous and diffuse and without a clear research aim. 2.6

B   APPROACH (73 percent)
1   The proposed tests, methods, or scientific procedures are unsuited to the stated objective. 34.7
2   The description of the approach is too nebulous, diffuse, and lacking in clarity to permit adequate evaluation. 28.8
3   The overall design of the study has not been carefully thought out. 14.7
4   The statistical aspects of the approach have not been given sufficient consideration. 8.1
5   The approach lacks scientific imagination. 7.4
6   Controls are either inadequately conceived or inadequately described. 6.8
7   The material the investigator proposes to use is unsuited to the objective of the study or is difficult to obtain. 3.8
8   The number of observations is unsuitable. 2.5
9   The equipment contemplated is outmoded or otherwise unsuitable. 1.0

C INVESTIGATOR (55 percent)

1	The investigator does not have adequate experience or training for this research.	32.6
2	The investigator appears to be unfamiliar with recent pertinent literature or methods.	13.7
3	The investigator's previously published work in this field does not inspire confidence.	12.6
4	The investigator proposes to rely too heavily on insufficiently experienced associates.	5.0
5	The investigator is spreading himself too thin; he will be more productive if he concentrates on fewer projects.	3.8
6	The investigator needs more liaison with colleagues in this field or in collateral fields.	1.7

D OTHER (16 percent)

1	The requirements for equipment or personnel are unrealistic.	10.1
2	It appears that other responsibilities would prevent devotion of sufficient time and attention to this research.	3.0
3	The institutional setting is unfavorable.	2.3
4	Research grants to the investigator, now in force, are adequate in scope and amount to cover the proposed research.	1.5

10.4

GETTING THE PROPOSAL APPROVED FOR SUBMISSION

Different organizations have different procedures for handling proposals. Often proposals to someone within the writer's organization can be submitted without much red tape. You may need to get your supervisor's approval before sending out a proposal asking that you work for another department or unit, but a verbal OK may be all the approval you need. (Note, however, that you should at least inform your supervisor if you propose to work on a project outside your unit.)

In contrast to the simple procedures for approving internal proposals, the procedures for approving proposals to funding agencies outside the proposer's organization may be quite involved. Organizations often insist on formally reviewing and approving any proposal to external funders. This allows the organization to monitor commitments made to outside organizations; it also allows the organization to eliminate undesirable competition between members of its own staff who may be applying for funding in competition with each other. The problem with this organizational review and approval—from the proposer's point of view—is that it takes extra time, and the proposer must plan this time into the writing process. In addition, the proposer may need to get special forms filled out and signed by administrative officers in the proposer's organization.

For instance, the sample approval form in Figure 10-5 for proposals

FIGURE 10-5   Sample approval form for a proposal seeking
external funds.

Proposal no. DRDA-81-2096-P1
Date 5/21/81

The University of Michigan

APPROVAL OF APPLICATION FOR GRANT OR CONTRACT
(See back of form for instructions.)

____X____ Research   _____ New
_____ Instruction   ____X____ Continuation of acct. no.: 014524
_____ Facilities   _____ Competing renewal
_____ Fellowship/traineeship   _____ Non-competing renewal
_____ Equipment/materials   ____X____ Supplement
_____ Service

To be submitted to: National Aeronautics and Space Administration
                              (proposed sponsor)

Project Director   Prof. John F. Meyer
Department or unit   Elec. & Comp. Eng.
                              (Limit to One)
Organization code   2160

1. Major fields of study to which project is related:
   Code nos.   11.80.50
                              (Please Limit to 3 Codes)

2. PROJECT TITLE (60 spaces max.)   Models and Techniques for Evaluating
   the Effectiveness of Aircraft Computing Systems

3. Proposed period (Must agree with period for which budget in Item 9 is
   given):
   From 7/1/81 through 6/30/82   Total months: 12   Total cost: $39,932

4. Does the proposed activity involve:
   a. Use of human sub- Yes _____ No __X__ If yes, date of committee approval ____
      jects
   b. Use of vertebrate   Yes _____ No __X__
      animals
   c. Lease or purchase Yes _____ No __X__
      of computer equip-
      ment
   d. Recombinant DNA Yes _____ No __X__ If yes, date of committee approval ____
   e. Classified research Yes _____ No __X__ If yes, date of committee approval ____

5. Participating faculty and students:
   a. The following faculty members other than the project director will partici-
      pate.

FIGURE 10-5 (*continued*)

Name	Rank and Unit	Nature of Participation
_____	_____	_____
_____	_____	_____
_____	_____	_____
_____	_____	_____

b. Number of students to participate: Postdoctoral: _____ Graduate: __2__
   Undergraduate: _____

6. University facilities required:
   a. Will adequate space be available for the period proposed?
      Yes __X__   No _____
      If yes, Bldg.: <u>E. Eng.</u>   Room: <u>2523, 2506, 1</u>   Approved: _____
      If no, space required: _____ sq. ft. Source: _____ Approved: _____
   b. Space renovation $_____   UM account no.: _____ Approved: _____
   c. Will acquisition of major equipment items require installation and building
      modification at cost to the University?
      Yes _____   No __X__.

7. Is additional office equipment required for this project? Yes _____ No __X__
   Is it provided for in the proposal? Yes _____   No _____
   If not, attach list and indicate University unit to provide: _____
   _____
      (approved)

8. Is work to be performed off University property? Yes _____ No __X__
   If yes, indicate location and duration: _____

9. Summary of proposed budget:

	UM sources:	Sponsor:	Total:	Acct. no. for UM sources:	Approved:
a. Salaries and Wages (S&W)	_____	21,405	21,405	_____	_____
b. Staff Benefits (SB)	_____	3,853	3,853	_____	_____
c. Fellowships/Stipends	_____	_____	_____	_____	_____
d. Tuition	_____	_____	_____	_____	_____
e. Consultants	_____	_____	_____	_____	_____
f. Travel	_____	3,100	3,100	_____	_____
g. Supplies/Materials	_____	_____	_____	_____	_____
h. Equipment*	_____	_____	_____	_____	_____
i. Computer costs	_____	1,000	1,000	_____	_____
j. Subcontracts	_____	_____	_____	_____	_____
k. Other (Itemize under notes)***	_____	600	600	_____	_____
l. Total Direct Costs (TDC)	_____	29,958	29,958	_____	_____
m. Indirect Costs**	_____	16,974	16,974	_____	_____

FIGURE 10-5 *(continued)*

```
    17.8% Computer = $178
    58% MTDC = $16,796
    Less carryover          ( -7,000)  ( -7,000)
  n. TOTAL                   $39,932    $39,932
```

10. NOTES:
    *Itemize items of equipment costing more than $10,000.
    **Explanation if other than negotiated rates.
    ***Publication costs/supplies.

_____	_____
(Requested by: Project Director)	(Fiscal/Legal review)
J. F. Meyer	

_____	_____
(Approved by: Dept. or Unit Head)	(Approved for DRDA by: Project Representative)
G. I. Haddad	Mary L. Egger

_____	_____
(Approved for School/College by: Dean)	(Approved for the University by: Vice-President)
M. J. Sinnott	C. G. Overberger

seeking external funds had to be filled out and signed by four administrative officers in addition to the proposer. Just finding these people can take some time, and theoretically each person should have some time to review the contents of the proposal he or she is approving.

If the proposer's organization requires such approval, the proposer should be aware of this fact long before the proposal's deadline and should plan to allow sufficient time for each reviewer.

EXERCISE 10-1

A   Compare the short formal proposal shown as Figure 10-6 to the proposal outlines given in Figures 10-1 and 10-2. See if you can identify the various sections outlined in Figure 10-2. Compare the treatment of these sections with the comparable treatment of sections in the short informal proposal in Figure 10-4 and then to that in the formal proposal in the Appendix.

B   Write a short formal or informal proposal for some project you want to do. Be sure to include all relevant information, and be prepared to justify the information you have included or excluded.

FIGURE 10-6 A short formal proposal.

THE DEVELOPMENT OF INSTRUCTIONAL MODULES IN LECTURE
COMPREHENSION FOR FOREIGN STUDENTS

A proposal submitted to the Center for Research on Learning
and Teaching in competition for a Faculty Development Fund
Teaching Grant

Project Director: Leslie A. Olsen, Associate Professor, Department of Hu-
                           manities, College of Engineering

Co-Principal Investigator: Thomas N. Huckin, Assistant Professor, Department of
                           Humanities, College of Engineering

Date of Submission: February 15, 1979

Proposed Starting Date: May 1, 1979

Proposed Duration: 1 year

Amount Requested: $5000

ABSTRACT

Many foreign students attending the University of Michigan have serious
problems understanding course lectures. The objective of this proposal is to improve the University English instruction for such foreign students by developing
expertise and instructional materials in lecture comprehension. Specifically, we
propose (1) to compile experimental evidence from several disciplines to provide a
solid explanatory base for teaching lecture comprehension and (2) to develop a set
of graded video tapes and handout materials on lecture comprehension aimed at
science and engineering students.

The results of this project, which will be class tested and reviewed by outside reviewers, could affect potentially all of the foreign science and engineering
students at the University who need instruction in lecture comprehension. It is
estimated that one third of the approximately 1630 foreign technical students
here now need such instruction and that about 300 of the students entering each
year would profit from this instruction when they first arrive. The instructional
materials will be made available for both class and individual use.

For years the English Language Institute (ELI), the Department of Linguistics, and the Department of Humanities in the Engineering College have conducted
a program to improve the English of foreign students at the University. The ELI
program has been oriented toward basic, sentence-level skills and spoken "survival" English, while the programs in the Linguistics and Humanities Departments have been oriented toward improving writing and conversational ability.

FIGURE 10-6   *(continued)*

Recently we in the Humanities Department have become aware of a strong need for another component in our programs for science and engineering students, a component on lecture comprehension. This need has been expressed by the Oster-Brown report on the state of English-as-a-second-language (ESL) instruction at the University, by specific requests from Dean Sussman at Rackham (stimulated by complaints from various professors), by complaints from professors in our college to us, and by comments and requests from our students, even those students who did well on the various English tests.

Science and engineering professors often complain that foreign students cannot follow lectures, especially the interpretive material added by the professor in the lecture. On the other hand, science and engineering students complain that they can't remember all of the material (they often try to write down and memorize a lecture word for word rather than to look for generalizations). Further, the students often complain that they can't tell what is important and what is unimportant material (they can't tell when an aside or a digression occurs in contrast to an important gloss or a shift in the line of argument) and they can't follow a lecturer who is disorganized, who mutters, or who digresses. In addition, we and other researchers have noted that students often do not know how to read visual aids presented to summarize or to illustrate a point, and they often don't see the redundancy we put in our speech to signal our various shifts in purpose, mood, or importance of subject. In short, there is a strong consensus that something needs to be done and done quickly.

We are thus requesting $5000 to develop some materials in lecture comprehension appropriate for the foreign science and engineering students at the University. (These students account for approximately 70% of the 2100 foreign students at the University of Michigan.) In the pages which follow, we would like to argue (1) that there is a need for new materials to be developed because there are no adequate materials available and (2) that we are qualified to develop such materials.

NEED FOR NEW MATERIALS

An important recent study from the University of Lancaster (England), one of the few studies of "real" lecture situations and the only serious in-depth study of the problems foreign students have with science and engineering lectures, has identified all of the problems we have just described. This study has also stressed the critical importance of using lecture comprehension materials (1) which incorporate a real lecture situation and (2) which really represent the total communication situation with all of its visuals, movements, digressions, etc. We agree completely with this conclusion.

Unfortunately, the materials currently available to teach listening comprehension are inadequate from both points of view. First, most materials are inadequate because they are too basic; they do not represent real lecture situations even at the end of an instructional sequence. (People seem to have assumed that a student living in this country would readily assimilate listening skills and thus would

FIGURE 10-6 *(continued)*

not need more advanced, real lecture materials. We have already tried to demonstrate that this assumption is false.) Second, the available materials are inadequate because they simply do not simulate the total communication situation in a classroom with its asides, glosses, hesitations, digressions, visual aids, blackboard notes, etc.; the available materials are usually audiotapes of texts carefully read from written or highly polished material.

Thus, if we are to develop adequate instructional modules in lecture comprehension, we need to prepare a series of videotapes of lectures of varying difficulty and to develop class materials to accompany these tapes. We also plan to survey the literature in cognitive psychology, psycholinguistics, and speech for further insights and experimental support for our current hypotheses on lecture comprehension for foreign students.

OBJECTIVE

The general objective of this proposal is to improve the Humanities Department's (and the University's) offerings in the area of foreign student instruction by developing expertise and instructional materials in the now neglected area of lecture comprehension. We have already done some basic reading devoted to lecture comprehension in scientific and engineering disciplines and have some ideas about those features of lectures in science and engineering which must receive special emphasis—such as kinesics (movement and how it is tied to the lecture's structure), visual aids, and ways of signaling transitions within a unit and transitions to peripheral or explanatory material. Thus, we propose for this project to

1) compile a body of experimental evidence from cognitive psychology, psycholinguistics, and speech and rhetoric to test and support or refine our analysis of those features of lectures needing special attention for foreign students. This will develop our expertise as well as provide a solid explanatory base for our teaching materials;

2) develop a set of graded videotapes and handout materials for a class unit on lecture comprehension.

ACCOMPLISHING THE OBJECTIVES

To accomplish the objectives listed above, we plan to

1) survey the literature in speech, psychology, and psycholinguistics for evidence describing the relative importance of various features in establishing coherence in lectures or in oral discourse;

2) define additional areas or features not covered in the literature above which we feel might be important in establishing (or breaking) coherence for foreign students;

FIGURE 10-6   *(continued)*

3) survey tapes of lectures already available to find at least some segments of lectures we could use to describe and teach the important features identified in steps 1 and 2 above;

4) secure permission to use appropriate sequences from prerecorded tapes for instruction in lecture comprehension;

5) produce tapes as necessary to provide a modularized series of 5 or 6 tapes (3 tapes 5-10 minutes long and 2 or 3 tapes 25-30 minutes long);

6) develop for each tape written handout materials and questions focusing on the important features of the tape.

Any materials developed would be made available to teachers in any unit of the University and would be placed in the Sight and Sound Room of the Undergraduate Library for individual use.

EVALUATING THE OUTCOME

In general, we have been successfully utilizing—and plan to continue utilizing—two different forms of evaluation for our project: testing of students and evaluation by peers. Student-testing is provided mainly through class assignments (written reports and papers) and subsequent classroom critique sessions and editing exercises; we observe the students' performance to see if our instruction has been learned and put to use. We also use class exercises aimed directly at the teaching points and solicit student response and suggestions on these class exercises. Peer evaluation will be provided as indicated by the following sources:

1) on a monthly basis by colleagues in the English Language Institute, the Linguistics Department, the Speech Department, and the communications group of the Humanities Department;

2) at the project's end by researchers at Carnegie-Mellon University working on discourse analysis from the perspectives of cognitive psychology (John Richard Hayes) and rhetoric (Richard E. Young and Linda Flower); and

3) by researchers at the University of Lancaster (England), Christopher Candlin and Dermot Murphy, who have done the most significant research to date on listening comprehension in technical discourse.

SCHEDULE FOR PROJECT COMPLETION

If our complete budget request is allocated, our proposed team could complete the pre-evaluation part of it within 12 weeks. We could begin as early as May 1 and complete this part by the end of July. Evaluation and subsequent refining would begin in early September and continue through the academic year.

FIGURE 10-6 *(continued)*

PROPOSED BUDGET

Since there are three of us working on this project, including a research associate, we are asking for the maximum amount of support the CRLT can provide, namely, $5000. This total is broken down as follows:

Partial salary support for the project director, Leslie Olsen, during May-July	$1550.00
Partial salary support for the co-principal investigator, Thomas Huckin, during May-July	1550.00
Research Associate II salary for Russell S. Tomlin: 200 hours @ $7.75	1550.00
Supplies, photocopying, typing, 2 videotapes	350.00
	$5000.00

We are especially requesting support to work on the project during the summer because we have heavy teaching loads and thus limited research time during the school year. (Our normal course load is 3 courses per term per instructor, and this winter term Olsen and Huckin are each teaching 4 courses.)

QUALIFICATIONS OF RESEARCHERS

Any teacher trying to develop materials on lecture comprehension for students in science and engineering should be able:

1) to apply the principles of discourse analysis developed in linguistics to analyze texts (in order to be able to adequately describe important features of the material students need to study);

2) to describe specifically the features of the English of science and technology (EST) and especially to describe the special ways in which EST establishes coherence and marks transitions;

3) to describe and analyze features of intonation and phonology and to describe such features to students having trouble with the sounds of words;

4) to explain the relationship between speech, movement, and gestures in a lecture (be able to explain why the speaker moved when and where he/she did);

5) to understand enough about the subjects chosen for the lectures to select appropriate material for the videotapes (this would be done in consultation with specialists in the field).

Our proposed team meets all of these qualifications. To address the first and second criteria, all three of us have done research and teaching in discourse analysis.

FIGURE 10-6 *(continued)*

Olsen and Huckin have been teaching scientific and technical writing for five and two years, respectively; have published in the area; and are finishing up a text on the subject. Further, all three team members are currently teaching a graduate-level course in the area. To address the third and fourth criteria, one of our team, Huckin, wrote his dissertation on phonology and intonation and all three members of our team have started studying the relationship between speech, movement, and gestures in videotaped lectures on technical subjects. Finally, to address the fifth criterion, Olsen and Huckin have had up to 12 years of experience with technical students and subjects and Olsen has worked as a technical writer and editor and has a degree in chemistry with significant work in math, physics, and biochemistry.

POTENTIAL IMPACT OF PROJECT

The expertise and materials developed in this project could potentially affect all of the foreign science and engineering students at the University who need instruction in lecture comprehension. We don't know exactly how many students there would be in this category, since none of the standard English tests for foreign students (including the Michigan Test and the TOEFL) measure lecture comprehension. However, from all of our informal measures, experience, and anecdotal evidence, we estimate that one third, or 550, of the approximately 1650 foreign technical students here at both the undergraduate and graduate levels need such instruction in lecture comprehension now and that about 300 students entering each year would profit from this instruction when they first arrive. The instructional materials developed on lecture comprehension will be made available for both class and individual use.

Further, the results of this project, including the materials and the evaluations of the materials, should increase our ability to secure funding from the National Science Foundation for a large, multidisciplinary research project on the discourse analysis and comprehension of lectures.

REFERENCE

1   *Proposal Writer's Guide,* Division of Research Development and Administration, University of Michigan, Ann Arbor, 1975. From E. M. Allen, *Science,* November 25, 1960, pp. 1532-1534.

11

THE BUSINESS LETTER

There are a number of activities which a business does frequently. These include requesting information or equipment, ordering supplies, praising or thanking someone for a job well done, complaining about a job badly done, and responding to someone else's request, order, praise, or complaint. While some of these activities may be accomplished orally or by filling out a form, many of them require writing a letter. A letter provides a record of the activity for someone's file, it allows the writer to provide more context or explanation than is usually possible on a form, and it helps the audience remember what is to be done. While letters may be written more frequently in some jobs than in others, all technical people should be able to write a good letter when it is needed.

Writing a good letter is an art. It basically requires a writer to produce a one-sided conversation with the reader. Of course, all writing does this to some degree in that the writer anticipates the reader's questions and provides answers to those questions where the reader might ask them. However, letters differ from most other forms of writing: they are often more personal, even emphasizing the reader/writer relationship with the generous use of such pronouns as *I, we,* and *you.*

Despite this frequent personal sense, letters share several of the organizational features already described for the less personal technical report. Like reports, letters need to first orient the reader to the topic at hand, then explain why the writer is writing, and then provide enough information so that the reader can easily understand what he or she is to do.

The orientation of the reader is especially important when the reader does not expect a letter or might not remember the subject the letter refers to. For instance, consider the letters in Figures 11-1 and 11-2. The first letter has no orientation to the topic being discussed. Its writer jumps right into the letter as if she were simply continuing the phone conversation which initiated it. Unfortunately, however, the conversation occurred almost 3 weeks before the letter was written, and the reader may well have a hard time remembering which shroud is being discussed in the first paragraph and what he needed to know about it. In contrast, the version of the letter in Figure 11-2 gives a good orientation to the topic; it reminds the reader of the conversation and the main questions raised. It also provides a rationale for the organization of the letter's details: they are grouped as answers to the three questions.

FIGURE 11-1 Sample business letter without a good orientation to the topic.

ADVANCED TECHNOLOGIES, INC.
40 Technology Park, Milford,
Massachusetts 01757
617-555-4553

October 4, 1981

Mr. Frank E. Lee
Equipment Department
Harrison Radiator, Inc.
1451 Murray Avenue
Pittsburgh, PA 15217

Dear Mr. Lee:

The shroud you called about on September 15 has the same shape, construction, and functions as the shroud previously purchased by Harrison Radiator Division from A.T.I. However, that machine was designed to balance a series of blower assemblies having only one fan. With the present balancer you wish to have the capability to balance a new blower assembly having two fans, one on each end of the motor shaft. This new blower assembly is longer than all previous single-fan blower assemblies. The only dimensional change needed to account for this is to widen the shroud from 10″ to 16-½″. Also, a new clamping fixture is needed to hold the double-fan assembly, necessitating that a 3″-diameter hole be added to allow a clamping cylinder to protrude. These changes are illustrated on the enclosed print.

I have found that the change of the focal point of the fan speed sensors, due to the new blower fan's position within the shroud, in relation to the sensors, is well within the limits of the adjustments designed into the original shroud, as is shown on the print. This should not present a problem.

The estimated weight of the shroud is 6-½ lbs., only ½ lb. heavier than the previous one. This results in a cost difference of an additional $5.75 for the new shroud.

I refer you to the enclosed print; it might answer additional questions.

Sincerely,

Barbara C. Benson

Barbara C. Benson
Engineering Department

BCB/rs

Enclosure

FIGURE 11-2 Sample business letter with a good orientation
to the topic.

ADVANCED TECHNOLOGIES, INC.
40 Technology Park, Milford,
Massachusetts 01757
617-555-4553

October 4, 1981

Mr. Frank E. Lee
Equipment Department
Harrison Radiator, Inc.
1451 Murray Avenue
Pittsburgh, PA 15217

Dear Mr. Lee:

In our phone conversation of 28 September 1981, you wanted to know:

--What the differences will be between the original shroud for Harrison Radiator Division's D-25-PS balancing machine and a similar one purchased
from A.T.I. on September 15, 1981 (Ref.: Sales Order #N5682).

--How these differences affect the cost of the shroud.

--If the introduction of our double-fan blower assemblies will create any
problems with the fan speed sensors.

I have listed this information below and enclosed a blueprint of our new shroud
for your reference (#25508).

Basically, the two shrouds have the same shape, construction, and functions. However, there are three major differences between the Harrison and A.T.I. shrouds:

1) The A.T.I. shroud has a new double-fan blower assembly.

2) The A.T.I. shroud is 6-½″ wider.

3) The A.T.I. shroud has an additional 3″-diameter hole in the top of the
shroud.

These changes will result in an additional cost of $5.75. Also, the new blower as-
semblies will not present any problems with regards to the fan speed sensors.

The original Harrison shroud was designed to work only with a series of single-
fan blower assemblies. The introduction of longer double-fan blower assemblies
(fans on both ends of the motor shaft) on the A.T.I. shroud requires two things.

FIGURE 11-2 *(Continued)*

Frank E. Lee
October 4, 1981
Page Two

The shroud must be lengthened from 10″ to 16-½″. Also, a new cradle fixture is required to hold the blower assemblies, resulting in the addition of a 3″-diameter hole in the tip of the shroud, to allow a clamping cylinder to protrude.

Since the two shrouds are the same construction--same material, same method of assembly--the only cost difference will result from the greater weight of the new shroud. Widening the shroud 6-½″ adds ½ lb. to the total weight. This translates to an additional cost of $5.75.

As can be seen on the enclosed print, the focal point of the new blower fan is well within the range of the electro-optical fan sensors, as provided for by the sensor adjustments designed into the original shroud.

I refer you to the enclosed print. It may answer additional questions.

                                                    Sincerely,

                                                    Barbara C. Benson

                                                    Barbara C. Benson
                                                    Engineering Department

BCB/rs

Enclosure

11.1

BASIC LETTER FORMATS

Even though letters do not have labels such as *To, From, Subject,* and *Date,* they do have conventional places where readers look for such background information. Figure 11-3 indicates these places, and Figure 11-4 gives an example of a short letter arranged in a conventional format for business letters. Notice that the subject of the letter is mentioned in the first sentence to orient the reader.

The conventional format illustrated in Figure 11-4 is only one of three quite common formats for business letters: the unblocked format, the semi-blocked format, and the blocked format. These formats are presented in skeletal form in Figures 11-5 to 11-7 for letters written on plain paper, that is, not on letterhead stationery. It should be noted, however, that when a writer is representing a company or organization, the writer should use the

FIGURE 11-3 Location of background information in a letter, unblocked format. (On letterhead stationery, the writer's address will already be printed.)

Writer's street address
Writer's city, state, zip code
Date

Reader's name
Job title
Reader's company/organization
Organization's street address
Organization's city, state, zip code

Subject:

Dear name of reader:

Subject of letter and references to past correspondence, if any

Sincerely yours,

Writer's signature

Writer's name typed

WRITER'S INITIALS: typist's initials
(if writer did not type letter)

Enclosure (if appropriate)

cc: Name of recipient of carbon copy of letter
    (if appropriate, or xc: Name of recipient of photocopy of letter)

FIGURE 11-4   Example of a short business letter in the unblocked format.

---

205 State Street
Livonia, Alabama 35851
June 20, 1982

Mr. R. B. Sparks
Sales Representative
Computer Connection
384 Grand Drive
Harrington, Georgia 30018

Subject: BASIC Software for Apple II System

Dear Mr. Sparks:

The Computing Club of Brewster High School will be sponsoring a series of computing activities during the fall of 1982. Since our current computing equipment is inadequate for our projected needs, we are considering buying an Apple II system to expand our equipment line.

Last week I wrote to you requesting information on the cost of the Apple II Plus system and its peripheral equipment as well as information on the PASCAL software available with the system.

I would now like to request additional information on

1)   the BASIC software currently available with the system, especially for plotting graphs such as that enclosed with this letter; and

2)   any BASIC software expected to be released in the next year for the APPLE system.

Thank you for your attention. We are looking forward to hearing from you in the near future.

Sincerely yours,

Ann Jones

Ann Jones
President, Computing Club

AJ:vg

Enclosure

cc: Mr. Peter Murray
    Ms. Mary Corvina

FIGURE 11-5 Unblocked format for business letter.

*(Number of Blank Lines
Between Sections)*

205 State Street
Livonia, Alabama 35851
June 20, 1982

(3)

Mr. R. B. Sparks
Sales Representative
Computer Connection
384 Grand Drive
Harrington, Georgia 30018

(1)

Subject: BASIC Software for Apple II Systems

(1)

Dear Mr. Sparks:

(1)

_____
_____
_____

(1)

_____
_____
_____

(1)

_____
_____
_____

(1)

Sincerely yours,

(3)

Ann Jones
President, Computing Club

(1)

AJ:vg

(1)

cc: Mr. Peter Murray
Ms. Mary Corvina

FIGURE 11-6 Semiblocked format for business letter.

*(Number of Blank Lines
Between Sections)*

                                                                205 State Street
                                                                Livonia, Alabama 35851
                                                                June 20, 1982

(3)

             Mr. R. B. Sparks
             Sales Representative
             Computer Connection
             384 Grand Drive
             Harrington, Georgia 30018
(1)

             Subject: BASIC Software for Apple II Systems
(1)

             Dear Mr. Sparks:
(1)

             _____
             _____
             _____
(1)
             _____
             _____
             _____
(1)
             _____
(1)

                                                      Sincerely yours,

(3)

                                                      Ann Jones
                                                      President, Computing Club

(1)

             Enclosure
(1)

             cc: Mr. Peter Murray
                  Ms. Mary Corvina

FIGURE 11-7 Blocked format for business letter.

*(Number of Blank Lines
Between Sections)*

205 State Street
Livonia, Alabama 35851
June 20, 1982

(3)

Mr. R. B. Sparks
Sales Representative
Computer Connection
384 Grand Drive
Harrington, Georgia 30018

(1)

Subject: BASIC Software for Apple II Systems

(1)

Dear Mr. Sparks:

(1)

_____
_____
_____

(1)

_____
_____
_____

(1)

_____

Sincerely yours,

(3)

Ann Jones
President, Computing Club

(1)

AJ:vg

(1)

Enclosure

(1)

cc: Mr. Peter Murray
      Ms. Mary Corvina

FIGURE 11-8   Sample continuation page, blocked format.

(Number of Blank Lines
Between Sections)

      Mr. R. B. Sparks   -2-   June 20, 1982

(3)

      ————————————————————————————————
      ————————————————————————————————
      ————————————————————————————————
      ——————————————————

(1)

      Sincerely yours,

(3)   *Ann Jones*

      Ann Jones
      President, Computing Club

(1)

      AJ: vg

(1)

      Enclosure

(1)

      cc: Mr. Peter Murray
           Ms. Mary Corvina

organization's letterhead stationery for any correspondence with people outside the organization. When such letterhead is used, the only change in the
formats presented in Figures 11-5 to 11-7 involves the location of the writer's
address, city, and state. These are usually given, along with the organization's
name and telephone number, in the letterhead printed at the top of the page,
as shown in Figure 11-1, for example.

    If a letter requires more than one page, the additional pages are called
continuation pages. They are typed on plain paper, not letterhead. A sample
continuation page appears in Figure 11-8. Notice that it contains enough
information at the top of the page to identify it if it gets separated from the
first page.

    In Figures 11-5 to 11-8 the number of blank lines separating the various
sections of the letter are indicated in the left margin. One blank line separates
most sections, but three blank lines separate the writer's address at the top

of the page from the reader's name and address and three blank lines separate the closing (*Sincerely yours*) from the typed version of the writer's
name. The latter spacing should leave plenty of room for the writer's signature.

The letters in Figures 11-1 to 11-8 all have some additional bits of
information the function of which may not be as obvious as that of other
parts of the letter. This information appears at the left of the page below the
typed version of the writer's name. Such lines should be added to a letter
whenever they are appropriate; if any one of these lines is omitted, the
remaining lines should be moved up, with one blank line separating them
from each other and from the typed version of the writer's name. The following information is included in these lines:

Initials line. The first line under the writer's name is the initials line
(*AJ:vg*), which indicates that someone other than the writer typed the
letter. The capital letters are the initials of the writer, the lower case
letters the initials of the typist. If the writer types the letter, the initials
line is omitted.

Enclosure line. The second line under the writer's name (*Enclosure*)
indicates that something has been enclosed with the letter and alerts
the reader to look for this item. The enclosure may be a whole document, a single page, a picture, a drawing, or anything else. If nothing
is enclosed, the Enclosure line is omitted.

Copy line. The third line under the writer's name (*cc: Mr. Peter Murray*)
indicates that a carbon copy or photocopy of the letter has been sent
to the person (or persons) whose name appears after the *cc:*. Sometimes *xc:* is used to indicate that a photocopy has been sent. If no one
has been sent a copy of the letter, the *cc:* line is omitted.

11.2

LETTER OF TRANSMITTAL

When a document of more than two or three pages—a report, article, or
proposal—is sent to some reader, it is often accompanied by a letter of
transmittal. The purpose of this letter is to identify the document and to orient
the reader to it. It may help the reader decide whether to read the document,
file it, or send it on to someone else. Thus, the letter of transmittal should
be short, no more than one page if possible, but it should identify the document being sent, the project about which it was written, and any especially
important points being made about the project which the reader does not
already know.

The form of the letter of transmittal may vary with the audience. For
instance, suppose you have been working with a partner on a proposal for

the city of Grand Rapids to develop an industrial waste pretreatment program and a modification to a facilities plan your company submitted to Grand Rapids last spring. You and your partner have written the proposal together, and you have agreed to incorporate some last-minute details into a final draft and have that typed. When the final draft is ready, you might put the following short letter on the report and send it to your partner:

Ann–

Here is the final draft of our Grand Rapids proposal. How do you like it?

Roger

Obviously, this short letter assumes that the reader knows which proposal you are talking about when you mention "our Grand Rapids proposal" and what points are especially important. Since the reader is a coauthor of the proposal, this is probably a safe assumption.

On the other hand, if you were sending this proposal to your supervisor for review and approval before you send it out to the city of Grand Rapids, your letter of transmittal might look like this:

Dear Phil:

You assigned Ann and me to prepare a proposal for the development of an industrial waste pretreatment program for Grand Rapids as an addendum to the facilities plan we submitted to Grand Rapids last spring. Here's a final draft of our proposal for both subjects. As you suggested earlier, this proposal is based on a cost-plus-fixed-fee-with-a-maximum contract. It does not include the city's involvement by force account in the work other than identifying equipment to be purchased by the city.

Think it's ready to go?

Roger

This letter is longer and provides more information than did the version to your partner. It assumes somewhat less knowledge on the part of the reader about the particular project and thus points out certain assumptions behind the proposal which would have been obvious to your partner. On the other hand, it assumes that the supervisor has a relatively high level of technical knowledge in waste pretreatment programs and facilities plans. It assumes, for instance, that the supervisor will see the appropriateness of basing a proposal of this sort on a cost-plus-fixed-fee-with-a-maximum contract and of not including the city's involvement in the work other than identifying the equipment it will purchase.

Finally, if you were sending the final proposal to the city of Grand Rapids, you would need to provide much more context and information than provided in the two previous letters. Those letters were written to people who had knowledge about the project, who understood the implications of important technical statements, and who were aware of the experience and credentials of your company in developing industrial waste programs for waste pretreatment and facilities modifications. Much of this knowledge needs to be spelled out in more detail for someone outside the company, someone less familiar with the company's credentials and the details of the development plans. Thus, a letter of transmittal accompanying the proposal might look like that presented in Figure 11-9. This letter, written to the city engineer but potentially read by anyone else to whom the document will be routed, identifies the proposal it accompanies and then spells out some of the information the city engineer and city planners might need to know about the proposers. It emphasizes the proposers' willingness to cooperate with the city and stresses their qualifications in the area of the proposed work. This is probably the most important thing to stress, since any community about to spend a large amount of money wants to believe that the money will be well spent, that the community's problem will be solved for the specified amount. This will most probably occur only if the company funded to do the work is knowledgeable, effective, and experienced.

Whatever the audience, the function of the transmittal letter is to orient the audience to the document. This can be done well only if the writer of the letter is sensitive to the needs, interests, and knowledge of the reader.

11.3

LETTER OF COMPLAINT

Sometimes in our dealings with individuals or businesses we find that we have not received what we were promised or what we expected. We may have received a part different from what we ordered, one that is defective, or one that does not perform as advertised. We may have been badly treated by someone, or we may be exasperated by a long-term dangerous situation. Whenever these sorts of things happen, whenever we feel cheated or victimized, especially in important ways, we want to complain. We want to get things "off our chests" and to correct unpleasant situations.

This desire to complain and correct gives rise to the letter of complaint. The goal of such a letter is usually to improve the situation about which one is complaining. To do this, the writer of a letter of complaint should politely but firmly

1   Identify the nature and seriousness of the problem

2   If possible, request or suggest a solution to the problem

This procedure is illustrated in Figure 11-10, a sample letter of complaint.

FIGURE 11-9 Sample letter of transmittal to an
audience outside the writer's organization.

June 15, 1981

Mr. John L. Hornbach, P.E.
City Engineer
300 Monroe Avenue, N.W.
Grand Rapids, MI 49503

RE: Request for Proposals

Dear Mr. Hornbach:

We are pleased to submit four copies of our proposal for the Development of
an Industrial Waste Pretreatment and Non-Domestic Users Program and the Preparation of a 201 Facilities Plan Addendum.

Our proposal is based on a cost plus fixed fee with a maximum type contract
and is our best estimate at this time to complete the work outlined in your request. We are prepared to make an oral presentation of its contents when requested, and, should we be selected as your engineer, we are prepared to discuss
our proposal in greater detail.

Our proposal does not include the city's involvement by force account in the
work other than identifying equipment that is to be purchased by the city. We believe that there will be considerable manpower saving on the city's part resulting
from our understanding of your needs; our team approach; our knowledge of your
wastewater collection system, your wastewater treatment facilities, and your
NPDES requirements; our ongoing working relationship with the Michigan DNR;
and the fact that we did prepare and had Michigan DNR approval of a Step 1
Grant Amendment for this work.

We are aware of the slow-down in the U.S. EPA grant program and the possibility of some of the work not qualifying for federal and state assistance. We are
prepared to discuss a scale-down of this work if so directed by the city.

We value our long association with the city of Grand Rapids and look forward to its continuance. Please contact us if you have any questions on our proposal.

Sincerely,
McNAMEE, PORTER AND SEELEY

By _____
Philip C. Youngs

FIGURE 11-10   Sample letter of complaint.

┌─────────────┐
| HARRISON |
| RADIATOR, | 1451
| INC. | Murray
└─────────────┘ Avenue
Pittsburgh, Pennsylvania
412-555-4356   November 12, 1981

Ms. Barbara C. Benson
Engineering Department
Advanced Technologies, Inc.
40 Technology Park
Milford, Massachusetts 01757

Dear Ms. Benson:

Two weeks ago we purchased a shroud #22508 from Advanced Technologies, Inc., for our Harrison Radiator Division's D-25-PS balancing machine (reference sales order #N5682). Your sales representative told us that your shroud would fit our Harrison balancing machine. Unfortunately, when we tried to install the shroud, it did not fit.

I am writing now to request instructions for adapting your shroud to our Harrison balancer. If you have such instructions, we would appreciate receiving them as soon as possible. We need to use our balancer within two weeks, if that is possible, and we cannot operate it until we have a suitable shroud.

If we cannot adapt your shroud to our Harrison balancer, we would like to return the shroud for full credit and find a suitable substitute from another supplier.

Thank you for your prompt attention to this matter.

                                                          Sincerely,

                                                          Frank Lee
                                                          Frank Lee

FL:aa

Notice that this letter of complaint asks the reader to provide time, effort, and perhaps money to address the complaint. Since this is typical of many letters of complaint, you can imagine that such letters should be polite but firm. They probably shouldn't threaten the reader, at least not on the first exchange of letters, since they shouldn't alienate the person who may be able to solve the problem. For the same reason, if the letter of complaint suggests a solution to some problem, the solution should be reasonable. From a purely practical standpoint, reasonable solutions have a better chance of being accepted than unreasonable ones. Further, they show the writer to be fair and helpful even in a difficult situation, and thus they may encourage people to do business with the writer in the future.

11.4

RESPONSE TO A LETTER OF COMPLAINT

Responding to a letter of complaint poses its own set of problems. The writer of such a response needs to look knowledgeable, helpful, concerned, and appropriately apologetic. As with other letters, the writer needs to

1 Identify the purpose of the letter, reminding the reader of the source of the complaint as well as of any suggestions the reader originally made for dealing with the complaint

2 Deal with the complaint, outlining whatever the writer can do to help the reader

3 Assure the reader of the writer's goodwill and attention to the problem

These characteristics are illustrated in Figure 11-11. Notice that the writer of this letter stresses the speed with which she is responding and then carefully provides the help the reader needs: she outlines the characteristics of the two shrouds and the few simple modifications which need to be made to the problematic shroud. Finally, the letter ends on a positive, helpful note. It does not grovel, but it stresses the company's commitment to satisfying its customers.

11.5

LETTER OF REQUEST

Writers sometimes need to request information or equipment. Since such requests usually cost the reader time or resources in responding, the writer should very clearly

1 Orient the reader to the topic of the letter

2 Indicate why the writer should be willing to respond

3 Indicate exactly what the writer is requesting the reader to provide

Orienting the reader to the topic was discussed earlier in this chapter and in Chapter 6 and so will not be treated again here. Indicating why the writer

FIGURE 11-11 Sample response to a letter of complaint.

ADVANCED TECHNOLOGIES, INC.
40 Technology Park, Milford,
Massachusetts 01757
617-555-4553   November 16, 1981

Mr. Frank E. Lee
Equipment Department
Harrison Radiator, Inc.
1451 Murray Avenue
Pittsburgh, PA 15217

Dear Mr. Lee:

On November 12, you wrote to me about a problem you were having fitting
our shroud #22508 on your Harrison balancing machine. You asked for instructions, as quickly as possible, which would allow you to adapt our shroud to your
balancer.

I have just received your letter and am sending by return mail the instructions you requested. I am also enclosing, for your reference, a blueprint of our
shroud #22508.

Basically, our shroud and the original Harrison shroud have the same construction, shape, and functions. However, there are three major differences between
our shroud and the Harrison model:

1) Our shroud has a new double-fan blower assembly.
2) Our shroud is 6-$\frac{1}{2}$ inches wider.
3) Our shroud has an additional 3-inch-diameter hole at the top of the
    shroud.

These differences are probably causing your problem with fit.

Fortunately, it will be very simple to accommodate our shroud to your Harrison balancer. The original Harrison shroud was designed to work only with a
series of single-fan blower assemblies. Our shroud introduces longer double-fan
blower assemblies (fans on both ends of the motor shaft), and this requires only
two small modifications.

1) Our shroud must be lengthened from 10″ to 16-$\frac{1}{2}$″.

2) Our shroud will require a new cradle fixture to hold the blower as-
    sembly. You can insert a cradle fixture by drilling an additional 3″-

FIGURE 11-11   *(continued)*

Mr. Frank E. Lee   - 2 -   November 16, 1981

 diameter hole in the top of the shroud, inserting the fixture, and allowing the clamping cylinder to protrude.

If you make these simple modifications, I am sure that your new shroud will fit perfectly.

 If there is anything else we can do to help you in this matter, please feel free to call us. We firmly believe that customer satisfaction is our most important goal.

Sincerely yours,

Barbara C. Benson

Barbara C. Benson
Engineering Department

BCB/rs

Enclosure

---

should be willing to respond may be done along with the orientation or in addition to it. An example of a letter which provides a motive for responding along with an orientation appears in Figure 11-12; one which provides a motive for responding in addition to the orientation appears in Figure 11-10, a letter of complaint as well as a letter of request.

11.6

RESPONSE TO A LETTER OF REQUEST

The response to a letter of request, like any of the other letters already described, should (1) orient the reader, (2) identify the purpose of the letter, reminding the reader of his or her request and any conditions upon it, and (3) provide the information requested. If the reader requested a physical object that will be sent in a separate package or at a later date, the writer should include any available information on the package's status and projected date of arrival. If the reader requested information which is now available, the letter writer should try to include all of the relevant information the reader requested without overwhelming the reader with irrelevant or unnecessary detail. A sample response to a letter of request appears in Figure 11-13. Notice that in form this letter resembles a very short technical report, minus the *To, From, Subject, Date* format and headings to indicate the Foreword and Summary.

FIGURE 11-12 Sample letter of request.

ANDERSON COLLEGE
400 Haggerty Road
San Francisco, California 94105
(415) 555-6400

May 28, 1982

Mr. R. B. Sparks
Sales Representative
Computer Connection
384 Grand Drive
Harrington, Georgia 30018

Dear Mr. Sparks:

The Computing Science Club of St. Mary's School will be sponsoring a new course in the PASCAL programming language for the Fall Term of 1982. Due to the over-crowded conditions of our present computer system, we are going to purchase several microcomputers rather than add PASCAL software to our present system.

We are considering purchasing ten of your Apple II Plus systems for our PASCAL course. However, before we can make a final decision, we must review the following:

1. Information on the cost of the Apple II Plus system and its peripheral equipment.

2. Information on the PASCAL software available with the system.

3. Specifications for the computer system and its microprocessor.

Could you please send us whatever information you have which addresses these questions?

Thank you for your attention. I am looking forward to hearing from you in the near future.

Very truly yours,

C. Beth Norther

C. Beth Norther
Instructor, Data Processing

CBN/vr

FIGURE 11-13   Sample response to a letter of request.

Pine Dunes Country Club
Orchard Dunes, New Jersey 07081

August 16, 1980

Mr. Thomas F. Wilson
Club Manager
Pine Dunes Country Club
Orchard Dunes, New Jersey 07081

Dear Mr. Wilson:

After receiving numerous complaints from some of the club members about the unsafe conditions of the wooden boat docks on the club's west shoreline, you asked me to investigate the condition of the docks. In particular, you asked me to find out what parts were broken or missing from the dock (if any) and what it would cost to make the necessary repairs. I am now writing to report my findings and actions.

I inspected the boat docks on August 12. In general, they seemed to be in good condition, but they did have a few minor trouble spots. Six crossarm brackets were missing bolts, and three stanchion braces had pulled away from the dock and required some minor welding.

After receiving authorization from Virginia Lette, the Beach Director, I bought the required bolts for $4.75, replaced the missing bolts in the crossarm brackets, and welded the three stanchion braces. The docks are now in good working order.

Details on Repairs

On August 13, the docks which were causing the problem were removed from the water. The missing bolts from the crossarm brackets were replaced with new bolts (stock #83443.7AG) that were bought from Commerce Dock Supply for $4.57. They were tightened to the recommended torque of 40 ft-lb. as stated in the installation instructions.

The three stanchion braces that needed welding were brought over to the Maintenance Department, where they were rewelded by Roger Sands. No problems were encountered during the welding, and the stanchion braces seemed to be completely repaired. There was no cost for the welding.

On August 14, the repaired dock sections were reinstalled and the dock system is now in good working order.

Sincerely,

Steven M. Berggruen

Steven M. Berggruen
Assistant Beach Director

EXERCISE 11-1

A Write a letter of request for information that can be provided by one of your classmates, friends, or colleagues. Then ask that classmate, friend, or colleague to write a letter responding to your letter of request.

B Write a letter of request or complaint to someone you do not know. Try to pick a situation about which you really need information or about which you really want to complain. Send your letter and see if it is effective.

C Write a letter of transmittal for a report you have produced, found in the library, or encountered on the job. If you are a student, you may choose a report you have produced in one of your technical classes or in your technical writing class. Define the audience and situation for your letter, and write at least one version that might accompany your report to total strangers.

11.7

LETTER OF APPLICATION FOR EMPLOYMENT OR ADMISSION AND RESUME

If you need to write a letter of application for employment or admission and a resume, it is probably important that you do them well. If they are impressive enough, you may get an interview with a prospective employer or admission to a desired graduate school or other training program. If they are poorly done, you will probably be screened out of the pool of applicants and into the ranks of those still looking. (Most companies and programs have more applicants than positions and thus are looking for ways to eliminate applicants; a bad letter or resume makes elimination easy.)

What Makes a Good Applicant?

If you were hiring an employee, you would probably want someone who filled a need in the company and who was (not necessarily in this order) well trained, technically competent, smart, hardworking, reliable, honest, loyal, well organized, disciplined, tactful, helpful, resourceful, and able to get along with others, to take initiative, and to communicate well orally and in writing. Therefore, when you are applying for a position your letter of application and resume need to say as many of these things as possible about you as an applicant.

This can be done by following a few general principles. First, you should understand that applications for jobs or training programs are usually read by at least several readers coming from different perspectives and having different "pet peeves," assumptions, and preferences. For instance, many readers believe that competent, educated, serious writers should not make grammatical errors; most readers are offended by form letters; many readers prefer a conservative look in a letter or applicant, while others prefer a "different" look. Thus, you should design your letter with multiple audiences in mind. Second, you should stress what you can do for the company or program, not what it can do for you. This means showing that you are aware of *its* needs, that you meet these needs, and that you want to be a part of *this* company or program, not just any one. Third, you should address the person having the authority to arrange an interview or to admit you into a program; don't waste your letter on someone who can't help you. Fourth, you should stress your accomplishments and responsibilities, since these will show off many of the desirable qualities listed above. Fifth, you should set up your letter and resume in expected formats, but still highlight your best qualifications.

Applying the first principle is easy: to be successful, you must write a letter and resume which are short (no more than one page each, if possible) and which please as many and offend as few readers as possible. We suggest that you look as competent, careful, and "proper" as possible: have each letter individually typed by a professional secretary with a professional-quality typewriter on good white bond paper. This may seem expensive, but it really isn't. You're preparing a $20,000 to $50,000 letter (your salary for a year or so if you get the job or the training program that prepares you for the job). Applying the next three principles is difficult unless you know what you want to do, what you can offer, what the company or program does, and what it offers. Thus, writing a good letter involves doing some serious thinking and research.

To find out what you want to do, you need to consider your short-term and long-term goals, your previous training and interests, and your financial, geographical, or personal constraints. Do you eventually want to be a technical person or a manager? (Managers are often technical people who, because of the demands of their jobs, have developed managerial skills but grown away from their technical skills.) What type of training and experience do you need to reach your goal? Is there a conflict between your short-term and long-term goals? If so, what is the best way to resolve the conflict? What are the constraints on your plans and goals? Be sure to consider financial and geographical constraints, as well as such personal constraints as limits on your time, attention, concentration, and energy. You may not be able to answer all of these questions without consulting other people, but if you need some outside advice, go get it. You are making important decisions and should do so with the best knowledge you can get.

To assess your previous training and interests, you might refer to the matrix in Table 11-1. Be thorough and immodest at this stage; you are generating information for yourself, not for an offendable outsider. And don't forget your special talents and accomplishments; if you speak a language besides English or play the violin, write that down. Most Americans do not have such skills, and what is unexceptional to you may be very exceptional

TABLE 11-1 A MATRIX TO HELP YOU ASSESS YOURSELF AND YOUR INTERESTS

RESULTS ACTIVITIES	PRODUCTS OR ACCOMPLISHMENTS	SKILLS	KNOWLEDGE	REWARDS OR RECOGNITION
Working				
Studying				
Playing				
Living				

and interesting to an employer. Note that not everything you write down in the matrix will appear in your letter and resume. You are writing to ensure that you don't leave out something important; you still have to select the information you will finally put in the letter. Be bold and thorough.

Designing a Good Letter of Application

To write a successful application letter, you need to know the conventions for such things. This is easy. The letter is ideally only one page long and has four main sections:

1   The heading, which includes the writer's address, the reader's address, and the date

2   A first paragraph, which introduces you (if possible by citing an impressive recommender) and then establishes the company's need and your ability to fill it (or your reason for writing and what you are requesting of the reader)

3   A second paragraph, or series of paragraphs, which establish your *most relevant* experience and qualifications and which stress your accomplishments, responsibilities, and work quality, not just your activities. This section should include some proof that you did a good job, by citing either a reference ("X will provide a letter of reference") or some objective measure of a job's quality. (If you have designed an oven as a senior engineering project, distinguish it from an oven an English major would design: "My oven maintained a constant temperature within a 1 percent error.") This section should be very selective and focus on only your *most relevant* qualifications.

4   A closing paragraph, which gives any other pertinent data, asks for an interview, provides your telephone number and the hours you may be reached, and thanks the readers for their attention.

Two sample letters are shown in Figures 11-14 and 11-15. The letter in Figure 11-14 was written by an engineering senior and has been nicely

FIGURE 11-14 Letter of application for employment and resume.

6601 Bursley-Lewis
1931 Duffield Street
Ann Arbor, MI 48109

May 16, 1980

Mr. Lewis R. Tassellski
Technical Group Coordinator
Pacific Power & Light Company
1634 S.W. Columbia
Portland, OR 97201

Dear Mr. Tassellski:

At the suggestion of Mr. S. Feldman, Instrumentation Supervisor, Fermi II Project Pioneer Corporation, I am writing to apply for a position in the Instrumentation and Control Group at the Astoria Point Power Project. My work experience in both instrumentation and start-up groups should be of value to the project, as should my engineering experience at the University of Michigan in automatic control and power plant systems. My credentials include:

Employment as an engineering aide with the Detroit Edison Company at the Enrico Fermi II Nuclear Power Project. While there, I independently researched in-service inspection problems of certain ASME Class II welds, reporting to the Production Department Lead Superintendent. I also worked on the initial rough draft of the In-Service Lubrication Manual for the plant. (Ref. R. J. Szcotnicki, Technical Group Supervisor)

Employment as a student engineer with Daniels International Corporation at the Enrico Fermi II Nuclear Power Project. I researched material for input into the Component Control System, checking data for authenticity and inputting the data to update the system. (Ref. T. R. Bietsch, I. & C. Lead Superintendent)

I expect to receive my bachelor of science in mechanical engineering from the University of Michigan in April, 1981, and my future goals include a desire to advance in the area of technical management. I am very interested in discussing my credentials with you at your convenience. If this is possible, I can be reached at (313) 555-1857 during the afternoon and evening.

Thank you for your consideration.

Sincerely,

Steven J. Kaercher
Steven J. Kaercher

PERSONAL DATA RESUME

Steven J. Kaercher
6601 Bursley-Lewis
Ann Arbor, Michigan 48109
(313) 555-1857

Career Plans and Objectives

I am seeking employment in the consumer power industry, especially in the area of instrumentation and control of power systems. I feel that my experience and education will enable me to contribute to such projects now; later I plan to advance into management.

Qualifications and Experience

Summer 1979	Engineering Aide (student) with Detroit Edison Company at Enrico Fermi II Nuclear Power Station. While working for the production department, I 1) researched potential problems in the In-Service Weld Inspection program and 2) worked on the Plant Lubrication Manual, documenting equipment and specified procedures.
Summer 1978	Engineering Aide (student) with Daniels International Corporation at Enrico Fermi II Nuclear Power Station. While there, I worked on a Component Control System for the Instrumentation and Control Group. This involved monitoring all sources of information and doing in-plant inspections to keep the system updated.
College Work	Employed by University of Michigan, Ann Arbor, during all terms in school. I worked at the School of Public Health as a coder one year and at the Bursley Hall Cafeteria for the last two years. This employment paid for twenty-five percent of my college costs.

Academic Experience

Education	Bachelor's Degree in Mechanical Engineering, College of Engineering, University of Michigan, expected in April, 1981. Emphasis on instrumentation and control of power system equipment.
Societies	American Society of Mechanical Engineers, student member.

References

References and transcript will be supplied upon request.

FIGURE 11-15 Letter of application for admission and resume.

717 Dartmoor
Ann Arbor, MI 48103
(313) 555-9872

February 15, 1978

Dean Arjay Miller
Graduate School of Business
Canston University
Bolton, Connecticut 06431

Dear Dean Miller:

At the suggestion of Mr. Richard E. Spaid, I am writing to request your support for my application to the Canston Graduate School of Business. My qualifications include a bachelor of science in engineering from the University of Michigan and the following abilities:

Ability to Solve Problems: I have concentrated in mechanics, a strongly problem-oriented field, while maintaining a diverse background in engineering and science. The human aspects of engineering have received the majority of my attention; I have independently researched current advances in the fields of stress-related anxiety, materials for implantation in humans, and nerve conduction velocities in the hand. (Ref. - Prof. D. A. Sonstegard)

Ability to Lead and Motivate Others: I had 24-hour direct counseling responsibility for 20 teenage boys at the National Music Camp at Interlochen (summer, 1976). I used my motivational and leadership skills to keep the boys interested in their musical and physical activities as well as to deal with homesickness, personality conflicts, group problems, and personal problems. Additionally, I supervised many of the waterfront, recreational, and evening activities. (Ref. - L. E. Dittmar)

Ability to Achieve Desired Goals: I have been able to simultaneously attain my goals in swimming and school over the past twelve years through self-motivation, my competitive spirit, and my organizational abilities. I stand in the top fifth of my class scholastically, and I have won three varsity letters as a member of the University of Michigan swim team. (Ref. - A. P. Stager)

Ability to Maintain Diverse Interests: I have kept my interests diversified through activities such as being Rush Chairman of my fraternity and joining Tau Beta Pi, National Engineering Honorary Society. I have been able to share my interests with others as a counselor, scholastic tutor, swim instructor, and music student.

It is essential that the manager master these abilities if he is to contribute to the business world. My perspective of these abilities differs from that of the typical applicant with a B.B.A., and I believe it is a perspective that would be valuable to the M.B.A. program at Canston. I would be happy to provide any further information you might wish. Thank you for your consideration.

Sincerely,

Brian Wylie

Brian D. Wylie

Brian D. Wylie
717 Dartmoor
Ann Arbor, MI 48103
(313) 555-9872

Career Goals

My career goal is to become an effective and conscientious corporate general manager. My strong engineering background, a projected M.B.A., and my diverse activities will have formed a solid foundation for this goal.

Education and Work Experience

Current	Bachelor's degree in engineering-science from the University of Michigan (expected completion: May 1978). Emphasis on bio-mechanics and industrial human performance. Independent studies on stress-related anxiety, materials used for human implantation, and nerve conduction velocities in the hand.
Summer 1977	Assistant Pool Manager, Travis Pointe Country Club, Saline, Michigan. Responsibilities included maintaining effective relations with the membership, organizing the physical plant, and teaching group and private swim lessons.
Summer 1976	Cabin Counselor, National Music Camp, Interlochen, Michigan. 24-hour counseling responsibility for 20 teenage boys with homesickness, personality conflicts, group problems, and personal problems. This job enabled me to use my motivational and leadership skills to keep the boys interested in their musical and physical activities. I also had responsibility for the waterfront, the recreation program, and special events.

Activities and Honors

Activities	Michigan swim team, 3 varsity letters (1974-78) Michigan varsity club water polo (1974-77) Psi Upsilon fraternity (1976-78), Rush Chairman (1978)
Honors	Tau Beta Pi National Engineering Honorary (1977-78) Dean's List (3 semesters); Regents' scholar (1974) Academic scholarship (1974-75, 1976-78) Athletic scholarship (1975-78)

References

Richard E. Spaid, President, Hillman Manufacturing Company
David M. Sonstegard, Ph.D., Assoc. Prof. of Applied Mech., U.M.
Larry E. Dittmar, All-State Boy's Director, National Music Camp
Augustus P. Stager, U.M. Varsity Men's Swim Coach

presented at many levels. It is addressed to the person in charge of screening applicants, written specifically to the company, and individually typed on a professional-quality typewriter. This gets the letter to the right reader(s) and shows that the writer cares enough about the job to prepare the letter carefully. It first cites a reference known and respected by the readers and then defines the job being sought and the writer's general qualifications for the job. The two indented paragraphs in the middle of the letter give the writer's main qualifications, focusing on his accomplishments and on proof that he did a good job. This section is relatively straightforward since the writer comes from a good school and has some obviously good qualifications for the job. Finally, it provides useful information if the employer wants an interview and then ends politely.

The letter in Figure 11-15 presents a more difficult problem. It is a letter of application to an excellent graduate program, written by a good student but one with apparently marginal qualifications for that program: swimming, work as a camp counselor, and an engineering degree as preparation for one of the most competitive business programs in the country. Since the letter was relatively successful in a difficult situation, it is useful to see why.

First, the student has a good reference in the first sentence. (If no one with an impressive title or reputation has offered to recommend you, then politely ask someone to do so; the worst you can get is "no.") Second, the student has assessed his qualifications and activities and reinterpreted them from the point of view of the reader. (What are the characteristics an admissions officer would look for in applicants to business school? In future managers? What kinds of things have I done to demonstrate these characteristics?) Obviously, a writer is limited by the facts of a given case, but she or he can organize those facts in more or less effective ways. This student has chosen a more effective way and highlighted it. He has focused on his qualifications by using indented paragraphs which draw the reader's eye. Further, he has used underlining and parallel construction (the repetition of *Ability to*) to highlight the qualities he wants the reader to note most carefully: Ability to Solve Problems, Ability to Lead and Motivate Others, Ability to Achieve Desired Goals, and Ability to Maintain Diverse Interests.

When you actually write such a letter, try to avoid having a lot of empty space at the bottom of the page. You don't want to suggest that your life has been a blank. Also, try to produce a balanced and visually attractive format. If you have only a little information, spread it out and use wide margins. Finally, make your letter and resume as concise as possible, a model of efficiency.

Designing a Good Resume

The resume is a summary of *all* your activities and experience. It will repeat some (or all) of the information in the letter of application, and it may be more inclusive than the letter. However, since the letter and resume may get separated, the resume should be as strong and impressive as the letter and arranged to highlight your most important qualifications.

What should the resume look like? First, it ought to be easy to read: not too long, not too much material crammed on a page, easily visible headings, and, if you prefer, short phrases rather than full sentences. Second, it ought to give your vital statistics: your name, address, and telephone number and, if necessary, a permanent address and telephone number if you intend to move soon. (Resumes used to include such information as height, weight, sex, and marital and military status, but this information is often irrelevant and prohibited by law, and so it is not included here. Obviously, you can include it if you particularly want to.) Third, the resume ought to suggest where you are headed professionally (perhaps in a section called "Career Goals") and where you have been (in a section called "Qualifications and Experience," including education and work experience, or two sections entitled "Work Experience" and "Education").

The ordering of information is important in the Qualifications and Experience section. Generally, you want to put your most relevant and impressive qualifications first. If you have a lot of relevant work experience, you should list that before your educational experience. If you have only a little work experience, you will have to emphasize your education and its special features. What makes you different from any other student with your degree? Have you had any research or design courses which simulate a job situation? Do you have a number of honors and extracurricular activities? You might want to highlight them in a separate section entitled "Honors and Activities," since such features show that you are organized enough to handle several activities at one time.

Finally, you need a section entitled "References," which either states that references are available on request or lists your references' names (and addresses) if these are particularly impressive or if you need to use up some extra space. (Before you list someone as a reference, ask that person if she or he is willing to write letters of recommendation for you. It is impolite and potentially disastrous to list someone who is annoyed at being taken for granted or unwilling to write a good letter.)

One note about the placement of information on the page. As with your letter of application, try to avoid a lot of empty space at the bottom of the page. Again, use wide margins if necessary, spread the information out evenly on the page, and give the full names and addresses of your references to use up more space.

Sample resumes appear in Figures 11-14 and 11-15, and other resumes for discussion are presented in the following pages as Figures 11-16 to 11-19 for Exercise 11-2 (page 274).

FIGURE 11-16 Letter of application for admission and resume: example 1 for analysis.

1320 South University
New Brunswick, New Jersey 07032
January 23, 1981

Prof. Walter H. Arnaz
Assistant Professor of Electrical
and Biological Engineering
University of Illinois, Urbana-Champaign
Urbana, Illinois 60537

Dear Professor Arnaz:

At the suggestion of Professor Lin of Rutgers University, I am writing to express my desire to become a part of the Bio-Engineering Department at the University of Illinois in the capacity of a teaching or research assistant. Because of my financial situation, it will be necessary for me to obtain an assistantship while working for my master's and possibly my doctorate degree. Considering the current research and educational activities at your University, including biological computation and instrumentation, I am sure that my educational background and activities would be of great value to the Bio-Engineering Department. My background includes:

Six months of employment in the Kubik Hydradrive Company of Troy, New Jersey. In the Design Department of this company my duties included the design of control mechanisms for hydraulic machinery. I was also assigned individual design projects in which I worked in the field without direct supervision.

My studies have given me an extensive background in instrumentation, biology, and chemistry. In May, I will receive a combined degree in Electrical Engineering-Instrumentation and Engineering Science-Biology.

Active participation in extracurricular activities, including serving as Vice President, Athletic Director, and Educational Programming Chairman for a University dormitory. I am currently an instructor for the New Jersey Heart Association in Cardio-Pulmonary Resuscitation. I am also a member of the Engineering Honor Society, Tau Beta Pi.

Ranking in the upper quarter of my class at Rutgers University.

I believe my background and abilities will make me an asset to the Bio-Engineering Department. I would appreciate your contacting me as to the projected needs of your department as they apply to my experience and background. I can be contacted at the above address or most evenings at the enclosed number.

Sincerely,

Patricia J. Lingri

Patricia J. Lingri

PERSONAL DATA RESUME

Patricia Jensen Lingri
1320 South University #4
New Brunswick, New Jersey 07032
(201) 555-9064

Career Plans and Objectives:

I am seeking employment as a University Research Assistant or closely related position. My particular area of interest is in the application of the rapidly increasing technology in instrumentation to the area of biological
measurements and diagnostics. My experience in the areas of teaching and
leadership will enable me to make significant contributions to the reputation
and quality of the Bio-Engineering Department.

Academic Summary:

September 1977 to present	Bachelor's degree in Engineering Science-Bio-Engineering (Rutgers University; expected completion: December 1981). In this engineering program I obtained an extensive background in most areas of engineering, chemistry, and biology.
September 1977 to present	Bachelor's degree in Electrical Engineering-Instrumentation (Rutgers University; expected completion: May 1982).

Qualifications and Experience:

May 1980 to January 1981	Employed by the Kubik Hydradrive Company of Troy, New Jersey, in the Design Department. My duties included the design of control mechanisms for hydraulic machinery. I was also assigned individual design projects in which I worked in the field without direct supervision.
September 1979 to present	Instructor for the New Jersey Heart Association in Cardio-Pulmonary Resuscitation. In this capacity, through the New Jersey Heart Association, I teach groups of individuals who request instruction in this life-saving technique.

Organizations

While attending Rutgers University	Vanner Hall House Council, Vice President 1980 Vanner Hall Educational Programming Committee, Chairperson 1980 Vanner Hall Athletic Director, 1979-1980 Tau Beta Pi, Rutgers University Chapter

References and transcript will be supplied upon request.

FIGURE 11-17 Letter of application for employment and resume: example 2 for analysis.

921 Second Street
Charleston Heights, South Carolina 29405
September 21, 1981

Mr. Donald Abelson
Senior Engineer-in-Charge: Works Engineering Activity
Fisher Body Central Engineering
General Motors Technical Center
Warren, Michigan 48019

Dear Mr. Abelson,

Clara Kinnie, who is employed in the Mechanical Engineering section of your department, has informed me that you require an engineering student to fill the temporary vacancies which will occur when vacations begin next summer. I will be graduating from the University of South Carolina this April with a bachelor of science degree in engineering, and I have prior training which would assist my performance in this position.

Three years ago, I attended General Motors Institute under the sponsorship of Fisher Body. While there I obtained work experience in many different central engineering activities which included:

Works Engineering Activity--Plant Layout and Mechanical Product Testing
Laboratories--Elastomers
Manufacturing and Development--Resistance Welding
Body Engineering Activity--1980 Detroit Pilot
Production Engineering Activity--Fixture Design
Product Research and Safety--Seats and Headliners
Seat and Wagon Load Floor--B-C-E Body Rear Seats

While assigned to the Works Engineering Activity I worked on the following projects:

Plant Layout--Worked with Hugh Reins at Fisher Body Fleetwood during the rework for the 1980 model change-over. Also interacted with the Architecture group to produce a collection of plant elevation cross sections to be utilized in layout work.

Mechanical--Aided Clara Kinnie in the preparation of a study concerning effluent guidelines on chrome plating operations, requested by the EPA. Drew the new drainage system for the plant in Syracuse, N.Y.

My schedule is very flexible, and I would be glad to come to Warren for an interview, at your convenience. I will be at the number listed in my resume on Monday and Friday afternoons and also most nights if you require any additional information.

Sincerely,

Alexander Herrod

Alexander Herrod

<u>PERSONAL DATA RESUME</u>

Alexander Herrod
921 Second Street
Charleston Heights, South Carolina 29405
(805) 555-5879

Career Plans and Objectives:

    I am seeking summer employment in a company that requires an engineer
    with a diverse background. My experience and training will enable me to
    make contributions in many different work situations.

Qualifications and Experience:

August 1978   Cooperative student, General Motors Institute, employed by
to 1980   Fisher Body Central Engineering, Warren, Michigan. While
                 there I was rotated for six-week periods throughout the many
                 different departments and acquired experience in Production
                 Engineering, Designing, Layout, Product Testing, and Product
                 Development. I saved the corporation $15,000 by locating a
                 design error in a plant drainage system construction project.
                 I also designed a screwless installation sunshade support
                 which would save an estimated $1 per car on installation
                 costs. This support was under patent investigation at the time
                 I left the company.

Academic Summary:

Education   College of Engineering, University of South Carolina (en-
                 rolled September 1980). BS Engineering degree expected April
                 1982, in interdisciplinary program, pre-biomedical-engineer-
                 ing option. This option is geared toward the obtaining of a
                 master's degree and involves engineering applications in the
                 medical field.

                 Engineering program, General Motors Institute, electrical
                 engineering option (enrolled August 1978-August 1980). Left
                 program to enter University of South Carolina.

                 Am applying for admission to Samuels School of Graduate
                 Studies, Yale University, where I plan to obtain a master's
                 degree in biomedical engineering.

Societies   Considered for Management Honor Society (GMI). Withdrew
Honors   name from consideration because of transfer to USC.
                 Dean's List (GMI).

References:

    Will be supplied upon request.

309 Packard Road
Embry, Ohio 44107
September 22, 1981

Dr. Rodger L. Krakau
Biomedical Engineering Research
Johns Hopkins University
Applied Physics Laboratory
8621 Georgia Avenue
Silver Spring, Maryland 20910

Dear Dr. Krakau:

Pursuant to the advice of Dr. Roberta Shofer, Department of Neurology, Einstein Medical College, New York, I am writing to you concerning potential employment. I will be graduating in May, 1982 from the University of Ohio, College of Engineering, with a degree in electrical engineering and with an emphasis in biological sciences and bio-engineering. I believe that I could be a valuable addition to your research staff. My qualifications and background include the following.

    For five summers I have been employed as a refrigeration and air conditioning mechanic and technician. I hold an unlimited Refrigeration Journeyman's License. While not directly related to my field of study and interest, I believe this employment (and the invitations to return each year) are indicative of my integrity and initiative in a commercial situation and my ability to solve technical problems through a logical diagnostic approach. I have letters of recommendation from my two employers, which I would be happy to forward should you so wish.

    During the course of my education at the University of Ohio, I have studied in both the College of Engineering (Department of Electrical and Computer Engineering) and the College of Literature, Science, and the Arts (Department of Zoology). I believe that my background in zoology, supplemented with course work in bio-engineering, gives me a solid and adequate basis for work in biomedical engineering.

    Outside of the classroom, I have been active on many fronts. I am currently serving as treasurer of the U of O chapter of IEEE. I served on the executive board of the local Democratic Party and did research on issues and policy for an unsuccessful bid for a Congressional seat (Ohio 2nd district) in 1974. I have also performed in Washington, D.C., for the President with an Ohio-sponsored vocal jazz ensemble.

I would like very much to discuss your research projects and my potential employment and am confident that a convenient date and time can be arranged. At present, I am planning to be in the Maryland area for the Thanksgiving and Christmas recesses, or I could arrange to come earlier should it prove necessary. As I am difficult to reach, I will contact your office shortly to arrange a suitable and convenient meeting.

I thank you for your time and consideration, and am looking forward to meeting you.

Sincerely yours,

Lisa C. Caron

Lisa C. Caron

PERSONAL DATA RESUME

Lisa C. Caron
309 Packard Road
Embry, Ohio 44107
(614) 555-8932

Career Plans and Objectives:

I am seeking employment in biomedical engineering research and development. My education and background in electrical engineering, biology, and
biophysics will enable me to contribute significantly to the growth of knowledge in this vast and growing field and thereby eventually help to improve
the quality and technology of medical care.

Work Experience:

Summer 1977, 1978, 1979, 1980
Bishop Equipment Company, Fairfax, Virginia

Summer 1981
The Alma Corporation, College Park, Maryland

I worked for these five summers, to finance my education, as an air-
conditioning mechanic, beginning as an apprentice and working my
way up to master mechanic. I worked in all phases of residential and
commercial air conditioning.

Education:

University of Ohio
College of Literature, Science, and the Arts
Department of Zoology 1977-1979

University of Ohio
College of Engineering
Department of Electrical and Computer Engineering 1979-1982
B.S.-E.E. May, 1982

Language fluency: French

Societies and Honors:

IEEE 1980-1982; presently serving as treasurer, U of O chapter
Invitation to join Eta Kappa Nu

Professional License:

Unlimited Refrigeration Journeyman's License

References will be supplied upon request.

FIGURE 11-19 Letter of application for employment and resume:
example 4 for analysis.

806 Packard
Ann Arbor, Michigan 48104
September 24, 1976

Mr. Edwin L. Russell
Manager, Planning Department
FMC Corporation
Prudential Plaza
Chicago, Illinois 60701

Dear Mr. Russell:

My job placement investigation of careers combining engineering and finance has led me to your Business Planning Department of FMC Corporation. I
believe that I would be of value to your department and firm, as my education,
work experience, and personal objectives closely match your preferred qualifications as outlined in your recent Business Planner job bulletin. My qualifications
include:

Over three years of work experience in the research and development divisions of General Motors Corporation. Functioning as a development engineer, I redesigned the GM rotary engine coolant flow distribution, directed
development tests on rotary engines, and worked with instrumentation and
computers on engines and suspensions.

My overall experience in many engineering departments has made me familiar with the functions of an R&D division within a major corporation.

Controllership of a college student body and a social fraternity, giving me
three years of practical experience in designing budgets, formulating financial policies, and resolving financial problems.

Graduate studies in the area of financial management, including managerial
accounting, management planning, a seminar in corporate financial policy,
and hopefully a seminar in management problems in large corporations.

Undergraduate studies in the area of automotive mechanical engineering.

I will graduate with an MBA from the University of Michigan in the beginning of May, 1977, and be available for employment by the end of the month. I
would like to meet with you and your department on the first Monday or Friday
convenient for you. If you have any questions, please call me any evening at the
number listed in my resume.

Sincerely,

Lincoln H. McGhee

LINCOLN H. McGHEE
806 Packard
Ann Arbor, Michigan 48104
(313) 555-6279

OBJECTIVE   A position in corporate financial management requiring the use of my engineering experience and analytical abilities in the formulation and solution of problems in the areas of financial, product, and corporate planning. Long-term goal is corporate management.

EDUCATION   Graduate School of Business Administration   The University of
1976 - 77   Michigan
MBA emphasizing financial management. Primary supportive coursework: managerial accounting, management planning and control systems, seminar in corporate financial policy.

1970 - 75   General Motors Institute (GMI)   Flint, Michigan
B.S. in Mechanical Engineering. Ranked in top five percent of class.

Offices: Comptroller, GMI student body (annual budget $125,000).
Comptroller, Phi Delta Theta Fraternity (annual budget $90,000).
Treasurer, national convention and local chapter of Tau Beta Pi.
President and Founder, Karate Club.

Honors: Sobey Scholar (school's highest honor). GMI Alumni travelship.
Robot Society (leadership). Tau Beta Pi (engineering honor society).
Listed in Who's Who in American Colleges.

EMPLOYMENT
1970 - 75   General Motors Corporation   Warren, Michigan
Co-operative student engineer accumulating 3.5 years' work experience at GM Technical Center, Milford Proving Grounds, and Chevrolet Flint Assembly.
Responsible for:

Redesign of the GM rotary engine coolant flow distribution.
Computer simulation of a high-performance rotary engine.
Test and evaluation of anti-drive on Monte Carlo front suspensions.

Exposed to testing, development, computer simulation and design of automotive engines, transmissions, and suspensions. Worked at European engineering facilities, summer 1974.

1968 - 80   Manual laborer for a construction company and assembler for a
(summers)   mobile home corporation.

FIGURE 11-19 *(continued)*

PERSONAL BACKGROUND	Born in Tokyo, Japan; raised in several locations in United States and Europe.
	Graduated top five percent in high school; member National Honor Society.
	Volunteer fire fighter and ambulance assistant for a hometown fire company (1968-70).
	Self-supporting since 1970.
	Interests: cars, golf, karate, skiing, camping, and tennis.
REFERENCES	Will be supplied upon request.

October 1976

EXERCISE 11-2 Evaluate and revise where necessary the letter and resume combinations of Figures 11-16 through 11-19. Note the order of categories and information in both the letters and the resumes. Do the most important qualifications come first? Are they appropriately highlighted? In each pair, are the letter and resume of equal strength? (If they get separated, you do not want someone important to see only the weaker unit.) Are they concise and smooth enough?

12

THE JOURNAL ARTICLE

A good journal article is like other good technical communications: it addresses a problem or issue important to its audience and presents its argument clearly and coherently. In this sense, it is not exempt from the principles of communication presented in this text. However, it differs somewhat from the other types of technical communication because it is written for a very different audience. The other types are written for a wide spectrum of readers with varied backgrounds, attention, interest, and time—audiences including nontechnical people, technical nonspecialists, and specialists. Thus, the other types have to provide features appropriate to this diversity of audience. In contrast, the journal article is written for a narrow audience: specialists in a field, who share assumptions, knowledge, and backgrounds and who have the need and interest to read carefully. This means that the journal article can eliminate some of the features provided primarily for nonexperts, in particular, the Foreword and Summary. However, the article should maintain the features providing clarity and ease of reading for experts, in particular, the movement from generalizations and claims to data and support. Unfortunately, this movement is sometimes reduced by writers relying perhaps too strongly on the expert reader's interest and determination to understand.

Articles in different fields and journals vary somewhat in style. If you are writing an article for a particular journal, you should read that journal's instructions on organization and format and then look at sample articles in several issues to become familiar with the journal's style. (Here, style includes such things as format for footnoting, ways of setting up headings and visual aids, length of paragraphs, and other special features.)

Although they may differ in style, most articles share a remarkably uniform purpose and structure. The purpose of an article in any field is to advance an argument of fact or policy: (1) an argument of fact that the results reported are valid, that previously reported results are supported (or not), that a given theory is supported (or not), that other observations are necessary to resolve some debate in the field; or (2) an argument of policy that previous results should be questioned or reinterpreted, that a given theory should be abandoned, recast, or extended. These arguments are made in a structure that is quite consistent over many fields and includes the following sections:

1   *Introduction,* which defines the problem and describes its importance

2   *Materials and Method,* which describes how the research arrived at the results

3   *Results,* which describes what was discovered

4   *Discussion,* which analyzes the importance of the results and their implication(s)

Let us examine these sections in more detail.

12.1

INTRODUCTION

The Introduction is an especially tricky part of the article since it must present a great deal of information and orientation in a short space. Robert Day, an experienced journal editor and author of a helpful and entertaining book on writing articles, has provided four rules for a good Introduction:

> (i) It should present first, with all possible clarity, the nature and scope of the problem investigated. (ii) To orient the reader, the pertinent literature should be reviewed. (iii) The method of the investigation should be stated. If deemed necessary, the reasons for the choice of a particular method should be stated. (iv) The principal results of the investigation should be stated. Do not keep the reader in suspense; let him follow the development of the evidence. An O. Henry surprise ending might make good literature, but it hardly fits the mold that we like to call the scientific method.[1]

Articles aimed at specialists may begin with an Introduction based on either a long-form or a short-form problem statement. (Since specialists share assumptions, methods, and knowledge of their field, a writer can often assume this shared experience and not state it.) In contrast to other types of Introductions, article Introductions aimed at specialists include technical details and a short review of previous work on the topic. These establish a context against which the author contrasts an observation or an inadequacy in theory or method. This is illustrated in the following example.

THE THERMAL CONDUCTIVITY AND SPECIFIC HEAT OF EPOXY-RESIN FROM 0.1-8.0 K

A term	The thermal properties of glassy materials at low temperatures are still not completely understood. The thermal conductivity has a plateau which is usually in the range 5 to 10 K, and below this temperature it has a temperature dependence which varies approximately as T^2. The specific heat below 4 K is much larger than that which would be expected from the Debye theory,
Amplification of or proof for A term	

(Unstated *however*) B term	and it often has an additional term which is proportional to T. Some progress has been made towards understanding the thermal behaviour by assuming that there is a cutoff in the phonon spectrum at high frequencies (Zaitlin and Anderson 1975 a,b) and that there is an additional system of low-lying two-level states (Anderson 1975, Phillips 1972).
Missing information	Nevertheless, more experimental data are required, and in particular it would seem desirable to make experiments on glassy samples whose properties can be varied slightly from one to the other. The present
Purpose	investigation reports attempts to do this by using various samples of the same epoxy-resin which have been subjected to different curing cycles. Measurements
Assignment	of the specific heat (or the diffusity) and the thermal conductivity have been taken in the temperature range 0.1 to 8.0 K for a set of specimens which covered up to nine different curing cycles.[2]

As suggested by this example, article Introductions and especially their problem statements can be quite complex and varied. They have been studied in some detail by John Swales, who has proposed the amplified structure for them presented in Table 12-1; this is based on 16 articles each from physics, biology/medicine, and the social sciences. In Swales's proposed structure, the problem statement extends into or through Move Three, the statement

TABLE 12-1   A POSSIBLE STRUCTURE FOR ARTICLE INTRODUCTIONS

THE FOUR MOVES				NUMBER OF OCCURRENCES IN 48 ARTICLE INTRODUCTIONS
Move One	Establishing the field			43
	A	Showing centrality		25
		i	By interest	6
		ii	By importance	6
		iii	By topic prominence	7
		iv	By standard procedure	6
	B	Stating current knowledge		11
	C	Ascribing key characteristics		7
Move Two	Summarizing previous research			48
Move Three	Preparing for present research			40
	A	Indicating a gap	20	
	B	Raising questions	14	
	C	Extending a finding	6	
Move Four	Introducing present research			46
	A	Giving the purpose	23	
	B	Describing present research	23	

SOURCE: John Swales, *Aspects of Article Introductions,* Aston ESP Research Report No. 1, University of Aston, Birmingham, England, 1981, p. 22a.

of missing information (the question or task) appears in Moves Three and Four, and—if it appears at all—the purpose is found in Move Four. (Note that a statement of purpose appears in only half of Swales's examples.) The sample article introduction analyzed above is repeated below, this time with Swales's divisions.

THE THERMAL CONDUCTIVITY AND SPECIFIC HEAT OF EPOXY-RESIN FROM 0.1-8.0 K

1 Establishing the field	The thermal properties of glassy materials at low temperatures are still not completely understood. The thermal conductivity has a plateau which is usually in the range 5 to 10 K, and below this temperature it has a temperature dependence which varies approximately as T^2. The specific heat below 4 K is much larger than that which would be expected from the Debye theory, and it often has an additional term which is proportional to T. Some progress has been made towards understanding the thermal behaviour by
2 Summarizing previous research	assuming that there is a cutoff in the phonon spectrum at high frequencies (Zaitlin and Anderson 1975 a,b) and that there is an additional system of low-lying two-level states (Anderson 1975, Phillips 1972).
3 Preparing the present research	Nevertheless, more experimental data are required, and in particular it would seem desirable to make experiments on glassy samples whose properties can be varied slightly from one to the other. The present
4 Introducing present research	investigation reports attempts to do this by using various samples of the same epoxy-resin which have been subjected to different curing cycles. Measurements of the specific heat (or the diffusity) and the thermal conductivity have been taken in the temperature range 0.1 to 8.0 K for a set of specimens which covered up to nine different curing cycles.[2]

12.2

MATERIALS AND METHOD

The Materials and Method section of any report you write is a critical part of the report's argument since it establishes the validity of your results and allows them to be taken seriously. It demonstrates that you have done everything "the right way": that you have been scrupulously careful and thorough, that you have used an accepted method, that you have made no technical mistakes.

This section also provides the mechanism by which the scientific community can repeat and verify your work. It must contain sufficient detail to allow any relatively experienced researcher in your field to reproduce your results exactly. This means that you must

1   Identify exactly what materials you used to conduct your research—reactants, enzymes, catalysts, organisms, experimental subjects (human or animal), etc. You should also identify your materials specifically enough that another researcher could use exactly what you specify to reproduce your results.

2   Identify any special conditions under which you conducted your research—special temperatures, irradiation with ultraviolet light, testing with unusually high current or voltage loads.

3   Identify any special criteria you used to select materials, subjects, test apparatus, or test method. (If you chose one gluing material or catalyst over another, why?)

4   Identify the specific method you used to conduct the research. If you followed a standard procedure, you may simply reference it. If you followed an unorthodox or new procedure, you need to describe it fully.

5   Justify, where necessary, any of your choices of criteria, materials, method, or conditions.

12.3

RESULTS

The Results section of an article presents (1) the major generalization(s) you are making about your data and (2), in compact form, the data supporting the generalization(s). The generalizations must be clearly and obviously stated: "This glue has successfully bonded stainless steel artificial joints to human bone" or "UV-irradiated DNA contains 5,6-dihydroxydihydrothymine as well as pyrimidine dimers in ratios dependent upon wavelength."[3] The supporting data must be presented fully enough that the reader can evaluate the strength of your generalizations but succinctly enough that the data do not overwhelm the generalizations.

For instance, consider the relationship between generalizations and data in the following example from an article entitled "The Effect of Graded Doses of Cadmium on Lead, Zinc, and Copper Content of Target and Indicator Organs in Rats." The subheadings in the Results section are "Zn," "Cu," "Pb," and "Correlations." Only the section on zinc (Zn) and one of the tables containing the data are presented here.

Zn

The Zn contents of the target and the indicator organs are presented in Tables I and II. The Tables also state significant differences ($p \leq 0.001$) between the control group and the experimental groups for each tissue. As can be observed several of the soft tissues had increased Zn levels when the Cd dose had been

TABLE I ZN CONTENT IN μG/G WET WEIGHT IN TARGET ORGANS OF RATS HAVING RECEIVED GRADED DOSES OF $CdCl_2$ IN THE DRINKING WATER FROM CONCEPTION UNTIL SACRIFICE AT THE AGE OF TEN WEEKS

CD DOSE IN PPM	KIDNEYS		LIVER		SPLEEN		HEART		MUSCLE		ADRENAL GLANDS		EPIPHYSES		DIAPHYSES	
	x̄	SD	x̄	SD	x̄	SD	x̄	SD	x̄	SD	x̄	SD	x̄	SD	x̄	SD
0	23.38	±1.53	27.71	±1.21	17.42	±0.98	14.54	±1.05	9.86	±2.51	13.79	±2.61	119.86	±10.90	121.73	±8.17
2.5	22.53	±5.26	22.98	±1.77	17.09	±0.54	13.68	±0.73	8.35	±1.96	11.25	±2.18	114.64	±11.87	111.45	±7.73
5	24.69	±1.30	30.48	±3.01	16.81	±0.95	14.40	±1.25	8.46	±1.12	14.61	±5.05	123.86	±12.89	123.67	±6.02
7.5	28.12	±1.14[a]	33.44	±2.42[a]	21.71	±0.83[a]	18.42	±1.21[a]	11.17	±1.27	20.31	±1.98[a]	125.13	±6.63	129.77	±6.45
10	25.04	±1.52	31.84	±3.33[a]	18.21	±1.20	15.31	±0.57	7.65	±1.04	13.38	±0.96	122.39	±4.65	123.93	±6.04
12.5	24.35	±1.15	32.35	±3.04[a]	19.91	±1.96[a]	17.90	±0.95[a]	10.02	±1.79	27.60	±13.18	106.66	±8.17[a]	112.52	±9.59
15	23.11	±1.62	26.43	±3.74	18.64	±1.24	16.66	±0.82[a]	9.79	±2.12	17.30	±2.38	102.50	±8.56[a]	114.80	±7.80
17.5	27.45	±0.99[a]	42.67	±4.85[a]	21.40	±3.47[a]	17.07	±0.97[a]	10.96	±1.32	17.63	±2.14[a]	104.63	±10.89[a]	107.85	±5.65[a]
20	27.20	±1.53[a]	35.13	±2.64[a]	20.30	±1.09[a]	16.96	±0.74[a]	10.44	±1.91	16.44	±3.06	106.41	±5.17[a]	110.54	±7.71
22.5	25.72	±2.07	32.67	±1.84[a]	19.05	±0.97[a]	15.83	±0.89	9.80	±1.29	13.43	±3.40	92.99	±9.11[a]	101.67	±6.65[a]
25	26.77	±1.65[a]	31.87	±3.37[a]	16.43	±1.11	14.63	±1.13	16.10	±3.81[a]	20.73	±3.62[a]	106.98	±5.09[a]	99.95	±6.01[a]

[a]The concentration is significantly higher/lower ($p \leq 0.001$) than in the control group of the same tissue.

above 5 ppm, whereas for Cd doses above 12.5 ppm significantly decreased Zn contents in epi's and dia's were registered (see Table I). In addition, Cd supply was associated with a significant decrease in Zn content in incisors and molars of the majority of the experimental groups (see Table II) [not included].

In none of the pooled left molar samples was the Zn content of enamel different from that of dentin. This was valid also for incisor samples through all groups.[4]

Notice that the table contains the condensed results of many experiments, presenting average values for the ten experiments conducted with each dose of cadmium (Cd).

A second example of the relationship between generalizations and supporting data in the Results section comes from an article on tensile and fatigue tests on human cortical bone specimens. The example presents only a part of the Results section, the results of the tensile tests; it does not include the two tables of supporting data referenced in the text.

TENSILE TESTS

The mean values of the physical characteristics of the test specimens are shown in Table 2. The mechanical parameters of ultimate stress, yield stress, yield strain and modulus of elasticity showed no statistically significant correlation with porosity, ash fraction, wet density, or dry density.

Statistically significant ($P < 0.05$) positive linear regressions were found between ultimate stress and elastic modulus and between yield stress and elastic modulus (Table 3). A weak negative linear regression (not statistically significant) was found between yield strain and elastic modulus. The influence of bone elastic modulus of the ultimate stress, yield stress, and yield strain are shown in Fig. 1 [see page 282]. The mean values ($\pm$ s.d.) were ultimate stress 140 ($\pm$12) MPa, yield stress 129 ($\pm$11) MPa, yield strain 0.0068 ($\pm$0.0004), and elastic modulus 17.5 ($\pm$1.9) GPa.[5]

12.4

DISCUSSION

The Discussion explains the implications of your results. It fits the results into the context of the field by relating your results to other work, both theoretical and experimental. Along with the Introduction, it explains why your work is important, how it contributes to the advancement of the field. It is critical that this be done carefully and thoroughly, for as Robert Day has noted,

Many papers are rejected by journal editors because of a faulty Discussion, even though the data of the paper might be both valid and interesting. Even more likely, the true meaning of the data may be completely obscured by the interpretation presented in the Discussion, again resulting in rejection.[6]

If you want to show how your work contributes to the advancement of your field, you might consider what circumstances create advancement.

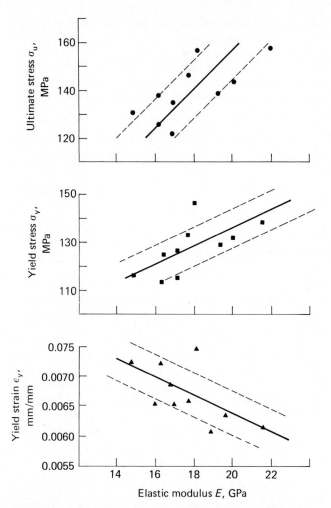

FIGURE 1   Influence of elastic modulus on the ultimate stress,
yield stress, and yield strain of the tensile specimens.

One philosopher, John Platt, believes that some fields advance much more
rapidly than others because of a rigorous intellectual methodology:

> These rapidly moving fields are fields where a particular method of doing
> scientific research is systematically used and taught, an accumulative method
> of inductive inference that is so effective that I think it could be given the name
> of "strong inference.". . . The steps are familiar to every college student and
> are practiced, off and on, by every scientist. The difference comes in their
> systematic application. *Strong inference consists of applying the following steps*
> *to every problem in science, formally and explicitly and regularly:*

1   *Devising alternative hypotheses;*

2   *Devising a crucial experiment* (or several of them), *with alternative possible outcomes,* each of which will, as nearly as possible, *exclude one or more of the hypotheses;*

3   *Carrying out the experiment so as to get a clean result; and*

4   *Recycling* the procedure, making subhypotheses or sequential hypotheses to refute the possibilities that remain; and so on.[7] [italics original]

The keys to this system, according to Platt, are logical rigor, the alternative hypotheses, the crucial experiments, and the disproving of hypotheses. These allow scientists in a field to identify those problems crucial to the advancement of the field, then to construct a series of logical options or lines of inquiry for each problem, and then to eliminate quickly unproductive hypotheses and lines of inquiry. This system reduces wasted time because it focuses the theoretical and experimental efforts of a field on the hypotheses and lines of inquiry likely to produce successful results.

Now, given this perspective, what do you say in an adequate Discussion? How do you establish the importance of your work for the advancement of your field? If you can, you argue that your work is a crucial experiment, disproving some hypothesis or supporting another. Notice that this view encourages the reporting and discussion of negative results. "Hypothesis A predicted that I would find B, but I did not. This result questions the validity of hypothesis A or indicates its inadequacy." Such a statement might save researchers and graduate students from wasting years of needless effort. It may also provide the best kind of statement scientists may make about nature:

> As the philosopher Karl Popper says today, there is no such thing as proof in science—because some later alternative explanation may be as good or better—so that science advances only by disproofs. There is no point in making hypotheses that are not falsifiable, because such hypotheses do not say anything; "it must be possible for an empirical scientific system to be refuted by experience."[8]

In arguing that your work is important, ideally because it is a crucial experiment, you typically consider some or all of the following questions.

1   Were your results expected? If not, why not?

2   What generalizations or claims are you making about your results? How do you interpret these generalizations?

3   Do your results contradict or support other experimental results?

4   Do they suggest other observations or experiments which could be done to confirm, refute, or extend your results?

5   Do your results support or contradict existing theory?

6   Do your results suggest that modifications or extensions need to be made to existing theory? What are they?

7   Could your results lead to any practical applications? What are they?

EXERCISE 12-1

A   If you expect to write technical articles, analyze the organization and style of articles in one journal to which you might submit an article. Compare the journal's features with those discussed above and with those presented in Chapter 9 for technical reports.

B   For a topic you are interested in, write an Introduction to an article on a research project you have conducted or would like to conduct.

C   Select a laboratory report from one of your technical courses and write a Materials and Method section for it. Then write a Results section. If you can, write an Introduction and then a Discussion. Try especially hard to produce a Discussion.

REFERENCES

1   Robert A. Day, *How To Write and Publish a Scientific Paper,* ISI Press, Philadelphia, 1979, p. 24.

2   John Swales, *Aspects of Article Introductions,* Aston ESP Research Report No. 1, University of Aston, Birmingham, England, 1981. p. 16.

3   Philip C. Hanawalt, Priscilla K. Cooper, Ann K. Ganesan, and Charles Allen Smith, "DNA Repair in Bacteria and Mammalian Cells," *Annual Review of Biochemistry* **48**:787 (1979).

4   G.B.R. Wesenberg, G. Fosse, and P. Rasmussen, "The Effect of Graded Doses of Cadmium on Lead, Zinc, and Copper Content of Target and Indicator Organs in Rats," *International Journal of Environmental Studies* **17**:193–194.

5   Dennis R. Carter and William E. Caler, "Uniaxial Fatigue of Human Cortical Bone: The Influence of Tissue Physical Characteristics," *Journal of Biomechanics* **14**:463–464 (1981).

6   Day, op. cit., p. 33.

7   John R. Platt, *The Step to Man,* Wiley, New York, 1967, p. 20.

8   Platt, op. cit., p. 27.

ADDITIONAL READING

American National Standards Institute, *American National Standard for the Preparation of Scientific Papers for Written or Oral Presentation,* ANSI Z39.16-1972, New York, 1972.

V. Booth, *Writing a Scientific Paper,* 4th ed., Biochemical Society, London, 1978.

CBE Style Manual Committee, *Council of Biology Editors Style Manual: A Guide for Authors, Editors, and Publishers in the Biological Sciences,* 4th ed., Council of Biology Editors, Washington, D.C., 1978.

Robert A. Day, *How To Write and Publish a Scientific Paper,* ISI Press, Philadelphia, 1979.

L. DeBakey, *Guidelines for Editors, Reviewers, and Authors,* C. V. Mosby, St. Louis, 1976.

J. H. Mitchell, *Writing for Professional and Technical Journals,* Wiley, New York, 1968.

John R. Platt, "Strong Inference," in *The Step to Man,* Wiley, New York, 1967, pp. 16–36.

John Swales, *Aspects of Article Introductions,* Aston ESP Research Report No. 1, University of Aston, Birmingham, England, 1981.

PART VI

Readability

13

MAKING YOUR WRITING READABLE: INFORMATION SELECTION

As emphasized in earlier chapters, most readers of scientific or technical writing do not have as much time for reading as they would like to have and therefore must read selectively. This is especially true of managers, supervisors, executives, senior scientists, and other busy decision makers, who often skim-read for main points and ideas. However, it is also true of professionals who often need to read more closely and slowly, for thorough understanding, and it is true of technicians, workers, and consumers who may need to read and follow operating instructions. These different types of readers are selective in different ways: the skim-reading decision maker may be looking for "bottom-line" cost figures and performance data; the professional may be looking for the main thread of an argument; the technician, worker, or consumer may need to use operating instructions only as a checklist.

For such readers, writing is readable to the extent that it provides the information they need, located where they can quickly find it, in a form in which they can easily use it. This takes considerable effort on the writer's part; as the old adage has it, "It's easy to make things difficult, but difficult to make things easy." Nonetheless, it's *worth* the effort. If you can make your writing readable, you'll greatly increase its chances of being read and used; in short, you'll greatly increase its *effectiveness*. Conversely, if you don't make the effort, your readers may not either.

How can you make your writing more readable? Unfortunately, there is no simple formula to follow. (The so-called readability formulas are not designed to guide the writing process, and so prescriptions based on such formulas, such as "Use short sentences and short words," are not very reliable.) There *are* steps you can take, however, that should be of some help, and these are laid out in the next three chapters. First, in this chapter, we make suggestions for selecting appropriate information and for making this information accessible to the reader. Then, in Chapters 14 and 15, we suggest a number of things you can do to make it easier for the reader to absorb details; the focus will be on sentences, phrases, and individual words.

BASIC PRINCIPLE
Put information that is new to the reader into a framework of information already known to the reader.

Communication involves the transmitting of information to someone who does not already possess that information, in other words, the transmitting of new information. When new information is absent from a message, the message does not communicate anything: it is noninformative. To most of us, the statement "The Pope is a devout Catholic" is noninformative: it tells us nothing we don't already know. In general, new information is a necessary ingredient of meaningful communication.

New information is not the only ingredient in successful communication, however. If statements consisted of nothing but new information, they would be incomprehensible. Try this, for example: "On big-wall Grade VI's, the second usually jumars all off-widths and overhangs." Got that—or did it go sailing right over your head? It's filled with technical rockclimbing jargon, and so unless you're familiar with rockclimbing, you probably found it totally incomprehensible. In general, for new information to be comprehensible to someone, it must be couched in a meaningful framework of information already known to that person, i.e., a framework of given information. Given information is the background knowledge that we call upon in trying to make sense of new information.

Most of what we know about the world, however, is stored in distant recesses of our mind, in what psychologists call long-term memory; it is not consciously on our mind. Thus, before this background knowledge can be put to use in interpreting new information, it must first (to use another psychological term) be activated, that is, brought to conscious awareness. Fortunately, our background knowledge exists not in isolated bits and pieces but rather in clusters and networks of associated concepts. Thus, mention of even a single word may trigger in the reader's mind a whole host of related images, facts, beliefs, etc. Some of these related concepts may be useful in helping the reader interpret a particular piece of new information; others may not be.

While there is almost no limit to the amount of knowledge we can store in long-term memory, there are very severe limits to the amount of knowledge we can maintain in active consciousness. We can only think about one thing at a time, as the saying goes. So, it's important for you as a writer to activate the *right kind* of given information in the reader's mind—the kind of given information that will help most in understanding the new information. It should pertain directly to the topic of discussion so that it fits into the context already established and allows the reader to anticipate what's coming next. The word or words you use to refer to this given information should be familiar, concise, and loaded with imagery so as to maximize the amount of background knowledge you can bring to the reader's conscious awareness. Consider the following negative example and its rewritten version:

NEGATIVE EXAMPLE

Modern cryptanalysis has attracted some of the most capable mathematical minds. In recent years the growing prospect that postal and diplomatic communication will soon be replaced by other forms of communication has furnished increased incentive for mathematicians and engineers to invent an

unbreakable cipher. Messages will be able to be typed, delivered, and quickly
typed on the other end. But unbreakable codes will be needed to protect
senders against snoopers.

No doubt you can understand the basic meaning of this passage. But is it
vividly memorable? Do you have a clear impression of what kind of unbreakable cipher or what kinds of snoopers the writer is talking about? If not, you
may find this rewritten version more to your liking:

REVISED VERSION

Modern cryptanalysis has attracted some of the most capable mathematical
minds. In recent years the growing prospect that postal and diplomatic communication will soon be replaced by electronic communication has furnished
increased incentive for mathematicians and engineers to invent an unbreakable
cipher. Electronic devices will permit messages to be telephoned, delivered,
and quickly typed on the other end. But unbreakable codes will be needed to
protect senders against electronic snoopers.[1]

The only difference between the two versions is the use of the word
electronic in the second one. This one word does not really add any new
information to the passage; someone reading the first version, for example,
could probably figure out that the phrase *other forms of communication* most
likely includes electronic communication. But this word does bring to mind
the kind of given information that's needed to fully understand what the
writer is talking about. The word *electronic* is familiar and concise. Even
more important, it activates in our minds a number of vivid images which
we can associate with words like *communication, devices,* and *snoopers,*
thus giving much greater meaning to these concepts. Notice, too, how the
writer has used the word repeatedly so as to keep it in the reader's active
consciousness and thus help maintain continuous focus on the topic of the
passage:

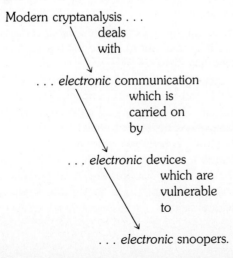

Modern cryptanalysis . . .
                    deals
                    with

. . . *electronic* communication
                    which is
                    carried on
                    by

. . . *electronic* devices
                    which are
                    vulnerable
                    to

. . . *electronic* snoopers.

As can be seen from this example, the use of key words can greatly enhance the coherence and memorability—and hence the readability—of a passage. Key words are most effective when they (1) trigger vivid imagery in the reader's mind, (2) are related in an obvious way to the topic of the passage, and (3) are related in an obvious way to the reader's purpose in reading the passage. If you select key words that satisfy these criteria, and if you put them in prominent positions in the text, you will be doing much to activate the right kind of given information in the reader's mind.

What are the prominent positions in a text? Basically, these are the places that are visually prominent by virtue of having more white space around them and being more easily located: titles, headings, subheadings, captions, labels, paragraph topic statements, and sentence subjects. Some of these text features have already been discussed (see Chapters 2 and 7 especially); the remainder of this chapter and Chapter 14 will go into further detail.

13.1

ESTABLISH YOUR TOPIC AND PURPOSE

Make it clear what the main topic of the report or the section is. Then state your purpose explicitly, so that your readers can anticipate how you will be dealing with the topic. Readers of scientific and technical writing are typically purpose-directed and pressed for time. So, rather than reading word for word and cover to cover, they often prefer to merely "consult" a document, looking only for the information they need. When you define your topic and state your purpose, you make it easier for the reader to determine right away how to process the document: whether to read it closely, skim-read it, pass it on to someone else, or disregard it. A clear statement of topic and purpose allows the reader to form certain expectations about the rest of the text, specifically, how the topic is likely to be developed. It is a well-known fact that we process information most quickly and efficiently when it accords with our preconceptions; this is why it's important to create the right preconceptions in the reader's mind in the first place.

Scientific and technical writing genres customarily have various features designed to announce the topic and set up initial expectations: titles, abstracts, summaries, overviews, etc. Use these to full advantage by loading them with key words and main ideas instead of vague phrases. If you're writing a report dealing with some problematic issue—as is the case with most reports—be sure to include a well-written problem statement at the beginning. Engineering and other applied sciences are fundamentally problem-oriented, and so, as discussed in Chapter 6, a good problem statement usually has important orientational value.

13.2

USE KEY WORDS PROMINENTLY

Build sections and paragraphs around key words related to the main topic. If possible, make these key words visually prominent by using them in headings, subheadings, topic statements, and sentence subjects. Once you've established a conceptual framework at the beginning of your text, you can turn your attention to filling it in with appropriate details. To make sure that your discussion is a coherent one, you should strive to link these details as directly as possible to the main topic. The best way to do this is to establish a hierarchy of intermediate topics and subtopics for the various units and subunits of your text, with each being directly related to the immediately higher topic or subtopic. (If you've taken the time to outline your text before writing the draft, this should be a fairly straightforward matter.) These intermediate topics and subtopics should consist of appropriate key words, as discussed above.

Figure 13-1 is an excerpt from a scientific article illustrating this technique. Notice how key words (circled) are used to weld the headings, subheadings, and topic sentences together into a hierarchy of topics and subtopics; notice, too, how the writer has given these key words prominence by putting them in the subject position of the sentences.

A well-structured discussion is highly functional in at least two respects. First, it builds on the basic framework established at the beginning of the text, allowing for easier interpretation and promoting greater coherence at the same time. As new information is progressively added to the initial framework, it is interpreted in terms of this framework and integrated into it. As such, this new information is transformed into given information and can then be used to help interpret succeeding pieces of new information. Second, a hierarchically structured text facilitates selective reading. Since the sections and subsections are arranged in a general-to-specific order, the reader can quite easily zero in on desired levels of detail—especially if the respective topics of these sections and subsections are made visually prominent through the use of headings and subheadings. (Paragraph-level topics can be made visually prominent by introducing them in topic statements at the beginnings of paragraphs, as discussed in Chapter 3.)

13.3

EXPLAIN IMPORTANT CONCEPTS WHEN WRITING FOR NONSPECIALIST READERS

When writing for nonspecialists, be sure to clarify the important technical concepts in your text by using examples, analogies, visual aids, or other forms of verbal or visual illustration. Research by information theorists in the past few decades suggests that communication proceeds best when there is a

FIGURE 13-1   Building sections and paragraphs around key words related to the main topic. The key words are circled. [From Ben Patrusky, "The Social Cell," *Mosaic* **12**(4):9 (1981).]

Reach out and touch

Cell/cell junctions come in a variety of shapes and sizes.

When cells come together—and stay together—they do more than just touch. Contiguous cells actually form highly specialized, highly organized interconnective structures. And with the development of a technique called freeze fracture, a method for separating the inner face of a membrane from its outer face, scientists have managed to extract many details about the micro-architecture of these fine intercellular contacts or junctions.

One of the earliest known of the intercellular junctions is the revet- or weld-like desmosome. Adjacent cell membranes linked by way of desmosomes don't actually touch; the neighboring surfaces are separated by a space of about 200 angstroms. (One angstrom equals one hundred-millionth—10^{-3}—of a centimeter.) And the space is packed with dense, extracellular material, which serves as binder.

Desmosomes have been found primarily between the cells of sheetlike epithelial tissue, which covers the surface of the body and the internal surfaces of most organs. A variation of the desmosome, the half-desmosome, does not link cells to other cells but anchors individual cells to the underlying mesh of connective tissue, between the epithelium and the underlying organ. According to Norton B. Gilula of Rockefeller University, an authority on intercellular junctions, the function of desmosomes is now thought to be strictly that of adhesion—to keep tissues and organs intact despite mechanical stress. That belief is prompted in part by the fact that cells normally subjected to severe, repeated stress—heart muscle, for instance, or the cervix and uterine walls—possess unusually large numbers of desmosomes.

Tight junctions

Epithelial tissue is especially rich in another kind of distinctive intercellular connection called the tight junction, clearly characterized by Marilyn Farquhar and George Palade of Rockefeller Institute (now Rockefeller University) in 1967. Unlike the desmosome, with its 200-angstrom separation between cell surfaces, the tight junction, as the name implies, seems to represent a true contact between cells. No intervening space is evident. It's as if the outer surfaces of the two bilayers of neighboring cell membranes had melted into each other.

Found at the apex, or shoulder, of epithelial cells, tight junctions act as sealants, occlusions, or barriers that enable an organism or an organ to maintain an internal environment that is chemically distinct from its surroundings, explains the California Institute of Technology's Jean-Paul Revel, another leading figure in intercellular junction research. Tight junctions he says, abound in places where a sharp physical separation between two compartments—or from the outside world—is essential. It's no surprise, then, that tight junctions have been discovered in embryonic cells at the very early stages of development.

For all that, tight junctions may not be totally impermeable. Freeze-fracture studies by Philippa Claude and Daniel A. Goodenough of Harvard Medical School show that different kinds of tissue have structural differences in their tight junction that appear to correlate with varying degrees of leakiness: the bigger the junction, the less the leakage.

[Patrusky, Ben, "The Social Cell," MOSAIC. 12:4, July/August 1981, National Science Foundation, p. 9] fairly even balance between given information and new information. This is what you should strive for in your own writing. This means that you must have some idea of who your readers are and what sort of background knowledge they have; as illustrated by the rockclimbing and Pope examples, "givenness" and newness are partly functions of the knowledge a reader brings to a test—and this can vary from reader to reader and text to text. To a rockclimber, the sentence about big-wall Grade VI's would be perfectly comprehensible; to anyone else, it wouldn't be. Thus, if for some reason you had to communicate that kind of technical information to a nonspecialist reader, you would have to insert some background information more familiar to the reader to provide a proper framework for interpreting the new information. In so doing, you would be creating a better balance between given and new.

In technical writing, it frequently happens that the writer feels it necessary to introduce key concepts that may be unfamiliar to the reader. Sometimes these key concepts even occupy topic roles: topics of paragraphs, topics of sections, perhaps even the topic of the entire text. In general, it's important to define such concepts, not necessarily with a formal definition but rather with some kind of *illustration*. How is the concept used? What is it similar to? What does it look like? If technical terminology is used, what is a nontechnical way of saying more or less the same thing? Not only will answering such questions with the reader's needs in mind help the reader understand that particular concept but, more important—especially if the concept is a topical one—it will enrich and sharpen the reader's interpretation of the text as a whole. It will provide some of the given information that a specialist reader would automatically and implicitly associate with that particular concept but which a nonspecialist reader would not.

There are several ways to illustrate and explain unfamiliar concepts for the nonspecialist reader. *Visual aids*, of course, should be used whenever the concept is suited to visual representation (Chapter 7). Often, however, a concept is too abstract to be represented visually. In such cases, specific *examples* of the concept are usually the most powerful means you can use to help the nonspecialist reader. Research by cognitive psychologists indicates that readers confronted with an unfamiliar abstract concept will often try to construct a concrete example they can relate to; in some cases, they will actually build a scenario with themselves as the principal actor acting out the concept. As a writer, you can sometimes save the reader this effort by providing an example, or even a scenario, yourself. This will also prevent the reader from constructing a misleading example or scenario. *Analogies* help explain an unfamiliar concept by showing that it is similar in certain ways to a familiar concept; they are useful in situations where the concept is so unfamiliar that you simply cannot think of any ordinary examples of it. *Paraphrases*, on the other hand, are useful in precisely the opposite situation:

where the concept is familiar to the reader but only if restated in more recognizable terms. Paraphrases have a distinct advantage over examples and analogies in that they usually take up less space; sometimes even a one-word paraphrase will accomplish the purpose. *Definitions,* of course, are a familiar way of explicating new concepts. They can be combined with some of the techniques mentioned above to form *extended definitions.*

Here is an example of an extended definition, explaining what the technical term *Remrak coefficient* means:

THE REMRAK COEFFICIENT

	In the production of powered detergents, spray drying is the technique used to evaporate the solvent from the liquid reaction mixture and physically form the finished powder product. In spray drying, the liquid is sprayed into the top of a tall tower and allowed to fall freely to the bottom of the tower, where it is removed as a dry powder. The solvent evaporates during the course of the fall. Particles dried in this
Analogy	fashion have an unusual shape, *like that of a saddle (or a Pringle's potato chip),* and consequently fall through the air in an unusual manner. Rather than falling in a vertical path, the
Paraphrase	particles fall in a helical *(spiral)* path. The shape of the helical path is described by the Remrak coefficient, *which is the ratio*
Definition	*of the diameter of the helix to the height required for one passage of the particle around the perimeter of the helix.* The coefficient, which is a function of drying conditions, is sought to be maximized, *so that the length of flight of the particle is*
Paraphrase	*made much greater than the actual height of the spray-drying tower.* [italics added]

The writer of this passage has obviously gone to considerable lengths to help us understand what the Remrak coefficient is. Among other things, most of the standard forms of verbal illustration have been used: analogy, paraphrase, simple definition, and extended definition. Has it been worth the trouble? Yes, if we didn't already know the concept and were interested in finding out; no, if we already knew or weren't interested in knowing. In general, verbal and visual illustrations are powerful devices, but they work only under the following conditions:

1   The concept is not already familiar to the reader.

2   The information used to illustrate the concept *is* familiar to the reader.

3   The concept being illustrated is an important one in that particular context.

4   The information used to illustrate the concept focuses on features of that concept that are relevant to that particular context.

Do take advantage of the power of illustrations in explaining unfamiliar technical terms to nonspecialist readers. When used correctly, illustrations can clarify things in an instant.

13.4
USE STANDARD TERMINOLOGY WHEN WRITING FOR SPECIALIST READERS

When writing for specialists, on the other hand, do not overexplain. That is, do not exemplify, define, illustrate, paraphrase, or otherwise explain concepts the reader is likely to already be familiar with. Instead, simply refer to such concepts with the standard terminology of the field. Part of what it means to be a specialist in a given field is to know the standard technical terminology of that field. Technical terms permit efficient and precise communication between specialists who know the concepts that such terms refer to. They should be used for that purpose, and used freely, even if they appear to be incomprehensible jargon to an outsider. When used among specialists, standard technical terms are not only comprehensible but are often "information-rich" in the sense that they may trigger a host of associated concepts in the reader's memory. These associated concepts then become part of the "given information" in the message. Thus, in the jargon-laden writing of specialists for other specialists, there is usually more than enough given information. Adding more given information in the form of examples, analogies, etc., would only produce a disproportionate and inefficient given/new ratio for that type of reader.

What do you do, though, if you are writing to a *mixed* audience of specialists and nonspecialists? This is always a very challenging—sometimes impossible!—situation, but there are a few things you can do. First, you might divide and conquer: produce two separate pieces of writing, or a single piece with two parts to it, so that each group of readers can be addressed with appropriate terminology. Alternatively, you might stick to a single text but briefly define the technical terms as you go along. The least objectionable way of doing this, usually, is to insert a short familiar paraphrase immediately after each technical term; in the Remrak coefficient example, for instance, notice how the writer has inserted the paraphrase *(spiral)* after the less familiar term *helical.*

13.5
STRUCTURE YOUR TEXT TO EMPHASIZE IMPORTANT INFORMATION

Structure the different parts of the text so as to give greatest prominence to the information you expect the reader to pay most attention to. For main ideas, use a hierarchical (general-to-particular) structure; for details, use a listing (coordinate) structure. As discussed above, a hierarchical text structure allows the reader to move quickly through a text, seeing what the main ideas are, how they're linked together, and what kind of detailed support they have. Many readers, expecially busy decision makers, habitually read this way. Thus, if you are writing for that type of reader, you should try to organize and present your information in a highly hierarchical pattern, with many levels of subordination.

On the other hand, if you are writing for a reader who will be focusing more on details, try to use a more coordinate structure—that is, with the details arranged in lists. A listlike structure, whether it's formatted as a list or not, draws the reader's attention to all of the items making up the list. Instead of one statement being subordinated to another, as in a hierarchical structure, the statements in a list are all on the same level and thus share equal prominence. Perhaps the most obvious examples of this phenomenon are lists of instructions, which are expected to be read and followed step by step. The same phenomenon can also be seen in carefully reasoned arguments and explanations, which are often cast in the form of a listlike sequence of cause-and-effect statements. Chronological sequences, too, as found in descriptions of test procedures or in progress reports, are often presented as lists. (As Chapter 3 emphasizes, remember to use parallelism for all lists, formatted or not.)

13.6

CONSTRUCT WELL-DESIGNED PARAGRAPHS

Make sure that each paragraph has a good topic statement and a clear pattern of organization. The paragraph is a basic and highly functional unit of discourse in scientific and technical writing. By definition, a paragraph is a group of sentences focusing on one main idea. If you use a topic statement to capture the main idea and a clear pattern of organization to develop it, you make it easy for the reader to either read the paragraph in detail or read it selectively. The topic statement, of course, should be presented within the first two sentences of the paragraph, and it should contain one or more key words for readers to focus their attention on. The pattern of organization you select for the remaining sentences in the paragraph should (1) be consistent with expectations likely to be raised by the topic statement, (2) be appropriate to the subject matter, and, most important, (3) be appropriate to the anticipated use of the paragraph by the reader (see, for example, Section 13.5). If you adhere to these principles with all your paragraphs, you will greatly enhance the overall readability of your writing.

(For a thorough discussion and review of the principles of paragraph writing, see Chapter 2.)

13.7

FIELD TEST YOUR WRITING

Field test your manuscript with its intended users or with representative sub-stitutes. Up to this point, you've had to make guesses about whether or not you're providing your readers with a proper mix of given information and new information for their purposes. Your decisions about what kind of ter-minology to use, what kind of structure to use, when to use verbal or visual illustrations, and so on, have been made on the basis of guesswork about your readers' background knowledge and the reasons they will have for reading your writing. Educated guesswork, perhaps, but guesswork none-theless.

This is why field testing with actual users (or representative users) is an essential part of making any manuscript maximally useful. Field testing allows you to see whether the assumptions you have made about your readers are accurate or not. This is so important that you should not put it off until the final editing stage (though you might want to do a second round of field testing at that time). As soon as you've finished writing a good first or second draft, try it out on a few intended users. Have them read it as if it were the final draft submitted for actual use. Tell them to mark it up, raise questions about it, criticize it. Talk to them about it, ask them for their comments. Does it leave anything out? Does it mislead them? Does it raise unanswered ques-tions? If they're using it for reference purposes, can they easily find what they need? If they're skimming it for main points, can they easily locate and understand them?

If your manuscript contains any instructions or any other material de-signed to be acted on, have your readers read and try to follow these instruc-tions while you observe. Take note of places where they have to pause and ponder. Take note also of places where they go astray. Do not interfere with their efforts—let the manuscript speak for itself. Later, do a postmortem with your readers. Ask them where they had trouble. Did something confuse them? Did something seem to be missing? Would they prefer to have the material written in some other way?

This is also a good stage at which to consult experts, to make sure that nothing you have written is substantively wrong. If you're writing a research proposal or article, for example, you might want to show your draft to other researchers in that area, so as to guard against the possibility that you've overlooked something important or misrepresented someone else's research. If you're writing a progress report for a group project, this would be a good time to show it to other members of the team.

In all of this testing, both with intended users and with experts, take note of *all* trouble spots—even those that do not seem to pertain to inade-quate selection or structuring of information. Some trouble spots will prob-ably be due to sentence-level problems: problems of phrasing, wordiness, diction, sentence structure, lack of emphasis, etc. These will be addressed in Chapters 14 and 15.

EXERCISE 13-1

A   Take a highly technical term or concept from your field and define it in standard terminology so that another specialist in the same field can quickly and easily understand it. Then check your definition against that found in a technical dictionary.

B   Take the same term or concept you chose for part A and write an extended definition of it so that a nonspecialist can understand it. Then test your definition by having a nonspecialist read it; if he or she has any trouble understanding it, make appropriate modifications.

C   The following paragraph is taken from a student report on two proto-type electric-powered cars, an AMC Pacer and a Ford Pinto. Reorganize and rewrite it in two different versions: (1) one that emphasizes a single main point and (2) one that emphasizes details.

> The fuel system integrity test is concerned with fuel leakage from the fuel lines in crash and rollover tests. The rollover test involves turning the car on its side at a 90 degree angle to the ground. Fuel leakage in electric cars is primarily concerned with the gas used in the heating system. The Ford Pinto had no problem in either test. The AMC Pacer had no leakage in the crash test, but it had excessive leakage in exceeding the standard in the rollover test. In the rollover test, the problem area was where the gas lines connected to the holding tank. Redesign of this setup could alleviate this problem. The fuel system integrity test can be met easily for electric cars.

D   Take a full-length sample of your own writing and field test it with one or more intended users (or representative substitutes). Then revise it according to the guidelines laid down in this chapter.

REFERENCES

1   Julie Wei, "Pure Mathematics: Problems and Prospects in Number Theory," *Research News* **30**(3):20 (1979).

ADDITIONAL READING

R. C. Anderson, R. E. Reynolds, D. L. Schallert, and E. T. Goetz, "Frameworks for Comprehending Discourse," *American Educational Research Journal* **14**:367–381 (1977).

Marshall Atlas, "The User Edit: Making Manuals Easier To Use," *IEEE Transactions on Professional Communication* **PC-24** (1):28–29 (1981).

F. C. Bartlett, *Remembering,* Cambridge University Press, London, 1932.

Walter Kintsch and Douglas Vipond, "Reading Comprehension and Readability in Educational Practice and Psychological Theory," in *Perspectives in Memory Research,* L. G. Nilsson (ed.), Lawrence Erlbaum, Hillsdale, N.J., 1979, pp. 329–365.

James R. Miller and Walter Kintsch, "Knowledge-Based Aspects of Prose Comprehension and Readability," *Text* **1**(3):215–232 (1981).

J. M. Royer and G. W. Cable, "Illustrations, Analogies, and Facilitative Transfer in Prose Learning," *Journal of Educational Psychology* **68**:205–209 (1976).

D. E. Rumelhart and A. Ortony, "The Representation of Knowledge in Memory," in *Schooling and the Acquisition of Knowledge,* Lawrence Erlbaum, Hillsdale, N.J., 1977.

D. Schallert, "Improving Memory for Prose: The Relationship between Depth of Processing and Context," *Journal of Verbal Learning and Verbal Behavior* **15**:621–632 (1976).

M.N.K. Schwarz and A. Flammer, "Text Structure and Title: Effects on Comprehension and Recall," *Journal of Verbal Learning and Verbal Behavior* **20**:61–66 (1981).

G. J. Spilich, G. T. Vesonder, H. L. Chiesi, and J. F. Voss, "Text Processing of Domain-Related Information for Individuals with High and Low Domain Knowledge," *Journal of Verbal Learning and Verbal Behavior* **18**:275–290 (1979).

A. B. Tenenbaum, "Task-Dependent Effects of Organization and Context upon Comprehension of Prose," *Journal of Educational Psychology* **69**:528–536 (1977).

Perry Thorndyke and Barbara Hayes-Roth, "The Use of Schemata in the Acquisition and Transfer of Knowledge," *Cognitive Psychology* **11**:82–106 (1979).

14

MAKING YOUR WRITING READABLE: INFORMATION ORDERING

One of the most important parts of speech in scientific and technical writing is the *noun phrase* (NP), which can be defined as any noun or noun-plus–modifier combination (or any pronoun) that can function as the subject or object of a sentence. Some examples are *tables, water, we, a potential buyer, the growing demand for asphalt,* and *strict limitations on the size of plates that can be handled.*

Notice that each of these NPs can serve as the subject of a sentence:

Tables usually have four legs.

Water can be dangerous.

We have an emergency.

A potential buyer has arrived.

The growing demand for asphalt is obvious.

Strict limitations on the size of plates that can be handled have been established.

By contrast, a singular countable noun, such as *table,* is not an NP because it cannot function by itself as the subject or object of a sentence. We cannot say

Table usually has four legs.

Instead, we would have to say

A table usually has four legs.

or

The table has four legs.

or

John's table has four legs.

NPs are important on the grammatical level because they account for most of the principal parts of a sentence: subject, indirect object, direct object, and object of a preposition. Not all of these parts are found in all sentences, of course, but here is an example that does have all four:

<u>We</u>   are sending <u>your company</u> <u>the new TK-140 model</u>
SUBJECT   INDIRECT OBJECT   DIRECT OBJECT

on <u>the assumption that you need an all-purpose sorter.</u>
                    OBJECT OF PREPOSITION

NPs are important on the functional, or communicative, level because they carry the main information of the sentence. In the example just cited, for instance, notice how much information is contained in the NPs (italicized)—almost all of it.

14.1

OPTIMAL ORDERING OF NOUN PHRASES

In English, NPs are expected to occur in certain orderings according to grammatical and functional criteria. These will be discussed in order of importance, beginning with the most important.

Put Given Information Before New Information

As with all languages, English sentences typically contain a mixture of given information and new information. That is, some NPs in a sentence refer to concepts or objects that have already been discussed or that are presumed to be understood from the context; this is *given information*. Other NPs refer to concepts or objects that have not yet been discussed and are not presumed to be understood from the context; this is *new information*. The optimal ordering of NPs within a sentence according to this view is *given* before *new* because the given information can serve as a frame within which the new information can be understood. Let us consider a specific example of this ordering:

> *The 5-year plan* does not indicate a clearly defined commitment to *long-range environmental research.* For instance, where *the plan* does address *long-range research, it* discusses the development of techniques rather than the identification of important long-range issues.

The key NPs in both sentences are in italics. By the time the first sentence has been read and understood, the phrases *the 5-year plan* and *long-range environmental research* have been mentioned and are part of the given information possessed by the reader. Notice that these words—the given information—come at the beginning of the second sentence and that the new noun phrases—the new information—come at the end of the second sentence. This ordering of given before new is desirable because the given information of the second sentence serves as a kind of glue between the information presented in the first sentence and the new information presented in the second sentence. Such an ordering allows a reader to more easily fit the new information into a meaningful context and to see the connection between the two sentences.

To see how much easier it is to understand sentences which put given before new information, compare the passage just discussed with the following one (which has the opposite ordering):

> The 5-year plan does not indicate a clearly defined commitment to long-range environmental research. For instance, the development of techniques rather than the identification and definition of important long-range issues is the subject of the plan where it does address long-range research.

Notice how hard it is to place the new information of the second sentence in an appropriate context.

Put Topical Information in Subject Position

Often, more than one NP in a sentence carries given information. In that case, which of these NPs should be promoted to subject position? Ideally (all other things being equal), the NP that carries information most closely related to the paragraph topic—call it "topical information"—should go there. Consider this example:

> Not all investors will benefit from All-Savers Certificates. Investors exceeding a deposit of $7931 ($15,861 joint return) would have an after-tax yield far lower than with alternative investments, such as money market funds or Treasury bills. Alternative investments would also yield better after-tax yields and no penalty if the certificate was redeemed within the one-year maturity period.

The last sentence in this paragraph has three overt NPs which contain given information: *Alternative investments, after-tax yields,* and *the certificate.* Of these, the last seems to come closest to being thought of as topical information; the word *Certificate,* after all, does appear in the topic statement. But what is the *real* topic of this paragraph? Isn't it *different kinds of investors?* Notice, for example, that the word *investors* appears not only in the topic statement but in the subject position of the next sentence. Notice also that investors are referred to by implication as the delegated agent of the passive main verb: *was redeemed (by investors)*. . . . Ideally, then, we should try to insert the word *investors* in the subject position of the third sentence, too, if it is at all possible. Indeed it is:

> Not all investors will benefit from All-Savers Certificates. *Investors* exceeding a deposit of $7931 ($15,861 joint return) would have an after-tax yield far lower than with alternative investments, such as money market funds or Treasury bills. *Investors* redeeming their certificates within the one-year maturity period would also have a lower after-tax yield and would pay a penalty besides.

Not only does this rewritten version keep the focus on the topic of the paragraph—and thus contribute to paragraph unity—it also establishes parallelism between the second and third sentences, thus making it much clearer to the reader that we are talking about two different classes of investors: those who exceed a deposit of $7931 ($15,861 joint return) and those who redeem their certificates early. These two classes do not necessarily intersect. The original version, by contrast, links the third sentence to the second through the repeated use of the term *alternative investments,* with the second use appearing in subject position. The third sentence might therefore easily be misunderstood as simply modifying or adding detail to the second, and the paragraph as a whole might be misinterpreted as having a general-to-particular (not listing) pattern.

Put "Light" NPs Before "Heavy" NPs

As can be seen in the examples at the beginning of this chapter, NPs vary considerably in length, complexity, preciseness, etc. If we use the word *heavy* to describe NPs which are long and complex and the word *light* for NPs which are short and simple, the preferred stylistic ordering is light NPs before heavy NPs. For instance, consider the following passage:

> We have received and acted upon requests for equipment from several branch offices. We have sent the research, development, and testing office in Chicago a gas analyzer.

The second sentence of this passage is awkward and difficult to read. It has a very heavy indirect object—*the research, development, and testing office in Chicago*—and a very light direct object—*a gas analyzer.* Thus, the ordering of NPs in this sentence, as it stands, is heavy . . . light. A more readable version of the second sentence and thus a better version would order the NPs light . . . heavy as follows:

> We have sent <u>*a gas analyzer*</u>
>   DIRECT OBJECT
>
>   to <u>*the research, development, and testing office in Chicago.*</u>
>   OBJECT OF PREPOSITION

Notice that in moving the heavy NP to the end, we have to insert the preposition *to.*

Sentences are more readable if they are ordered light . . . heavy because of important limitations on the capacity of the mind to process information. These limitations have been described by George Miller, a famous psychologist, who demonstrated that the mind can usually process a maximum of nine separate items of information at a time. This means that as the mind reads through a sentence, it must try to make sense out of the sentence as quickly as possible. Ideally, the mind would see the entire structure of the sentence—subject, verb, indirect object, direct object—all within the nineword processing limit. Unfortunately, this often does not happen since the main constituents of a sentence often span more than nine words.

However, even if you cannot put *all* of the main sentence constituents into a span of nine words or less, you can try to put *most* of them into such a span. One way to do this is to put light units before heavy ones. Let us consider our heavy . . . light sample sentence again:

1	2	3
We	*have*	*sent*
SUBJECT	VERB	

4	5	6	7	8	9	10	11
the	*research,*	*development,*	*and*	*testing*	*office*	*in*	*Chicago*
		INDIRECT OBJECT					

12	13	14
a	*gas*	*analyzer.*
DIRECT OBJECT		

In this version of the sentence, the reader sees only the subject, the verb, and part of the indirect object before reaching the nine-word limit. However, in the light . . . heavy version shown below, the reader sees the subject, verb, direct object, and part of the indirect object (now the object of the preposition *to*) before reaching the nine-word limit:

1	2	3	4	5	6
We	*have*	*sent*	*a*	*gas*	*analyzer*
SUBJECT	VERB		DIRECT OBJECT		

7	8	9	10	11	12	13	14	15
to	*the*	*research,*	*development,*	*and*	*testing*	*office*	*in*	*Chicago.*
			OBJECT OF PREPOSITION					

In the light . . . heavy version, then, by the ninth word the reader has a much fuller sense of the structure of the whole sentence and thus has a better chance of understanding the sentence with a minimum amount of effort.

From this discussion, it should be obvious that a heavy NP in the subject position would cause even more trouble for a reader than a heavy NP in the indirect object position. In a sentence with a heavy subject NP, the reader may not even get to the main verb before reaching the nine-word memory limit. Consider the following:

> The idea of designing and producing an economical AM/FM receiver that is both affordable for the average consumer and profitable for the company was introduced.

Here, all of the sentence before *was introduced* is the subject. This is a 23-word subject and poses serious comprehension problems for the reader, especially since the head noun, *idea,* must be held in memory while the reader processes 21 words before reaching the main verb, *was introduced.*
    Note how much easier the sentence is to read if it is rewritten as follows:

> It was suggested that we design and produce an economical AM/FM receiver that is both affordable for the average consumer and profitable for the company.

In this rewritten version the reader has a much fuller sense of the sentence's structure earlier on and thus a better chance of understanding the sentence with a minimum amount of effort. This greater chance of easy understanding is the main reason writers try to order their NPs light . . . heavy.
    One other situation should be considered before leaving this discussion of light . . . heavy ordering. That situation is the processing and "counting" of long introductory units, as in the following sentence:

> At the September and October meetings of all company engineers, it was suggested that we design and produce an economical AM/FM receiver that was both affordable for the average consumer and profitable for the company.

From what has been said so far, we would expect this example to be harder to read than the previous example because it begins with a 10-word phrase, *At the September and October meetings of all company engineers,* which comes before the subject and verb in the main clause. However, it doesn't seem to be any harder to read. This is because when readers process a sentence, they first process introductory clauses or phrases as a unit and then start over again with the main clause. Thus, in ordering units light . . . heavy, a writer should first order the units within the introductory clause or phrase and then order the units within the main clause:

> Light . . . heavy, light . . . heavy.

For ease of reading, the writer should also try to see that the introductory clause or phrase is no longer or not much longer than the nine words we have defined as the limit for easy remembering.

Sometimes a long subject is followed by a very short unit after the verb:

That we are in the midst of an energy crisis is	*obvious.*
SUBJECT	PREDICATE ADJECTIVE

Such a construction sounds strange to most people because it violates the criterion of light before heavy. In such cases, a writer might shift the heavy subject NP to the end of the sentence and replace it with the pronoun *it* (this last step satisfies the requirement that all English sentences have a subject NP):

It is obvious that we are in the midst of an energy crisis.

Of course, if the unit after the verb is quite long and complex itself, then the need to shift the subject is not so strong:

That we are in the midst of an energy crisis is
SUBJECT

obvious to everyone in the country except the big politicians.

This sentence is more balanced than the first version, does not violate our light . . . heavy rule, and sounds better to most people.

The three situations we have discussed are (where S stands for *subject* and V for *verb*):

S __ V _____.	preferred
S _____ V _____.	acceptable if subject not too long
S _____ V __.	not preferred

14.2

WAYS OF SATISFYING THESE CRITERIA

In most cases, sentences can be constructed so that at least the first two criteria, if not all three, can be satisfied. Sometimes this procedure is straightforward. At other times, however, writers may find themselves boxed in, having written a sentence that satisfies the light-heavy criterion, say, but not the other two. The purpose of this section is to describe several of the most common ways that good writers get themselves out of such a situation.

Passive-Active Alternation

We often have a sentence such as the following, consisting of a subject NP, an active verb, and an object NP:

<u>Christopher Sholes</u> invented <u>*the typewriter.*</u>
      SUBJECT   DIRECT OBJECT

We can create a passive form of such a sentence by (1) interchanging the two NPs, (2) inserting an appropriate form of the verb *to be,* and (3) inserting the preposition *by:*

<u>*The typewriter*</u> was invented by <u>*Christopher Sholes.*</u>
     SUBJECT   OBJECT OF PREPOSITION

Although these two sentences have the same meaning, they differ in word order and thus should be used in different contexts. For example, suppose the topic of discussion is *typewriters* and no mention has been made of *Christopher Sholes:*

> A typewriter is a machine that prints alphabetic characters, numbers, and other symbols. It [the typewriter] was invented by Christopher Sholes.

In this case, the passive form used in the second sentence is appropriate because the NP representing given information *(the typewriter)* comes before the NP representing new information *(Christopher Sholes).* On the other hand, if the topic of discussion is *Christopher Sholes,* as in the passage below, then reference to him would constitute given information and the active form of the second sentence would be preferred.

> Christopher Sholes was a nineteenth-century inventor who made the life of the secretary much easier. He [Christopher Sholes] invented the typewriter.

Equative Shift

Equative sentences consist essentially of two NPs connected by a linking verb (usually some form of the verb *to be*). Furthermore, as the name implies, the NPs are interchangeable without any change of meaning:

> Air pollution is one major form of pollution.
> One major form of pollution is air pollution.

Here again the choice of which form to use depends mainly on given-new considerations. If *air pollution* is given information and we want to add new information about it (namely, that it's a major form of pollution), then the first version is preferred:

There are several things the average person should know about air pollution. *Air pollution is one major form of pollution.* It exists in all large cities and most small ones. And it is a secondary cause of death among the sick and elderly.

If we have already been talking about different forms of pollution, however, and now want to introduce air pollution as one such form, as in the next example, then *air pollution* represents new information and should come after the verb:

There are several types of pollution with which the average person should be familiar. *One major form of pollution is air pollution.* Other major forms of pollution are water pollution and noise pollution.

Caution! Do not mistake a nonequative sentence for an equative one! Equative shift, which we have just been describing, cannot be applied to nonequative sentences without changing the meaning! For example, consider this sentence:

Football players are good athletes.

This implies that all football players are good athletes, which is true. However, if the NPs are interchanged, we have

Good athletes are football players.

This implies that all good athletes are football players, which is not true. Clearly, these two sentences do not mean the same thing. Thus, they cannot undergo equative shift.

Indirect Object Shift

As was mentioned earlier, the relatively fixed word order of English sentences requires indirect objects to precede direct objects. But then what do you do if, in a given situation, the indirect object NP is either heavier or newer (or both) than the direct object NP? The answer is that you shift the indirect object and convert it into the object of a preposition, following the direct object NP. Consider this sentence, for example:

We are sending <u>*your branch office in Longview*</u>   <u>*a copy.*</u>
                      INDIRECT OBJECT   DIRECT OBJECT

Suppose this sentence were to occur in a context where a report is being discussed:

We have finished our final report on the new product line. We are sending your branch office in Longview a copy.

In this case, *a copy* (of the report) would not constitute much new information since reports are customarily made in copies, but the location where one of the copies is to be sent probably does constitute new information. Furthermore, the location is a longer, more complex (i.e., heavier) element in this case. Given these considerations, a preferred version of the sentence would shift the heavier and newer information to the end of the sentence:

We have finished our final report on the new product line.

We are sending ____*a copy*____
                  DIRECT OBJECT

to *your branch office in Longview.*
       OBJECT OF PREPOSITION

14.3

A PROCEDURE FOR PRODUCING MORE READABLE SENTENCES AND PARAGRAPHS

The purpose of this section is to summarize the material presented in this chapter and to outline a procedure for applying it. The chapter provides a number of criteria and suggestions for producing functional and readable sentences. By applying the criteria given in this chapter—in the order they are given—to each sentence as it appears in a paragraph or text, a writer should consistently produce more readable prose.

A summary of these criteria and a procedure for applying them are given in the flow chart in Figure 14-1. To illustrate the use of this chart, let us consider this paragraph:

> The sound reproduction from a radio can be quite inferior to that from a record or tape. One of the causes of radio's inferior sound reproduction, and in some environments perhaps the major cause, is multipath distortion. This type of distortion occurs when there are tall structures or buildings in the receiver's environment. Most of the radio signal is reflected and diffracted by these tall structures before it reaches the radio's antenna. When the signal is received by the radio, the radio interprets it as distorted and can only produce an inferior sound from it. In contrast, the sound on a record or tape suffers none of this distortion.

To evaluate this paragraph and revise it where necessary, let us follow the procedure outlined in Figure 14-1. Step 1 asks us to read the paragraph, and then step 2 asks us to decide if the topic sentence (the first sentence of the passage) is adequate. The answer is "yes" here because the topic sen-

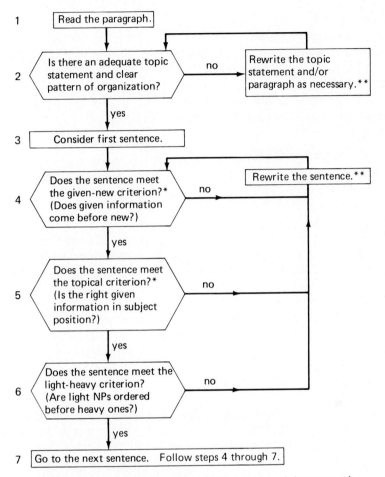

1 Read the paragraph.

2 Is there an adequate topic statement and clear pattern of organization? → no → Rewrite the topic statement and/or paragraph as necessary.**

yes

3 Consider first sentence.

4 Does the sentence meet the given-new criterion?* (Does given information come before new?) no Rewrite the sentence.**

yes

5 Does the sentence meet the topical criterion?* (Is the right given information in subject position?) no

yes

6 Does the sentence meet the light-heavy criterion? (Are light NPs ordered before heavy ones?) no

yes

7 Go to the next sentence. Follow steps 4 through 7.

*If the sentence being analyzed is the first sentence of the paragraph, this question may not be relevant.

**Sometimes it will not be possible to rewrite a sentence to meet all of the desired criteria.

FIGURE 14-1 Flow chart for editing sentences in paragraphs.

tence introduces the main idea, that *sound reproduction from a radio can be quite inferior to that from a record or tape,* and the rest of the paragraph goes on to describe why this is so.

SENTENCE 1 We then consider whether or not the first sentence meets the criteria outlined in steps 4 to 6.

STEPS 4 AND 5

It does not violate given-new or topical criteria since there is no previous given or topical information to consider here.

STEP 6

It meets light-heavy criteria because the subject NP *(the sound repro-duction from a radio)* is not significantly heavier that the object NP *(that from a record or tape)*.

SENTENCE 2 Since the first sentence does not require any revision, we move on to step 7 and decide whether the second sentence meets the criteria outlined in steps 4 to 6.

STEPS 4 AND 5

It does not really violate given-new or topical criteria, although it would be desirable to get the given and topical information *(radio's inferior sound reproduction)* closer to the beginning of the sentence. We could rewrite the sentence to read

One cause of radio's inferior sound reproduction, and in some environments perhaps the major cause, is multipath distortion.

This sentence meets given-new and topical criteria.

STEP 6

Revised sentence 2 does not meet light-heavy criteria at all. It has a 15-word subject NP (everything before *is*) and only two words after *is*. Thus, to meet light-heavy criteria, we must rewrite the sentence again. Three possible revisions which meet all of the criteria in steps 4 to 6 are:

1   One cause of this inferior reproduction is multipath distortion; in some environments, this is perhaps the major cause.

2   Radio's inferior reproduction is partly caused by multipath distortion; in some environments, this is perhaps the major cause.

3   This inferior reproduction is partly caused by multipath distortion; in some environments, this is perhaps the major cause.

Since all of these revisions are quite similar, we shall choose the first for our revision.

SENTENCE 3 Since the second sentence does not require any more re-vision, we move on to step 7 and consider whether the third sentence meets the criteria outlined in steps 4 to 6.

STEPS 4 AND 5

It meets given-new and topical criteria because the first NP *(This type of distortion)* is given and topical information. However, since we have moved *multipath distortion* away from the end of sentence 2, we may wish to repeat *multipath distortion* at the beginning of sentence 3:

Multipath distortion occurs when there are tall structures or buildings in the receiver's environment.

This new version meets given-new and topical criteria.

STEP 6

It also meets light-heavy criteria because the subject NP (now *multipath distortion*) is lighter than the other NP *(tall structures or buildings in the receiver's environment)*.

SENTENCE 4 Since sentence 3 does not require any more revision, we move on to decide whether sentence 4 meets the criteria outlined in steps 4 to 6.

STEPS 4 AND 5

It violates given-new and topical criteria since the given and topical information *(these tall structures)* occurs after significant new information *(Most of the radio signal is reflected and diffracted)*. We could revise the sentence to meet given-new and topical criteria as follows:

These tall structures reflect and diffract most of the radio signal before it reaches the radio's antenna.

This meets given-new and topical criteria.

STEP 6

The rewritten sentence also meets light-heavy criteria.

SENTENCE 5 Since sentence 4 does not need any more revision, we move on to consider sentence 5.

STEP 4

It violates given-new criteria in the main clause because the given information *(it,* meaning *the signal)* follows the new information *(the radio)*. We could revise the sentence to meet given-new criteria as follows:

When the signal is received by the radio, it is seen as distorted and produces an inferior sound.

STEP 5
The revised sentence also meets topical criteria since the previous sentence talks about the *radio signal:*

These tall structures reflect and diffract most of *the radio signal* before *it* reaches the radio's antenna. When *the signal* is received by the radio, *it* is seen as distorted and produces an inferior sound.

If we rewrite sentence 5 with *the radio* as the subject, it would violate topical criteria:

These tall structures reflect and diffract most of *the radio signal* before *it* reaches the radio's antenna. When *the radio* receives the signal, *the radio* sees it as distorted and produces an inferior sound from it.

While *the radio* is mentioned in sentence 1 and thus is given information, sentence 4 introduces the topic of *the radio signal,* not *the radio.* Thus, to satisfy the topical criterion set up in sentence 4, sentence 5 must place *the radio signal* (or *the signal*) in the subject position:

These tall structures reflect and diffract most of *the radio signal* before *it* reaches the radio's antenna. When *the signal* is received by the radio, *it* is seen as distorted and produces an inferior sound.

STEP 6
Sentence 5 now also meets light-heavy criteria.

SENTENCE 6   Since sentence 5 does not require any more revision, we move on to consider sentence 6. It meets all of the criteria outlined in steps 4 to 6. The revised paragraph follows.

The sound reproduction from a radio can be quite inferior to that from a record or tape. One cause of this inferior reproduction is multipath distortion; in some environments, this is perhaps the major cause. Multipath distortion occurs when there are tall structures or buildings in the receiver's environment. These tall structures reflect and diffract most of the radio signal before it reaches the radio's antenna. When the signal is received by the radio, it is seen as distorted and produces an inferior sound. In contrast, the sound on a record or tape suffers none of this distortion.

Once we have arrived at this reordered version, we can combine and cut the individual sentences to produce a more concise version of the paragraph:

The sound reproduction from a radio can be quite inferior to that from a record or tape. One cause of this is multipath distortion; in some environments it is perhaps the major cause. This distortion occurs when there are tall structures in the receiver's environment which reflect and diffract the signal before it reaches the radio's antenna. The signal is thus distorted and can be used only to produce an inferior sound. In contrast, the sound on a record or tape suffers none of this distortion.

Procedures for effectively combining and cutting sentences are discussed in Chapter 15.

EXERCISE 14-1 Where necessary, reorder the noun phrases in the following passages to create more readable sentences.

A   The labor movement has existed in the United States for about 180 years. In the past 40 years, labor unions have become an integral part of the workplace. Over 19 million workers belong to labor unions. The wage rates, working hours, and working conditions of both unionized and nonunionized industries are influenced and determined by organized labor. Clearly, unions are permanent and powerful institutions in America.

B   *Advanced Component Technology for the Ford Pinto*   A number of systems, devices, and technologies to work with in meeting your specified requirements are available. One such device is the aneroid control. This device is used to limit the fuel while the turbocharger is building up manifold pressure during acceleration. The pressure-reducing valve is another such device. It reduces the pressure of fuel entering the carburetor. We have also developed other devices that we think will be useful in the future.

C   Several recent developments have encouraged a re-evaluation of wind-powered vessels for the American merchant fleet. The sharp rise in the cost of energy required to drive a fuel-powered ship is the most significant of these developments. A second development is the improvement of technology associated with wind-powered ships.

D   *General Appraisal of the EPA Plan*   The plan prepared by the Environmental Protection Agency (EPA) lacks several essential characteristics. It does not clearly outline program priorities nor does it relate priorities to overall program goals. The planning process is vague, and no guidelines are offered for future updates of the plan. It is difficult to see a rationale for the major thrusts suggested in the budget. For example, the plan offers no basis for the dominant expenditure on developing control technology over the 5-year period covered by the plan.

E   *Accomplishments of the Massachusetts Institute of Technology's Sea Grant Program*   The Sea Grant Program estimated the environmental effects which would be associated with oil exploration on the Georges Bank, off the New England coast. Among other matters, we reviewed what is now known about the biological effect of oil, particularly its toxicity. We analyzed how long a Georges Bank oil spill would be likely to stay on the Bank and where it would be likely to go. Available statistics on oil spills were reviewed, and the amounts of oil which might be spilled in New England under a number of developmental hypotheses were estimated. Also the effect of offshore drilling and of possible oil spills on the fishing industry were estimated. Finally, we made a preliminary analysis of the loss in regional income which would be associated with oil spills as a result of cleanup costs.

F   *[From a report written by a geologist to a township council]*

FOREWORD
Within this township, the greatest industrial and residential development has occurred in areas with limited groundwater resources. This development has already created water shortages and a need to deepen existing wells; more seriously, the development of new industry has been greatly handicapped or prevented. Thus, the township council authorized Moody and Associates to locate, if possible, several high-yield water wells to provide a sufficient municipal water supply. The purpose of this report is to describe the successful location of sand and gravel materials with excellent groundwater yielding capability.

SUMMARY
We conducted studies of six test drillings to locate an area in which a reliable, high-quality, municipal water well field could be developed. We have located such a site.

Results: Four of the six wells drilled were productive and penetrated sand and gravel suitable for development of municipal wells. Each of the four productive wells showed very good water-producing capability, yielding 300 to 500 gallons per minute. Further, generally excellent-quality water, though with a potentially high but treatable concentration of manganese, was produced. One test well showed manganese concentration now high enough to require treatment, and there is a concern that manganese levels in all the wells might increase to such a level. However, this would not prevent development of the well field, and additional study will be required to determine if manganese removal is necessary at all.

G   *The Future of Solar Energy*   Various studies have shown that solar energy could supply between 10 and 35 percent of U.S. energy by the turn of the century. The 20 percent estimate is the most given figure, which is approximately equal to current imported oil levels. Much of the 20 percent supplied would be used for space heating and for heating of hot water. Shown below is a graph which gives the results of a Stanford Energy Institute Research Project. It states three different scenarios and shows the effect of each on the percentage of solar energy in the United States.

H   In comparing the costs of electric cars to the gas car, the major factors to be considered are the price of the car and the total fuel cost. Maintenance costs do not have to be considered because they are about the same for both cars. The price of the cars considered here is based on a standard model with minimum options. The total fuel cost is composed of battery replacement and electricity consumption for the electric car, whereas the cost of gas is the total fuel cost for the gas car. The following results were used for comparison.

I   Industry will be needed to produce the electric car. The present U.S. auto companies will be the ones who will have the capabilities to produce an electric car. All four U.S. auto companies (General Motors, Ford, Chrysler, and American Motors) are encouraged about the development and production of electric cars. These companies are all doing research in the area of electric cars, with General Motors leading the way.

ADDITIONAL READING

Herbert Clark and Susan Haviland, "Comprehension and the Given-New Contract," *Discourse Production and Comprehension,* R. Freedle (ed.), Ablex, Norwood, N.J., 1977.

Barbara Hayes-Roth and Perry W. Thorndyke, "Integration of Knowledge from Text," *Journal of Verbal Learning and Verbal Behavior* **18**:91–108 (1979).

Michael P. Jordan, "Some Associated Nominals in Technical Writing," *Journal of Technical Writing and Communication* **11**(3):251–264 (1981).

Ronald Langacker, "Movement Rules in Functional Perspective," *Language* **50**:629–664 (1974).

Charles Li (ed.), *Subject and Topic,* Academic Press, New York, 1976.

C. A. Perfetti and S. R. Goldman, "Thematicization and Sentence Retrieval," *Journal of Verbal Learning and Verbal Behavior* **13**:7–79 (1974).

15

MAKING YOUR WRITING READABLE: EDITING FOR EMPHASIS

Although some readers may prefer to skim-read, others have to read more closely and thoroughly, concentrating on details. For these readers, there is a danger of getting *lost* in the details, of overlooking main points and "not seeing the forest for the trees," so to speak. Consequently, the details themselves begin to lose significance: the reader cannot see exactly how they fit into the larger picture and thus cannot evaluate their importance. The reading process as a whole bogs down at this point, and the reader is forced to stop and start over.

When readers get bogged down in detail like this, it's often the writer's fault. Many writers make little effort to organize details in a coherent, unified way, preferring instead to have the reader do all the work. But this invites the kind of failure just described. Readers are often pressed for time, or tired, or have other things on their mind. Many readers lack the kind of background knowledge the writer has. Still others have poor reading techniques and are unable to decipher poor writing, no matter how hard they work at it. In general, readers are at the mercy of the writer; they depend on the writer to present details in such a way that the role of these details in support of main points is readily apparent. If the writer fails to do this, there is little the reader can do except try to figure things out.

It thus falls on the writer to mold the details of a text so that they reinforce main points in unified fashion. This is somewhat similar, actually, to the demands made on a speaker engaged in serious conversation. Face-to-face conversation is an intensive form of communication in which the speaker is acutely aware of the listener and vice versa. Because of this close speaker-listener relationship, conversations are governed by certain unwritten rules: say what you mean, don't beat around the bush, get to the point, be honest, etc. If the speaker violates any of these rules, the conversation will begin to break down unless the listener rescues it with a corrective comment, such as "I don't see what you're driving at" or "What's your point?" The possibility of such immediate feedback from the listener forces the speaker to make *every* detail relevant to the conversation; most listeners are simply intolerant of irrelevant details and will either intervene or break the conversation off if the speaker strays too far from the topic of discussion.

Good conversationalists, of course, are aware of such constraints and employ various techniques to make it clear to the listener that they are observing the rules. For one thing, they use emphatic intonation, physical gestures, inverted sentence structure, intensifiers, and other devices to signal important words—key words, topical words, words carrying new information. Conversely, they use none of these devices for the less important words—those that carry given information or redundant information. As for empty, meaningless words that serve no communicative purpose at all, they are simply omitted. In general, both by giving prominence to important words and by subordinating or omitting unimportant ones, good conversationalists emphasize those aspects of a detailed discussion that link the discussion to the main point or purpose of the conversation. As a result, the listener not only absorbs those details but also sees just how they support the main point.

Writers should do the same kinds of things as good conversationalists. They may not be in as close touch with their audience as speakers are and so they may not have such immediate demands placed on them, and they cannot, of course, use intonation and gestures in their writing. But writers do have an audience, and this audience needs to know, just as listeners do, how the details of a discussion are related to the main points. Furthermore, writers have as many devices as speakers do for helping the reader see how details support main points. In short, the use of emphasis is as appropriate and indeed necessary to good writing as it is to good conversation.

This chapter describes the most common and useful devices used by good writers to create emphasis within individual sentences. These fall into three categories: devices used to highlight important words and phrases, devices used to subordinate relatively unimportant words and phrases, and devices used to eliminate unnecessary words and phrases. This chapter thus complements several earlier chapters (2, 13, and 14), which also suggest ways of emphasizing details, but with more of a paragraph orientation.

15.1

COMBINE CLOSELY RELATED SENTENCES

Combine closely related sentences unless there is a compelling reason not to (such as maintaining independent steps in a list of instructions or avoiding extreme sentence length); put main ideas in main clauses. Many inexperienced writers have a tendency to use nothing but short, simple sentences, producing a very choppy style of writing which irritates the reader with its sing-song rhythm and, worse, fails to put emphasis on important ideas. This tendency derives, probably, from two principal sources: (1) an overemphasis in many quarters on the need to avoid dangling modifiers, comma splices, and other problems associated with complex sentence structures and (2) an erroneous belief, promoted by readability formulas, that short sentences make reading easier. Dangling modifiers, comma splices, and other errors of sentence structure and punctuation should, of course, be avoided—but not at the expense of emphasis, unity, and coherence. And although a short sentence by itself may be easier to read than a long sentence, the repeated use of short sentences may have just the opposite effect.

The best approach to take regarding sentence length is to *let the form reflect the content.* If an idea is complex enough to require qualification, the best way to qualify it may be with a relative clause, an adverbial phrase, or some other complex modifier. On the other hand, if an idea is simple and straightforward, a simple sentence may be the best way to represent it. Often, these choices can be made properly only within the context of an entire paragraph. For example, consider the following paragraph from a student report:

ORIGINAL VERSION

At the present time electric car utilization is not possible. The problems holding it back are satisfactory performance and costs. Performance problems of lack of speed, short mileage range, and lack of acceleration are present. Cost problems are the price of battery replacement and the base price of the electric car. It is possible though, with research and development, that these problems can be solved in the future.

Each of the first two sentences, taken in isolation, is grammatically correct and easy to read. When you look at them together, however, you notice that there is excessive overlap between them: sentence 2, in other words, contains too much given information *(The problems holding it back).* This unnecessary redundancy can be eliminated by combining these sentences:

At the present time electric car utilization is not possible because of performance and cost problems.

Not only does this move reduce the wordiness of the first two sentences, it also creates a better topic statement: it's more unified and emphatic, and it introduces the key terms *performance problems* and *cost problems.* (Notice how these terms are the subjects of the next two sentences.) If we also change sentence 3 to satisfy given-new and light-heavy criteria, we can reduce the wordiness of the paragraph and increase its readability still further. The overall result is this:

FIRST REWRITE

At the present time electric car utilization is not possible because of performance and cost problems. The performance problems are lack of speed, short mileage range, and lack of acceleration. The cost problems are the price of battery replacement and the base price of the car. It is possible, though, with research and development, that these problems can be solved in the future.

This is a significant improvement, but we have other options that might improve it even more. For example, now that we've converted the original sentence 2 into a prepositional phrase, we can shift it into presubject position in place of the time adverbial originally there:

> Because of performance and cost problems, electric car utilization is not possible at the present time.

This puts more focus on the key terms *performance problems* and *cost problems* and less focus on the less important time adverbial. Another change we could make, though not as compelling a one as those just described, would be to combine the two sentences in the middle with a semicolon. These two sentences are closely related in function; linking them formally would reflect this relatedness.

FINAL VERSION
Because of performance and cost problems, electric car utilization is not possible at the present time. The performance problems are lack of speed, short mileage range, and lack of acceleration; the cost problems are the price of battery replacement and the base price of the car. It is possible, though, with research and development, that these problems can be solved in the future.

We might entertain (though ultimately reject) one other change: substituting *its* for *the* in sentence 2 so as to have *Its performance problems* and *its cost problems.* This might be desirable from the point of view of reiterating the paragraph topic *(electric car utilization).* Unfortunately, it would create excessive repetition of the word *it* in the subject position: notice that the last sentence of the paragraph also begins with this word. Using the same pronoun to lead off all three sentences would constitute misleading parallelism, since *it* does not refer to the same thing in all three cases.

One must often make trade-offs like this in deciding how best to structure sentences and express ideas. This is simply inevitable. Language is highly complex and can trigger highly complex perceptual processes in the reader's mind. Expressing an idea in a certain way may have not only some benefits but also some undesirable side effects. Expressing the same idea in a different way may have fewer (or more) benefits but also fewer (or more) undesirable side effects. In such cases, the writer should strive for optimization, not perfection.

In the example just discussed, for instance, we could retain the two *its* in sentence 2 and avoid the misleading parallelism by simply changing the form of the final sentence into something like *Research and development, however, may be able to solve these problems in the future.* This would give us the benefits of repeated reference to the paragraph topic without the undesirable side effect. However, in doing so we would be sacrificing a number of benefits associated with the original form of the final sentence. For one thing, the original form begins with *It is possible,* which contrasts nicely with *is not possible* in the topic statement; this emphasizes the fact that these two sentences have contrasting functions. The revised alternative does not contain this phrase. Second, the original form further emphasizes this contrast with the topic statement by having two "possibility phrases": *is possible* and *can be;* the revised alternative has only one *(may be able)*. Finally, the original form gives more prominence to the key word *problems* by putting it in subject position in the *that* clause; the alternative form, by contrast, has it in object position.

There are, of course, many other alternatives that could be considered (though we might not want to, given the usual time constraints!). Let us assume, however, that the second rewrite is the best version we can devise, and now let us take stock of the benefits we've gained. First and foremost, we've produced a paragraph whose form reflects its content:

Because of performance and cost problems, electric car utili-zation is not possible at the present time.	MAIN IDEA
The performance problems are lack of speed, short mileage range, and lack of acceleration; the cost problems are the price of the battery replacement and the base price of the car. It is	DETAILS OF PROBLEM
possible, though, with research and development, that these problems can be solved in the future.	POSSIBLE SOLUTION

The ideas in this rewritten version are grouped so that there is a one-to-one correspondence between form and content: each sentence has a particular, identifiable function. Most important, we now have an effective topic state-ment. Second, we've put more prominence on the key words and phrases of the paragraph by placing them at the beginning of the topic statement. They are then reinforced in the following sentences, where they appear in subject and initial positions. Finally, we've eliminated a little wordiness (64 words instead of 68).

In general, combining sentences is often a good way to create emphasis in your writing. By making it easy for your readers to see the relatedness of ideas, you make it easier for them to absorb these ideas. You can also show explicitly that one idea is logically subordinate to another by putting the more important idea in the main clause of the sentence and the less important idea in a subordinate clause. For example, suppose you wanted to combine the two sentences in italics in the following paragraph:

NEGATIVE EXAMPLE
Electric cars must be able to meet the same safety standards that gasoline cars must meet as set up by the Department of Transportation. *These standards are derived from an established crash test. In the crash test, the car is propelled against a solid wall at 30 mph.* The data obtained from the crash test are analyzed for fuel spillage, fuel system integrity, windshield retention, and zone intrusion.

In combining the two italicized sentences, we could subordinate the more detailed second sentence to the more general first one:

> These standards are derived from an established crash test, in which the car is propelled against a solid wall at 30 mph.

Alternatively, we could maintain prominence on the details and subordinate instead the idea that the crash test is an established one:

> These standards are derived from propelling the car against a solid wall at 30 mph, which is an established crash test.

Clearly, the first option is the more appropriate one in this context: the fact that the crash test is an established one underscores the main idea of the paragraph, as stated in the topic sentence.

REVISED VERSION
Electric cars must be able to meet the same safety standards that gasoline cars must meet as set up by the Department of Transportation. These standards are derived from an established crash test, in which the car is propelled against a solid wall at 30 mph. The data obtained from the crash test are analyzed for fuel spillage, fuel system integrity, windshield retention, and zone intrusion.

There are times when it is best *not* to combine sentences. For example, if you are giving a list of instructions and want to emphasize independent steps in accordance with how the user might carry out the instructions, you might want to state these steps in independent sentences. To see how this might apply in a specific case, consider the following set of instructions for replacing a brake line in an automobile:

1   Disconnect the union nuts at both ends.

2   Unclip the line from the chassis.

3   Pull the line out.

4   Install the new line in the chassis clips.

5   Moisten the ends in brake fluid.

6   Tighten the union nuts.

You could leave this set of instructions as is, in the form of a formatted list. Or you could combine some of the steps (2 with 3, 5 with 6) to create a more realistic four-step sequence of disconnect-remove-install-reconnect, as is done in this excerpt from a repair manual:

> To replace a brake line, disconnect the union nuts at both ends. Unclip the line from the chassis and pull it out. Install the new line in the chassis clips. Moisten the ends in brake fluid, then tighten the union nuts.[1]

To combine sentences beyond this, however, would be a mistake because it would destroy the emphasis we want to maintain on certain individual steps. For example, if we were to combine sentences 2 and 3 in the repair manual version, this would be the result:

NEGATIVE EXAMPLE
To replace a brake line, disconnect the union nuts at both ends. Unclip the line from the chassis, pull it out, and install the new line in the chassis clips. Moisten the ends in brake fluid, then tighten the union nuts.

By lumping together the remove and install steps like this *(Unclip the line from the chassis, pull it out, and install the new line in the chassis clips),* we would be creating an imbalance in the sequence: no mechanic would consider this to be a single step, as the form of the description implies.

It's also best not to combine sentences when the result would be too long a sentence. Suppose, for example, you have been writing a proposal for a computer-aided design system and have included this paragraph in your summary:

> The proposed system is required to alleviate the increase in demand. The system will do that by removing the burden of data entry from the present system, CADDS. This is accomplished by utilizing the microcomputer as a stand-alone data entry system. The microcomputer has all of the graphics and software capabilities required to implement this concept.

As it stands, this paragraph is a nicely written one, with an adequate topic statement, a clear general-to-specific pattern of development, and properly constructed sentences satisfying the given-new, light-heavy, and topical criteria discussed in Chapter 14. The result is a highly readable paragraph with appropriate emphasis on the main ideas and key words. If you were to combine the sentences into one, on the other hand, much of this emphasis would be destroyed:

NEGATIVE EXAMPLE
The proposed system is required to alleviate the increase in demand by utilizing the microcomputer as a stand-alone entry system with all the necessary graphics and software capabilities to remove the burden of data entry from the present system, CADDS.

This is a more economical version, no doubt, insofar as it contains 16 fewer words than the original. But is it more readable? Absolutely not! In fact, it's a perfect example of the kind of incomprehensible gobbledygook that so many readers of technical writing complain about. The lesson to be learned from this example, then, is this: do not combine sentences just for the sake of doing so; do it only when it serves a purpose.

EXERCISE 15-1   Each of the following paragraphs lacks flow and emphasis because the sentences are all short and choppy. Break up this choppiness by combining those sentences that are most closely related. Be sure to put main ideas in main clauses.

A   The Cleveland Engine Plant will begin making diesel engines in 1982. The diesel engines will be produced along with the gas engines already in production. However, there is only one area for machining pistons. Both gas pistons and diesel pistons must be machined there.

B   I recommend using the cart system to reduce the costly existing production layout for machining gas pistons and diesel pistons. The cart system is the least expensive of the three systems. Also, the cart system meets the desired production rate for all probable product mixes. The cart system is the most compact of the three systems and fits in the existing piston machining area. The cart system layout should be used in the piston machining area.

C   The maximum speeds of the electric cars are much lower than those of the gas cars. The maximum speeds of the electric cars are barely fast enough to maintain freeway speeds. This causes problems with utilization of electric cars on freeways, and it almost limits their use to residential streets. The maximum speed must be increased to give electric cars a better chance for utilization.

D   Operation of the ammonia process has been only marginally profitable. Deterioration of the hydrogen recovery unit to unacceptably low efficiency has contributed to the high operating costs of this process. Process Engineering is preparing recommendations for replacing the hydrogen process unit with a similar unit. However, an alternative process is available. The alternative is to recover argon from the synthesis loop purge gas. On my own initiative, I have investigated the applicability of this process to the Des Plaines River Ammonia Plant. In this report, I present the results of my study of product outputs, utility requirements, and capital cost. I conclude that this process could improve considerably the ammonia process profitability. I seek approval from management to investigate this option in more detail.

15.2

USE INTENSIFIERS AND SIGNAL WORDS

All main points of a discussion should be supported, of course, by details; conversely, all details used should support main points. When this relationship between main points and details is established in an obvious way, emphasis is created. On the other hand, when it is left up to the reader to figure out how the details and main points are related, emphasis is lost. This means that you should take pains to show how details and main points are related. Make it obvious! If this means taking sides in an issue, so be it! In most situations, readers won't expect you to be totally neutral; indeed, they will want you to cast information in a certain light so as to more easily give it a single, unambiguous interpretation. This does not mean that you should forsake your objectivity. Not at all. It means simply that once you have used your objectivity to reach a certain conclusion, you should then make every effort to orient your writing toward that conclusion. Don't make the mistake of doing what many inexperienced writers do, just presenting detail after detail in a flat, colorless, pseudoobjective way and forcing the reader to see what purpose these details are supposed to have; you will only be wasting the reader's time and risking misinterpretation as well. Instead, take care to present details in such a way that the reader can easily see how they support your main points.

How can you do this? In addition to a number of techniques already discussed—such as using good topic statements, putting topical words in the subject position, and combining sentences—you can use appropriate signal words and phrases at the beginning of sentences: conjunctive adverbs like *specifically, in general, conversely, furthermore, on the other hand, however, therefore* and subordinating conjunctions like *since, because, although, despite.* (Many of these have already been discussed in connection with the paragraph patterns described in Chapter 2.) You can also use short adverbial phrases like *for this purpose, with this in mind, before taking any further steps, in order to determine the market for this product* at the beginning of sentences.

To see the effect of using signal words and phrases, consider first this negative example:

NEGATIVE EXAMPLE

The more-complete-expansion cycle is more efficient than the conventional diesel cycle. It uses the same four strokes but a relatively smaller combustion chamber. The intake valve closes when the piston is at some point A between TDC and BDC in its intake stroke. This draws in less air-fuel mixture. The compression, power, and exhaust strokes are identical to those of the diesel cycle. The increase in volume of the air-fuel mixture due to the piston moving to BDC is the same. The temperature and pressure of the gases after the power stroke is over are lower. More of their energy has been utilized. This increases the efficiency of the cycle. The smaller combustion chamber is necessary. The temperature and pressure at TDC in the compression stroke must be high enough to cause combustion of the mixture. The mixture must be squeezed into a smaller volume to get the same temperature and pressure.

It's pretty hard to make sense of this, isn't it? Certainly there's little emphasis on main points. Notice what happens, however, when we insert a few conjunctive adverbs, subordinating conjunctions, and adverbial phrases:

REVISED VERSION
The more-complete-expansion cycle is more efficient than the conventional diesel cycle. It uses the same four strokes but a relatively smaller combustion chamber. *In its intake stroke,* the intake valve closes when the piston is at some point A between TDC and BDC, *thus* drawing in less air-fuel mixture. The compression, power, and exhaust strokes are identical to those of the diesel cycle. *However, since there was less mixture to start with and since* the increase in volume due to the piston moving to BDC is the same, the temperature and pressure of the gases after the power stroke is over are lower. More of their energy has been utilized, *thus* increasing the efficiency of the cycle. The smaller combustion chamber is necessary *because* the temperature and pressure at TDC in the compression stroke must be high enough to cause combustion of the mixture. *Since there is less mixture in the chamber,* the mixture must be squeezed into a smaller volume to get the same temperature and pressure.

Clearly, this rewritten version is far more comprehensible than the original. The signal words and phrases help us see the logic of the paragraph—how the sentences work together to explain how the more-complete-expansion cycle is more efficient than the diesel cycle.

Another way you can highlight details and support main points is to use intensifiers, that is, words that amplify the meaning or connotation of other words. For example, if you say that the loss of certain equipment would impede progress on a certain construction project, that's one thing; if you say that it would *greatly* impede progress, that's quite another. The word *greatly* is an intensifier: it intensifies the meaning of an accompanying verb (in this case, *impede*) and thus emphasizes that meaning. Some other commonly used intensifiers in scientific and technical writing are *significant, significantly, considerable, considerably, obvious, obviously, easy, easily, much, many, complete, completely,* and *very.*

The use of intensifiers is particularly important when you are presenting numerical quantities whose significance is not immediately apparent to the reader. Don't just say "The present worth of the hydrogen process is $4.14 million" and expect the reader to know how to evaluate that figure. Is $4.14 million a lot for such a process, in the circumstances given, or isn't it? Many readers wouldn't know. If you insert the word *only,* however, you can give the reader a big assist: "The present worth of the hydrogen process is *only* $4.14 million." This immediately indicates that $4.14 million is a relatively small figure for that set of circumstances.

Intensifiers should be used even when the significance of a numerical quantity is fairly apparent—simply for emphasis. For example, instead of writing,

> The present worth of the argon process is $10.25 million. The present worth of the hydrogen process is $4.14 million.

it would be more effective to write

> The present worth of the argon process is $10.25 million whereas that of the hydrogen process is *only* $4.14 million.

This use of *only* may not be necessary for clarity, but it provides emphasis and thus underscores the fact that the argon process is more valuable than the hydrogen process. It is often bits and pieces of emphasis like this that add up and have a significant cumulative impact on the readability of a passage as a whole.

Some other intensifiers that can be used to emphasize numerical values are *more than, at least, almost, less than, not even, barely more than, well under,* and *at most.* These usually put a positive or negative slant on values and so should be used judiciously, that is, only in cases where the reader expects you to take more of an advocate's role, as when making a proposal or making recommendations. Here is an example, taken from a report making recommendations to meet EPA clean air requirements:

> Repairing the flyash returns is the optimal solution because of economic and downtime considerations. The three return units can *easily* be repaired by three of our employees with a downtime of *no more than* two days and a cost of *less than* $1,375, including labor. To have the units replaced, *on the other hand,* would take *at least* two months' downtime and, because of our limited staff, would have to be done by outside contractors at a cost of $70,000 *or more.*

Leaving these intensifiers out would not change the meaning of the paragraph significantly but would definitely reduce its effectiveness:

> Repairing the flyash returns is the optimal solution because of economic and downtime considerations. The three return units can be repaired by three of our employees with a downtime of two days and a cost of $1,375, including labor. To have the units replaced would take two months' downtime and, because of our limited staff, would have to be done by outside contractors at a cost of $70,000.

There are many situations, of course, where facts and figures are expected to be given in a neutral, unbiased fashion, e.g., in reporting test results to other specialists. In such cases, it is presumed that the readers are very knowledgeable about the subject matter and can easily interpret the data themselves. There are many other situations, however, where the readers are not so knowledgeable and need a helping hand. Although you should never use intensifiers to create a false impression, you should use them to help readers gain a true impression.

EXERCISE 15-2   Pretend that you are the fleet manager of a taxicab company. Your profit margin was much lower than expected last year because your costs soared while revenues remained about the same. In light of this problem, the company president several months ago had you purchase and field test several new front-wheel-drive automobiles. You kept careful records of each car's performance, as summarized in the following table:

CAR	FUEL COST PER MILE	TURNING CIRCLE (FEET)	EPA VOLUME INDEX*	MAINT. COST PER MILE	PURCHASE PRICE
Old fleet average	10.0¢	41.6	150	3.6¢	$6800
Buick Skylark	6.8	33.5	135	3.3	7600
Plymouth Horizon	6.4	32.8	121	3.8	7200
Volkswagen Dasher	5.9	33.2	129	3.3	9100

*The EPA volume index is a measure of relative roominess and cargo capacity.

Write the Summary section of a short informal report to the company president describing these test results and recommending that the company replace its present fleet with one of these new models. Be sure to use signal words and intensifiers to emphasize your point of view.

EXERCISE 15-3   The following samples of technical writing fail to give proper emphasis to important points. Correct this deficiency by providing topic statements or summarizing statements, by combining sentences, by inserting intensifiers or signal words, by eliminating unnecessary words, and/or by making other appropriate changes.

A   The fuel cost for the electric car is much lower than that for the gas car. When the price of battery replacement is added, the fuel cost rises considerably. For an average distance of 10,000 miles/yr, the total fuel cost of the electric car is about 33% greater than that of the gas car. This amount may seem to be significant, but gas prices are rising faster than electricity prices. Also, the price of battery replacement should go down slightly as batteries improve. The total fuel cost poses a problem for utilization at the present time, but it could be alleviated in the near future.

B   *Labor Cost*   The same number of workers is not used for each layout, but each layout uses six machine operators. Four extra workers are needed to operate the forklifts in the process layout. The workers make $10 per hour. Based on a 40-hour work week, 50 weeks a year, the extra workers cost $20,000 per worker per year. No extra workers are needed in the product layout, but its production rate is well below that needed to meet forecasted demand. More worker-hours will be needed to meet production. This would require a second shift or overtime to be used. Overtime is less costly than running a second shift to meet required demand (see Appendix B for the calculations). There is no extra labor cost in the cart system since it is fully automated. The labor costs associated with each layout are given below. All three include the labor cost of the six machine operators.

LAYOUT	LABOR COST
Product Layout	$228,000
Cart System	120,000
Process Layout	200,000

15.3

BE CONCISE

Be concise. While the more important words and phrases of a text should be highlighted, the less important ones should be subordinated—or perhaps even eliminated altogether. Unnecessary words and phrases will only detract from the emphasis you've carefully tried to build up through the use of combined sentences, signal words, and intensifiers. A bloated, wordy style can submerge your readers in a sea of empty terms, making it next to impossible for them to follow your main points and be persuaded to your point of view. In fact, foggy language is more likely than not to turn readers against you. Recall the MacIntosh survey cited in Chapter 1: of the 182 company executives, senior scientists, project leaders, and other prominent scientific and technical people who were asked to evaluate the writing they read, all 182 complained about "generally foggy language." More specific complaints were "failure to connect information to the point at issue" (169 respondents), "wordiness" (164), and "lack of stressing important points" (163).

Inexperienced writers sometimes think that they *must* use a wordy, bloated style of writing in order to create a certain professional image. They seem to believe that by using pretentious language they will enhance their image as experts in their field. Actually, what evidence there is suggests just the opposite: pretentious, wordy language is less likely to promote one's credibility as an expert than is concise, direct, simple language. For example, in a recent survey[2] taken at a conference of the British Ecological Society, 74 scientists were asked to read two versions of the same information, one version (attributed to a person named "Brown") being noticeably wordier than the other (by a person named "Smith"):

BROWN'S VERSION

In the first experiment of the series using mice it was discovered that total removal of the adrenal glands effects reduction of aggressiveness and that aggressiveness in adrenalectomised mice is restorable to the level of intact mice by treatment with corticosterone. These results point to the indispensability of the adrenals for the full expression of aggression. Nevertheless, since adrenalectomy is followed by an increase in the release of adrenocorticotrophic hormone (ACTH), and since ACTH has been reported (P. Brain, 1972) to decrease the aggressiveness of intact mice, it is possible that the effects of adrenalectomy on aggressiveness are a function of the concurrent increased levels of ACTH. However, high levels of ACTH, in addition to causing increases in glucocorticoids (which possibly accounts for the depression of aggression in intact mice by ACTH), also result in decreased androgen levels. In view of the fact that animals with low androgen levels are characterized by decreased aggressiveness the possibility exists that adrenalectomy, rather than affecting aggression directly, has the effect of reducing aggressiveness by producing an ACTH-mediated condition of decreased androgen levels.[2]

SMITH'S VERSION

The first experiment in our series with mice showed that total removal of the adrenal glands reduces aggressiveness. Moreover, when treated with corticosterone, mice that had their adrenals taken out became as aggressive as intact animals again. These findings suggest that the adrenals are necessary for animals to show full aggressiveness.

But removal of the adrenals raises the levels of adrenocorticotrophic hormone (ACTH), and P. Brain found that ACTH lowers the aggressiveness of intact mice. Thus the reduction of aggressiveness after this operation might be due to the higher levels of ACTH which accompany it.

However, high levels of ACTH have two effects. First, the levels of glucocorticoids rise, which might account for P. Brain's results. Second, the levels of androgen fall. Since animals with low levels of androgen are less aggressive, it is possible that removal of the adrenals reduces aggressiveness only indirectly: by raising the levels of ACTH it causes androgen levels to drop.[2]

Not surprisingly, the vast majority of these scientists (88 percent) found Smith's version easier to read. They also found Smith's style "more appropriate" (58 percent versus 8 percent for Brown, with 33 percent finding both styles equally appropriate). What is most significant, however, is that those who were willing to venture an opinion about the authors' probable scientific competence on the basis of their writing styles clearly favored Smith, as can be seen in Table 15-1. A larger sample, with 338 other respondents, was taken elsewhere, with similar results.

TABLE 15-1   RESPONSES FROM 74 SCIENTISTS JUDGING
THE SMITH AND BROWN ABSTRACTS

	YES, SMITH (%)	YES, BROWN (%)	NO DIFFERENCE (%)
Does one author give the impression of being more competent?	31	12	57
Does one author seem to have a better organized mind?	73	3	23
Does one author seem to be more objective?	24	8	68
Does one author inspire more confidence in what is being said?	44	15	41
Does one author seem to have a more dynamic personality?	46	7	47

SOURCE: Turk, Reference 2.

The more concise style of Smith's abstract (155 words versus 179 for Brown's) was definitely preferred by these scientists, to the point even of crediting Smith with probably being a better scientist. What is it, aside from the better formatting, that makes Smith's style so appealing? For one thing, it's not so "noun-heavy": it *has a higher percentage of verbs and adjectives* than does Brown's style. For example, instead of saying *effects reduction of,* it simply says *reduces.* Instead of *point to the indispensability of the adrenals,* it has *suggest that the adrenals are necessary.* Instead of *producing . . . a condition of decreased androgen levels,* it has *causes androgen levels to drop.* Second, the Smith style *has simpler sentence structure,* with fewer and shorter adverbial phrases before the sentence subject. This means that the reader reaches the main verb of the sentence sooner, making it easier to process the sentence as a whole (see Chapter 14 for a detailed explanation). For example, in the Brown version, the third sentence (beginning with *Nevertheless*) has 32 words before the main verb, and the sentence immediately following that one has 26 before the verb! In fact, the average number of words preceding the main verb in the Brown version is 17.6, versus only 6.8 for the Smith abstract. Thirdly, the Smith style *avoids unnecessary technical terms* in favor of more commonplace equivalents, even when it requires more words to make the substitution. In place of *adrenalectomised mice,* for example, Smith has *mice that had their adrenals taken out;* instead of *are a function of,* there is *are due to.* Finally, Smith's style *uses more pronouns and demonstrative adjectives: their* in sentence 2, *these* in sentence 3, *this* and *it* in sentence 5, and *it* in the last part of sentence 9. By contrast, Brown's style has only the one demonstrative *These,* leading off sentence 2. Pronouns and demonstrative adjectives, in general, help make a text more cohesive—provided, of course, that it's clear to the reader what they refer to.

This last point deserves some discussion before we end this chapter. Scientists, engineers, and other technical people sometimes use full noun phrases repeatedly to avoid being "imprecise." They've heard of cases, perhaps, where a single misinterpretation of a pronoun by a single reader has led to some accident or mishap, which in turn has led to the writer's company being sued for damages. Therefore they tend to avoid pronouns and demonstratives altogether, preferring instead to repeat full noun phrases over and over. This strategy is certainly a safe one, and indeed it should be used in appropriate circumstances (such as when writing operating instructions for a potentially hazardous machine or when writing a legally binding contract). There are many circumstances, however, where such caution is uncalled for, and where in fact it simply disrupts the coherence of the text. Consider this example:

NEGATIVE EXAMPLE
In order to keep from delaying the construction phase of the Parkway Office Building, the Geotechnical Division needs to know the loads that will be placed upon the footings. I have investigated the proposed use of the structure and various flooring systems to determine the loads that will be placed upon the footings. This report gives the loads on the footings and explains how these loads were derived.

There is no reason to *describe* the loads every time they're *referred* to! Pronouns and demonstratives can be used instead without any real risk of misinterpretation, and the result will be a more coherent and more concise text:

REVISED VERSION
In order to keep from delaying the construction phase of the Parkway Office Building, the Geotechnical Division needs to know the loads that will be placed upon the footings. I have investigated the proposed use of the structure and various flooring systems to determine these loads. This report gives the loads and explains how they were derived.

In general, when you have to refer repeatedly to some object or concept that has first been introduced with a long noun phrase, you can usually use a shortened version of this noun phrase and a demonstrative adjective or definite article without much, if any, risk of ambiguity. Here is a typical example:

Damper position indicators shall be provided on all damper units. *These indicators* shall be located a minimum of 1'-0" from the breeching to allow for installation of thermal insulation.

It would also be safe to substitute *The indicators* or perhaps even the pronoun *They* in this context. (Note, however, that the demonstrative pronoun *These* by itself could be misinterpreted as referring to the damper units.) It would not be necessary to repeat the full noun phrase *damper position indicators*.

Sometimes, however, it *is* necessary to repeat full noun phrases to assure proper interpretation. Consider this example:

NEGATIVE EXAMPLE

The two dampers installed in the flues shall be tied with the inlet dampers of the fans located at the exit side of the electrostatic precipitators. *These dampers* will open automatically in the event of any failure in the electrostatic precipitator system. Contractor shall provide proper instrument and control for the actuation of *these dampers.*

Now, what does *these dampers* refer to? The flue dampers? The inlet dampers? Or both? In cases like this, the writer has little choice but to repeat the full NP. In fact, it is often important in such cases for the writer to repeat the full NP *with exactly the same wording,* especially if the paragraph is long and complex. Look at the following badly written example and see if you can keep track of the different orders involved:

NEGATIVE EXAMPLE

On July 8, 1981, I was given the assignment of determining a method to reduce the problems associated with *lost unit down orders* and *special orders.* The misplacement of *special orders* in the shipping and receiving area has caused many customers to become unsatisfied with our services. This situation has also caused our partsmen to spend many hours each day locating *lost orders.* These problems have arisen from the increase in the number of *special orders* awaiting to be picked up in our Shipping and Receiving Department. Currently, there are between 200 and 300 *orders* awaiting customer pickup. *These parts* are stored on shelves, on the Shipping and Receiving room floor, and outside of the building. The Shipping and Receiving Department has no records at this time concerning which *special orders* have arrived. To correct this situation a system is needed which reduces employee work. This system must also minimize the number of *lost orders* at a low cost.

Do you know for sure what the term *lost orders* refers to in the third sentence? (Is it just lost *special* orders or is it *all* lost orders?) How about the *orders* in sentence 5? Or *These parts* in sentence 6? We didn't when we first read it, and we still don't.

EXERCISE 15-4   The verb phrases in column A are representative of those that characterize a wordy, noun-heavy style of writing. Use column B to provide one-word equivalents. (The first two have been done for you.)

A	B
effect a reduction in	reduce
accomplish a modification of	modify
put emphasis on	_____
come to the conclusion that	_____
provide with information	_____
increase by a factor of two	_____
give an explanation of	_____
have a deleterious effect upon	_____
create an improvement in	_____
do an analysis of	_____
make a recommendation that	_____
conduct an investigation of	_____

EXERCISE 15-5   The writing samples given below can all be made much more concise. Rewrite them, eliminating all unnecessary words.

A   At this point in time we are engaged in a re-evaluation of the budget.

B   The design that is recommended for adoption would accomplish the removal of particulates with the desired degree of efficiency.

C   According to our estimates, the cost of the system would be on the order of approximately $1.2 million.

D   It is the purpose of this report to provide you with information about the current status of the proteolipid experiment.

E   In our judgment, the optimal course of action would be to maintain continued use of the double-alkali reactant.

F   My telephone number is 221-3600 if you would like to contact me about any questions you might have pertaining to the contents of this report.

G   At our meeting of 9/4/82 we had a discussion about the fact that people have been making numerous complaints concerning vertical movement in the Hudson wing. The problem at hand was the excessive wait time people have been experiencing while waiting for the Hudson elevator.

H   The causes for this change in consumption were then determined. The nature of the causes indicated the most probable recommendations. If they were easily corrected, it was suggested that the corrections be made. If, however, the problem was unavoidable by using the existing units, new units were designed.

I   For electrical generation it is desirable to have a windmill that has a rotational speed that is higher than that of the wind. The ratio of the rotational speed of the windmill to the wind speed is called the speed-tip ratio. High speed-tip ratios are therefore desired for windmills that are to be used for electrical generation.

J   Industry has shown interest in the electric car through research. This research has revealed that electric car use is possible. Industry will also be needed to produce the electric car. It is of necessity that industry, the present auto companies, be interested in producing electric cars because they are the ones who have the capabilities. With the auto companies showing interest, electric car usage is possible.

K   In order for electric cars to become of practical use in the United States, the government must have a direct hand in encouraging electric car research and development. Making it attractive to industry to produce electric cars is of grave importance. Thus, the government and industry becoming involved would be a major step forward in the production of electric cars.

In order for electric cars to operate in the United States, they must be able to pass safety standards set up by the Department of Transportation. It is very important that these standards be met not only from the standpoint of safety but also for consumer acceptance. Consumers are not receptive to anything that will put them in danger. The electric car must be safe because consumer acceptance will be hard to obtain, and if the car is not safe, it will be impossible.

In order for electric cars to be used in the United States, it must be shown that they have satisfactory performance and costs as compared to a gasoline car. If the electric car cannot perform up to the standards of a consumer's present gasoline car, the consumer will have no use for the electric car. The costs for electric cars must be competitive with gas cars in order to fit the consumer's budget.

REFERENCES

1   *Volkswagen Service-Repair Handbook,* Clymer, Los Angeles, 1971, p. 235.

2   Christopher Turk, "Do You Write Impressively?" *Bulletin of the British Ecological Society* **9**(3):5–10 (1978).

ADDITIONAL READING

Jack Selzer, "What Constitutes a 'Readable' Technical Style?" *New Essays in Technical and Scientific Communication: Theory, Research, and Criticism,* P. Anderson, C. Miller, and J. Brockmann (eds.), Baywood Press, Farmingdale, N.Y., 1982.

Joseph Williams, *Style: Ten Lessons in Clarity and Grace,* Scott, Foresman, Glenview, Ill., 1981.

16

PROOFREADING

The last step in preparing any piece of writing—be it a letter, a memo, a formal report, or whatever—is to proofread it for errors. Proofreading doesn't take much time, and it can really pay off in terms of conveying both accurate information and an image of credibility. Don't forget that readers will often judge you by the quality of your writing. If your writing is full of typographical errors, misspellings, and other kinds of easily noticeable mistakes, your readers may well judge you to be not only a careless writer but a careless person in general. This may lead them to question the quality of your technical work: the accuracy of your measurements, the preciseness of your calculations, the soundness of your judgment. In short, your professional image can be damaged if you allow your writing to circulate without first checking it over for errors.

Even more serious than errors of form are errors of substance: miscalculations, misrepresentations of data, misleading claims. Errors of this type may easily go unnoticed by your readers and thus may genuinely misinform them, sometimes to the point of having very serious consequences. Needless to say, such errors of substance must be eradicated from any piece of writing.

Given the fact that there are several different kinds of errors to look out for, it makes sense to divide the proofreading process into different phases so that you can "sweep" the manuscript, concentrating on one type of error at a time. We recommend the following sequence of steps: first, proofread for substantive errors; second, proofread for sequencing errors; finally, proofread for spelling errors.

SUBSTANTIVE ERRORS Are your calculations correct? Are your data correct? Have you plotted your curves accurately? Have you got the right dates, times, places, job titles, model numbers? Remember, this kind of information may be of critical importance to your readers—so make sure it's right!

SEQUENCING ERRORS There may be one or more numerical or alphabetical sequences in your manuscript: page numbering, references, illustrations, footnotes, report sections, lists, figures, steps in a procedure, etc. Make sure these sequences are all correct and complete.

MISSPELLINGS For many writers, this is where proofreading is particularly fruitful. The English spelling system is notoriously irregular, causing writers no end of trouble. Despite this, many readers—including many readers in supervisory positions—attach great importance to spelling, seeing it as a sign of how careful and even of how well educated the writer is. So don't take any chances! Correct all spelling errors!

Most spelling errors occur with ordinary words that can be found in any pocket-size dictionary: words like *foreword, receive,* and *vacuum.* So, as you proofread for spelling, keep a small dictionary near you and use it for any words you're not sure of. If you're using a word processor with a text editor, it may have a built-in dictionary and spelling program; you can save considerable time by using that. Another way to save time, of course, is to ask someone who's known to be a good speller to proofread your manuscript for you.

Although any of these stratagems will work for you, try not to become overly dependent on them; after all, you may someday have to write a report in a hurry and may not have a text editor or friend available to help you out, or the time to look up every other word in a dictionary. Also, no text editor or ordinary dictionary is likely to be of much help when it comes to technical terms or proper names. So try to learn the correct spelling of words as you encounter them, especially those words that you use often in your writing. Practice writing them down, so that you can see them in your own handwriting as well as in print. Make a list of words that give you particular trouble, and then test yourself on them from time to time.

EXERCISE 16-1   Below is a list of some of the most frequently misspelled words in scientific and technical writing. Study the list; then have someone dictate the words to you while you write them down. Repeat the exercise until you've spelled all the words correctly.

accidentally	commission
accommodate	definitely
achieved	dependent
acoustic	desirable
address	develop
alignment	development
allot	different
allotted	environment
analyze	exaggerate
apparent	exceed
appropriate	excel
argument	exhaust
auxiliary	existence
basically	feasibility
beneficial	February
calendar	fluoride
ceiling	foreign

foreword	parallel
government	personnel
irrelevant	precedent
judgment	probably
knowledgeable	quantity
laboratory	receive
likelihood	recommend
lose	refer
maintenance	reservoir
manageable	retrieve
mathematics	schedule
misspell	seize
necessary	separate
occasionally	therefore
occurred	truly
omit	unnecessary
omitted	vacuum

Many writers have trouble with homophones—words that have the same pronunciation but are spelled differently and have different meanings. Probably the most frequently misused homophones in English are *its* and *it's*. Are you sure which is which? (Answer: *its* is a possessive form; *it's* = *it is*.) It's especially important to learn the difference between homophones because a text editor generally cannot. That is to say, if a writer uses *it's* where *its* is appropriate, most word processor spelling editors will fail to notice the error because they don't take usage into account; they look only at the form of the word and see if that form is found in the dictionary. So, if you are using a word processor to check your spelling, don't expect it to pick up spelling mistakes when homophones are involved.

Here are some other homophones that often cause writers trouble:

affect   (The new policy will not affect us.)

effect   (We decided to effect a change.)
(The extra dosage had no effect.)

all ready   (They are all ready to begin.)

already   (She has already made her decision.)

complement   (The illustrations nicely complement the written text of the report.)

compliment   (Linda's supervisor complimented her for having devised such a workable system.)

foreword   (Many technical reports begin with a foreword.)

forward   (We look forward to your visit.)

lead   (That idea went over like a lead balloon.)

led   (past tense of the verb *to lead:* Our company has led the way in this field.)

plain   (Use plain English whenever possible.)

plane   (The two lines formed a plane angle.)

principal   (The principal feature of this approach is its simplicity.)

principle   (The basic principle in this design is the conversion of heat energy to work.)

stationary   (A land-based missile system would provide a stationary target for the enemy.)

stationery   (Please don't use company stationery except for official business.)

their   (We would like to try some of their products.)

there   (There are too many terminals already in place there.)

they're   (What they're trying to do is capture the market.)

weather   (We'll launch at 700 hours, provided the weather's OK.)

whether   (Whether we succeed or not probably depends on our marketing strategy.)

who's   (Who's there?)

whose   (Whose jacket is this?)

your   (Do you have your calculator handy?)

you're   (When you're ready to begin, give the signal.)

EXERCISE 16-2

A   Keep a list of any words you misspell and review them from time to time. After you've accumulated 20 of them, have someone test you as in Exercise 16-1.

B   Proofread the letter of application for employment shown as Figure 16-1 for misspellings and other errors.

FIGURE 16-1   Letter of application for employment, to be
proofread for misspellings and other errors.

225 Nugent Hall
Brookfield Polytechnic Institute
Ames, NY 12181
(518) 270-4391

Febuary 25, 1982
FEBRUARY

Mr. Bill Dahlen
Sperry Univac Semiconductor Divisions
U2X26, P.O. Box 3525
St. Paul, MN 55165

Dear Mr. Dahlen:

In response to your advertisement in IEEE Spectrum (August 1982), I am
writing to apply for the position of Bi-polar Developement Engineer. After considering your challenging requirements; I believe that my experience and educational
background would enable to make significant contributions to your expanding division. My backround includes:
backgdound   *Physics*   *BRookfield*

Employment by the Electron Physisc Laboratory, Brookfeld Polytechnic Institute. I participated in designing an innovative computer modeling process
of a MOS submicron devise. I was responsible for designing the numerical
analysis subroutines of the computer program.

Completion of a senior desing project in designing and fabricating bi-polar
device chips using state-of-the-art technics. I was also responsable for testing and characterizing the device chips.
these

Completion of several graduate courses in semi-conductor devices, inter-
grated circiuts and digital logic circuit design.
CiRCuits

I will recieve my B.S.E.E. from BPI in August 1982 and would appreciate
being considered for permanant employment starting therafter. I will call on you
next Wed. during your office hours to set up an interview at your convenience. If
you desire additional information, I can be reached at (518) 270-4381 during the
mornings.

Thank you for your kind consideration.

Yours truely,

Perry C. Culbert

Perry C. Culbert

APPENDIXES

A

PUNCTUATION

SENTENCE PUNCTUATION

Punctuation marks serve as useful "road signs" to a reader. They help the reader *see* the grammatical structure of a sentence and thus more easily grasp its meaning. Punctuation can also help the reader *see* the rhetorical relationships between sentence elements.

The most commonly used—and *misused*—punctuation marks in scientific and technical writing are:

The comma	,
The semicolon	;
The colon	:
Parentheses	( )
Dashes	— —

These punctuation marks should be studied and used effectively by every technical writer. Unfortunately, they often are not. Many inexperienced writers believe—mistakenly—that punctuation marks are supposed to indicate the sound pattern of a sentence, that is, the pauses one would make if reading the sentence aloud. It is true that vocal pauses and punctuation marks often coincide, but this is not always the case, and even where they do coincide, a vocal pause by itself is hardly a reliable guide as to which punctuation mark to use.

In short, punctuation marks do not indicate sound patterning (at least not directly); instead, they indicate *grammatical structure, meaning,* and *rhetorical relationships.*

Independent Clauses

Three marks of punctuation are used to show a close relationship between independent clauses:

1   Use a *comma* if the two independent clauses are coordinate to each other, and the second clause begins with a coordinating conjunction (*and, but, or, so,* etc.):

The pressure in a gas or in a liquid is different at points that differ in height, and properties that depend on the pressure are also different.

Computer-linked scanning devices have been developed for bubble chamber films, but they cannot function without human helpers.

2   Use a *semicolon* if the two independent clauses are coordinate to each other, but there is no coordinating conjunction:

The structure of a simple liquid such as argon or methane is determined primarily by the repulsive forces between molecules; the attractive forces act only as a "glue" that holds the fluid together.

There are many geometries that are different from Euclidean; however, only the elliptic and hyperbolic geometries are called *non*-Euclidean.

3   Use a *colon* if the second clause elaborates on or explains the first:

All chemical equations should be balanced: they should show the same number of atoms of a given kind and the same net charge on each side.

The concept of randomness can best be explained by the use of examples: the sequence of heads and tails in tossing a perfect coin is said to be random.

EXAMPLE

Technology never exists in a social vacuum: it is embodied in products, processes, and people. By the same token, technology does not "transfer," or circulate, in a social void; it circulates throughout diverse institutional channels and mechanisms. . . . Thus social values cannot be separated either from technology itself, or from its mode of transfer. When these values, which are usually determined by the "parent" culture, conflict with those of developing countries, the technology transferred can deter—rather than promote—those countries' social objectives

    The most significant sources of technology transfer are consulting firms and the transnational corporations (TNCs) for manufacturing, extraction, and service. TNCs disseminate product and process technologies. But

STATEMENT FOLLOWED BY EXPLANATION

STATEMENT FOLLOWED BY CONTRASTING STATEMENT

"decisional technologies" exist as well, and these are vital: ⎤ CLOSELY RELATED
they comprise the know-how required to diagnose ⎬ COORDINATE CLAUSES
complex problems and formulate solutions. As Argentine ⎦
physicist Jorge Sabato puts it: "The ability to conduct a ⎤ STATEMENT FOLLOWED
feasibility study with its own means is the touchstone ⎬ BY EXPLANATION
revealing when a country has conquered technological ⎦
autonomy."[1]

Introductory Subordinate Elements

Use a comma to separate introductory subordinate elements from the rest of the sentence (and thus make it easier for the reader to locate the sentence subject):

> Since most current plantings are transplants from the greenhouse and are several months old, they can tolerate stronger herbicides.

> However, the supply of rubber has so far been able to keep up with the slow increase in demand.

Parenthetical Information

Three different marks can be used to indicate parenthetical (nonessential) information:

1   Use *commas* in most cases:

> The so-called British system of weights and measures, which is now on its way out in Britain, has its roots in ancient history.

> Galileo originally intended to be a doctor, on his father's advice that there was much more money in healing ailments than in proving theorems.

2   Use *parentheses* for references, abbreviations, definitions, qualifications, and other subsidiary information that may be of interest only to certain readers:

> It has long been known that the vertebrate electroretinogram (ERG) in response to flashes of light consists of various components that have been ascribed to electrical sources located mainly in the receptor layer and in the inner nuclear layer of the retina.

> The nucleus of each atom is composed of protons and neutrons (except for the isotope of hydrogen of atomic weight 1, whose nucleus consists of a single proton).

3   Use *dashes* for emphasis (unless writing for conservative readers):

> Mixtures can be categorized as either homogeneous—uniform in properties throughout the sample—or heterogeneous.

> Many of the experimental methods of modern chemistry were developed originally by men and women regarded as physicists—for example, the methods of spectroscopy and those of structure determination by the diffraction of x-rays.

EXAMPLE

At the small end of the physical world, there is a realm in which it is difficult, *if not impossible,* to measure or manufacture anything. There, a stray cosmic ray can destroy a wire only a few atoms thick. There, the Heisenberg uncertainty principle prevents simultaneous measurement of both the velocity and position of a particle. There, the very act of examining a structure—*with the beam of an electron microscope, for example*—will degrade or destroy it.

} PARENTHETICAL INFORMATION

} PARENTHETICAL INFORMATION, EMPHASIZED

It is a realm too fascinating—*and promising*—to resist. When engineers explore it, they find a wonderland of faster switches, more efficient circuits, higher reliability, and lower costs. As scientists work on scales of atomic diameters and angstroms (*an angstrom equals a ten-billionth, 10^{-10}, of a meter*), they are discovering new and unexplored phenomena that increase their understanding of the physical world.[2] [italics added]

} PARENTHETICAL INFORMATION, EMPHASIZED

} PARENTHETICAL SUBSIDIARY INFORMATION

Lists

Unformatted lists—those not separated from the text—are set off either by a colon or by dashes or parentheses:

1   Use a *colon* if the list ends the sentence:

> The new technology is most widely known for its microelectronics: pocket calculators, microcomputers, and other products that have revolutionized information processing and communication.

> To produce 100,000 Btu's, one needs the following quantities of various fuels: $2\frac{1}{2}$ lb of natural gas, or 5 lb of petroleum, or 8 lb of coal, or 14 lb of wood.

2   Use *dashes* or *parentheses* if the list does not end the sentence:

> Every aspect of the peptide neurotransmitters—how they are made, what they do, where and how they act, how they are disposed

of, and even whether they are true transmitters or serve some other function—is being studied with great intensity.

The charge of an electron has been shown to occur in integral multiples (1, 2, 3, 4, etc.) of its basic value; therefore magnetic charges should also occur in integral multiples of some basic value.

To separate the items of a list:

1 Use *commas* in most cases:

The three parallel lines are as follows: $y = y_1$, $y = \frac{1}{2}(y_1 + y_2)$, and $y = y_2$.

There are many words in our language that indicate sets: *school* of fish, *swarm* of bees, *herd* of cattle, *squadron* of planes, etc.

2 Use *semicolons* if the items are long or are punctuated internally:

We have assembly plants in the following locations: Cedar Rapids, Iowa; Totowa, New Jersey; Edmonds, Washington; and Castle Rock, Colorado.

An efficient project cost system must accomplish three important objectives: (1) check actual and predicted costs of ongoing projects against the estimated cost; (2) obtain production rates for use in estimating of new work and create historical files; and (3) forecast the project final cost.

In formatted, displayed lists, that is, those set off from the body of the text, punctuation is optional:

The positions available are

Recovery plant supervisor
Management analyst
Administrative aide
Clerk-typist

The positions available are the following:

Recovery plant supervisor,
Management analyst,
Administrative aide, and
Clerk-typist.

A.2

WORD AND PHRASE PUNCTUATION

At the word and phrase level, the most frequently misused punctuation marks are:

The apostrophe	,
The hyphen	-
Quotation marks	" "
Underlining	_____

These are most often misused with regard to the functions discussed below.

Possession

1   Add an *apostrophe* and an *s* to all nouns except plurals ending in *s*:

Roberta's watch	the company's policy
my assistant's desk	the people's concern
the children's toys	the men's room

[Exception: Personal pronouns (*it, they, you*, etc.) never take the apostrophe to show possession. The correct possessive forms of these words are *its, their, your, our, my, her,* and *his* when they are used as adjectives and *its, theirs, yours, ours, mine, hers,* and *his* when they are used as pronouns.]

In cases of joint possession, attach '*s* only to the last noun: *Dun and Bradstreet's projections, John and Mary's sailboat.* When referring to combined cases of individual possession, on the other hand, attach '*s* to each of the nouns: *John's and Mary's birthdates.*

2   Add just an *apostrophe* (no *s*) to plural nouns ending in *s*:

the workers' pension plan	our parents' friends
the Joneses' house	the neighbors' garden

Occasionally you may want to show possession with a singular noun which ends in *s* or with an *s* sound. In such cases, follow rule 1 above unless the result would be hard to pronounce:

Bass's new hiking boots   but   Sears' new hiking boots

Alice's restaurant   but   Moses' restaurant

Noun Compounds

Nonspecialist readers may have difficulty interpreting long noun compounds. There are two simple ways to alleviate such problems:

1   Use an occasional *preposition.* Scientific and technical English abounds with terms made up of long strings of nouns: terms like *diffusion convection mass transfer resistance* and *4Pb0 · PbSO$_4$ lead acid battery paste formation.* Although such noun compounds may cause no trouble for specialist readers, they can be very confusing to nonspecialists, especially if there are not enough clues in the immediate context to serve as a guide to interpretation. Therefore, avoid using such terms out of context (for example, in titles or abstracts); instead of *3B pump maintenance recall rate,* try *rate of recalls for maintenance of the 3B pump* or *maintenance recall rate for the 3B pump.* It may take a little more space, but it enables the nonspecialist reader to grasp the meaning of the term much more easily and quickly.

2   Use an occasional *hyphen.* Another way of carving up long compound nouns into more digestible pieces is to pair off closely related words by means of hyphens:

> Diffusion-convection mass-transfer resistance
>
> High-frequency semiconductor layers
>
> Automatic score-keeping feature
>
> Dow coal-liquefaction process
>
> High-energy laser-beam weapons
>
> Manual materials-handling tasks

This grouping of related words helps prevent momentary misinterpretations on the part of a reader who is not familiar with such terms, especially when adjectives are involved. Notice, for instance, how *manual materials handling tasks* without the hyphen could momentarily be misconstrued as tasks involving the handling of manual materials. By inserting a hyphen between *materials* and *handling,* the writer alerts the reader to the fact that the adjective *manual* modifies *materials-handling* or *tasks,* not *materials.*

End-of-Line Hyphenation

To divide a word at the end of a line, use a *hyphen.* If you have started to write a word at the end of a line but find you don't have room to complete it, you can hyphenate it and carry a part over to the beginning of the next line. Bear in mind, though, that this device exacts a slight toll on the reader: it briefly interrupts the smooth flow of words and forces the reader to guess at the meaning of the hyphenated word before seeing the second part of it. So, do not overuse this technique and, when you do use it, try to make it as easy as possible for the reader to guess the meaning of the full word after seeing only part of it. The following rules are designed for that purpose.

1   *Never divide a syllable.* Syllables are the building blocks of words. You can often guess the meaning of an unfamiliar word just by putting together the "meanings" of the syllables it is made up of. If you see the syllable *chem-* leading off a word, you can pretty well guess what the rest of the word might be, but if you see only half the syllable (*ch-*), you have no idea what follows: *chart, chapter, chiropractic, chlorine, choices, chest,* to mention but a few. So be sure to maintain the integrity of syllables when you divide a word. If you don't have time to check the dictionary (all good dictionaries show the syllable divisions of words), you can use these two rules of thumb: (1) all syllables have at least one vowel sound; (2) all syllables can be pronounced easily and, if necessary, can be shouted easily to someone at a distance. *Chem-* satisfies these conditions, but *ch-* doesn't.

INCORRECT	CORRECT
ma-rket	mar-ket
repo-rt	re-port
thr-ough	through
proje-ct	pro-ject
dupl-ic-ate	du-pli-cate
persp-e-ctive	per-spec-tive
phl-ebi-ti-s	phle-bi-tis

2   *If there is a natural break in a word, try to divide it there.* Many words divide naturally into two parts: *work-shop, self-disciplined, turn-key, break-water, anti-toxin, electro-magnetic, thermo-meter,* etc. By dividing the word at the end of the line according to its natural parts, you make it easier for the reader to recognize the two parts. Notice how funny it looks, for example, to see a word divided like this: *solid-state.* It's much more natural and easier to read divided as *solid-state.*

Titles of Written Works

Underline (or *italicize* if you have access to a typewriter with an italics font) the names of books, journals, magazines, monographs, and other publications that are bound separately:

J. Galbraith, <u>Designing Complex Organizations</u>, Addison-Wesley Publishing Company, Menlo Park, Calif., 1973.

D. O'Brien et al., "Responsibility for Inspection," *Journal of the American Concrete Institute,* Vol. 69, No. 6, June 1972, pp. 320–333.

Put *quotation marks* around the names of articles, reports, memos, and other pieces of writing that are either not published or not bound separately:

J. D. Borcherding, "An Exploratory Study of Attitudes That Affect Human Resources in Building and Industrial Construction," Department of Civil Engineering, Stanford University, Stanford, Calif., 1972.

W. Larkin and D. Burns, "Sentence Comprehension and Memory for Embedded Structures," *Memory and Cognition,* 5 (1977), pp. 17–22.

M. J. Zakkak, "Cost Control Simulation," thesis presented to the University of Texas, Austin, TX, 1976, in partial fulfillment of the requirements for the degree of Doctor of Philosophy.

Note: Long formal technical reports are often cited with the descriptive name in quotation marks but the report number underlined or in italics:

"PCS, Project Control System/360," <u>GH 20-0376-3</u>, International Business Machines Corp., 1971.

S. H. Wearne, "Contractual Responsibilities for the Design of Engineering Plants," *Report No. TMR5,* School of Technological Management, University of Bradford, England, 1975.

A.3

CAPITALIZATION

When referring to something by its official name, *capitalize* all words in the name other than articles, short prepositions, and conjunctions:

Personal names:	Phyllis Gomez, Lawrence Van Vlack
Personal titles:	Project Officer, Professor of Mechanical Engineering, M.D.
Places:	115 South Cache Street Jackson, Wyoming United States of America

Organizations:	Dupont Chemical Company, Arizona Girls Leadership Club, Brooklyn College
Days, months, eras:	Monday, Memorial Day, Christmas, October, the Twentieth Century, the Eighties
Programs:	the Professionals-in-Training Program, the Stanford University Degree Program in Physics
Natural languages:	French, Spanish, Chinese
Computer languages:	BASIC or Basic, FORTRAN or Fortran, ALGOL, Pascal
Model names:	the IBM Selectric, the Ford Mustang

REFERENCES

1   Denis Goulet, "The Dynamics of International Technology Flow," *Technology Review,* May 1978, p. 32.

2   William J. Cromie, "Room at the Bottom," *Mosaic,* May-June 1981, p. 25.

B

LETTERS

McNamee, Porter and Seeley

3131 SOUTH STATE STREET · ANN ARBOR, MICHIGAN 48104 · (313) 665-6000

Consulting Engineers

January 4, 1982

Mr. Donald P. Ziemke
City Manager
City Hall
7800 Shaver Road
Portage, MI 49081

Re: Portage-Kalamazoo Rate Negotiations

Dear Mr. Ziemke:

As requested by Mr. Williams we have attempted to develop a rate that would be a fair compromise between the $531 per million gallons ($0.14/$M^3$) based on the 1967 agreement and the $946/mg ($0.25/M^3) requested by the City of Kalamazoo.

The 1967 agreement does not reference interest costs, and the City of Kalamazoo would be justified in asking that the interest costs associated with borrowing money be passed on to the system's customers. The data needed to develop an accurate interest cost for Portage are not readily available; however, assigning all of Kalamazoo's interest cost for 1983 to the treatment plant and calculating Portage's share based on design capacity at the plant would give a 1983 cost of $101,393 and an interim cost of $75,031. This equates to an additional commodity charge of $75,031/494 mg (1,871,100 M^3) = $152/mg, or $0.04/$M^3$. This rate would be expected to increase because of increases in actual interest rates over the rates originally estimated by Kalamazoo. A reallocation of interest costs to both the sewer system and the WWTP may also cause a change in the rate. As mentioned previously, the information to calculate an exact allocation is not currently available.

Based on the numbers as presented, the commodity rate for Portage should be $531 + $152, or $683, per million gallons. The existing compromise rate of $700 reflects this cost plus a small adjustment for increased interest costs or differences resulting from the revised allocation figures.

In the interest of the City, this should be changed to an OM&R rate of $239 per mg plus a fixed monthly capital charge of approximately $18,000 and a demand charge of $760 per month.

Mr. Donald P. Ziemke
Page Two
January 4, 1982

    If you have any questions, or comments, please advise.

                       Very truly yours,
                       McNAMEE, PORTER AND SEELEY

BY _____
                       Raymond J. Smit

RJS:dr
Enclosure

McNamee, Porter and Seeley

3131 SOUTH STATE STREET · ANN ARBOR, MICHIGAN 48104 · (313) 665-6000

Consulting Engineers

January 18, 1982

Mayor and City Council
City of Portage
7800 Shaver Road
Portage, MI 49081

Attention: Mr. Donald P. Ziemke, City Manager

Dear Mayor and Members of Council:

In accordance with your request on January 12, 1982, bids were received for constructing new production wells at Portage. A total of five bids were received and are summarized on the attached tabulation.

The lowest and best bid is that of Leo Riegler Corporation of Muskegon, Michigan, at an estimated cost for two wells of $48,090. We recommend that a contract be awarded to Riegler at the unit prices shown in their bid proposal.

Please advise us of your wishes regarding this recommendation, and we will prepare the necessary contract documents for signatures of the contractor.

In addition to the work of Leo Riegler we would propose to retain the services of L. M. Miller and Associates to review the well drilling operations and pumping tests. L. M. Miller and Associates would be working as a subcontractor to us. Their charges will range from $5,000 to $6,000, and in addition we may expect to incur an additional $3,000 expense in our work in reviewing the progress of construction. Accordingly, we would appreciate receiving your purchase order for services at our hourly rate basis at a cost not to exceed $9,000.

Should you have any questions regarding this project, please call on us.

Very truly yours,
McNAMEE, PORTER AND SEELEY

By _____
Raymond J. Smit

RJS:bh
Enclosure
cc: Lynn Miller

McNamee, Porter and Seeley

3131 SOUTH STATE STREET · ANN ARBOR, MICHIGAN 48104 · (313) 665-6000

Consulting Engineers

February 5, 1982

Mr. G. Dallas Williams, P.E.
City of Portage
7800 Shaver Road
Portage, Michigan 49081

Re: Master Stormwater Management Report

Dear Mr. Williams:

Pursuant to our phone conversation of February 4, 1982, we would like to request a 60-day time extension for the completion date for the Master Stormwater Management Report. Therefore, we are requesting April 29, 1982, as the completion date for this project. We understand you will contact Marcy Brooks of the Southcentral Michigan Planning Council with this request and obtain her agency's approval for the time extension. Please contact us when our signature is required on any forms due to the time extension request.

We will be submitting a Final Phase 1 report at the end of February 1982 and are currently proceeding with the Phase 2 portion of the report.

If you have any questions, or if there is any problem with our time extension request, please contact us.

Very truly yours,
McNAMEE, PORTER AND SEELEY

By _____
Raymond J. Smit
Partner

By _____
John S. Wise
Project Engineer

RJS:JSW:dr
cc: Lynn Miller

McNamee, Porter and Seeley

3131 SOUTH STATE STREET · ANN ARBOR, MICHIGAN 48104 · (313) 665-6000

Consulting Engineers

February 1, 1982

Newkirk Electric Associates, Inc.
1875 Roberts Street
Muskegon, MI 49442

Attn: Mr. Robert Mudget

      RE: City of Portage
         Improvement to Pump Controls

Dear Mr. Mudget:

Mr. Robert Kimmer has asked that MP&S come to the City of Portage job site and help solve the problems remaining with our Portage contract.

We have agreed that David Matthews from our office will be there at 9:00 a.m. Tuesday, February 9. Mr. Matthews called your office on January 29, 1982, and left word of this meeting with Joe Breckring. Mr. Matthews offered to help you solve the remaining problems and wanted to know if the above date was mutually acceptable. Mr. Matthews also suggested that certain parts and test equipment be brought to analyze and solve the problems in a methodical manner.

You called back at 4:30 stating you would not go back to the job site without a purchase order. May we point out in the specifications page 46-2 under "Responsibility and Coordination" -- "It shall be the responsibility of the contractor to furnish a complete and fully operable system. The contractor shall be responsible for all details which may be necessary to properly install, adjust and place in operation the complete installation."

Also, under "Service and Start-up" page 46-5 -- "The supplier's representative shall revisit the job site for 8 hours per day as often as necessary after installation until all trouble is corrected and the equipment has passed the acceptance test and is operating satisfactorily to the engineer."

The acceptance test as defined on page 46-5 has not yet taken place. The engineer to this date has not been requested to witness the operation for acceptability.

We are offering in good faith to help you solve the problems that remain and request that your technical people come prepared to solve these problems on February 9.

Newkirk Electric Associates, Inc.
February 1, 1982
Page Two

If you do not cooperate to complete this work, we will be forced to invoke the liquidated damages clause and such other measures at our disposal. May we receive your favorable response to completing this project beginning February 9.

Very truly yours,
McNAMEE, PORTER AND SEELEY

By _____
Raymond J. Smit
Partner

By _____
David R. Matthews
Associate

DRM:RJS:bh

cc: City of Portage
    Dallas Williams
    Robert Kimmer

McNamee, Porter and Seeley

CONSULTING ENGINEERS
3131 South State Street ● Ann Arbor, Michigan 48104 ● Phone (313) 665-6000

To: Leo Riegler Corporation   Date: 2/11/82
     2121 Glade Street   Project: City of Portage
     Muskegon, MI 49444   Water Works Improvements

Attn: David Riegler   Production Wells

                                  Contract No.:   81-W-1

       Gentlemen: Agency No.: _____

            ☒ We are enclosing   ☐ We are sending under separate cover

The following:

       ☐ Drawings   ☐ Specifications   ☐ Change Orders

       ☐ Pay Estimate   ☒ Contract Documents   ☐ _____

5 copies of contract documents on the above referenced.

       For:

       ☐ Your approval   ☐ Your corrections   ☐ Your comments

       ☒ Your execution   ☐ Your file   ☐ Your signature

Remarks: PLEASE SIGN BUT DO NOT FILL IN THE DATE on the following as indicated:

Agreement page No. 22; execute page 23 if it pertains to you. Performance Bond Page No. 24. Payment Bond Page No. 26. DATES WILL BE FILLED IN BY THE OWNER. After signing all copies, please return all copies to this office for further processing. Also return with the signed documents the necessary insurance papers as called for on page 7 of INSTRUCTIONS TO BIDDERS AND ARTICLE 8, GENERAL STIPULATIONS, pages 17, 18 and 19.

CC: Transmittal only/Owner   By: Richard G. Walterhouse/ml
     File (2)

C

ACE MANUFACTURING COMPANY
23200 DEARBORN AVENUE
DETROIT, MICHIGAN 48227

October 9, 1981

M Consulting Services
G. G. Brown Laboratories
Ann Arbor, Michigan 48105

Gentlemen:

We are planning to install several high-speed punch presses in each of our new plant locations. Both groups of presses will be supported by block foundations which rest directly upon a thick layer of sand. At one of the sites the sand is white, and at the other it is black. Samples of each type of sand have been sent to you under separate cover.

Our engineers have obtained some data on the in situ characteristics of each sand deposit. The natural field void ratio of the white sand is 0.65, and the dry unit weight of the black sand in situ is 118 lb/ft^3.

Our concern is the possibility that these machines will settle because of the vibratory loadings transferred from the machines to the underlying sand. Please send us your expert opinion as to whether these foundations will or will not settle when located on the white or black sands.

Very truly yours,

J. C. Starr
Chief Engineer

JCS/dd

M. Consulting Services
G. G. Brown Laboratories
Ann Arbor, Michigan 48105

October 16, 1981

ACE Manufacturing Company
23200 Dearborn Avenue
Detroit, Michigan 48227

Attention: Mr. J. C. Starr, Chief Engineer

Re: Soil Settlement Tests at New Plant Locations

Dear Mr. Starr:

In your letter of October 9, 1981, you were concerned with the possibility of soil
settlement, through vibratory loadings, if high-speed punch presses are installed at
your new plant locations; you sent sand samples from these locations for us to
inspect. Accordingly, we have run tests on these samples to determine if indeed
settlement is possible. The purpose of this report is to inform you of the test results,
which lead us to recommend that your company forego the installation of the
presses at either of the new plant locations.

Summary

We performed three tests on the sands; a specific gravity test, a minimum void ratio
test, and a maximum void ratio test. The relative density, a measure of looseness
or compactness of soil, was calculated for each sand using these test results along
with the information your engineers supplied. The results indicate that soil settlement is very possible, and therefore we would not recommend the installation of
the punch presses.

Test Descriptions

We determined the specific gravity of the sands by taking the ratio of two weights;
the weight, in air, of a given volume of soil versus the weight, in air, of an equal
volume of distilled water. We determined the minimum and maximum void ratios
using local standardized tests; the "air pliviation" method and the "elephant trunk"
method, respectively. The minimum void ratio test allows the soil particles to take
up a minimum amount of volume. The maximum void ratio test does just the opposite. Detailed test descriptions and calculations of specific gravity, minimum and

maximum void ratio, and relative density can be found in the attached appendix. The results are presented here:

	Ottawa sand	Turkish emery
Specific gravity:	2.65	3.99
Minimum void ratio:	.53	.78
Maximum void ratio:	.71	1.15
Relative density:	33.3%	10.8%

Conclusion/Recommendation

Soils having a relative density greater than 70% are quite dense or compact and would settle very little. Our tests indicate that the relative densities of the sands tested are considerably below this and therefore settlement is quite possible. After carefully studying the data and taking into consideration the limitations of the tests, we would not recommend the installation of the presses at either of the new plant locations.

Very truly yours,

Timothy J. Burns, PE

Attachments:

Data calculations
Test descriptions

Ford Research Center
Kennedy Avenue
Dearborn, Michigan 48500

Date: November 1, 1981

To: Dr. Daniel Smith
             Director of the Electronic Circuit Design Group

From: John Leong
             Numerical Analysis Group Director

Subject: Proposal for the implementation of the Sparse Matrix Algorithm as
             an efficient subroutine for solving large systems of linear equations

Ref: Sparse Matrices and Their Applications, by Rose (Plenum)
             Numerical Methods, by Dahlquist (Prentice-Hall)

Dist: Nick Mente, Research and Development Director
             David Hill, Computer System Director
             Cathy Hope, Budget Director

You stated in your memo of October 25, 1981, that your designs often involve large systems of linear equations (SLE), but our present subroutine for solving them is very inefficient, especially if there are many zero entries in the SLE.
In order to increase the efficiency, we have found a mathematical algorithm which
is especially suitable for solving large SLE efficiently; we call it the Sparse Matrix
Algorithm (SMA). In this report, we introduce the new algorithm and propose that
it be implemented as a subroutine in our computer system.

Summary

The Sparse Matrix Algorithm is highly recommended to be implemented as
a system subroutine for solving large SLE. The SMA is efficient because it operates
on entries of the SLE matrix selectively. It requires less computer storage area
and less computer processing time. Tests were performed which showed that the
SMA is more efficient than the present Gaussian Elimination Algorithm. It can
be implemented in the computer system easily, and considerable computer
processing expenditure can be saved after its implementation.

Introduction

The present subroutine for solving large SLE uses the Gaussian Elimination
Algorithm. A brief description of the subroutine is shown in Appendix A, and a
detailed description is found in Computer System Memo 451. This algorithm is inefficient in solving large SLE because it operates on zero entries and results non-
selectively. However, the designs of the Electronic Circuit Design Group often
involve large SLE which have many zero entries. The inefficiency increases expo-
nentially with larger SLE. Consequently, a new algorithm is needed, and the SMA
is found to be efficient in the application.

Sparse Matrix Algorithm

The algorithm is efficient in solving large SLE which have many zero entries because it operates on entries selectively. It does not operate on zero entries and does not store zero results. These features lead to efficient usage of computer storage area and shorter duration of computer processing time:

1. Efficient usage of computer storage area --
    The algorithm neglects all zero entries and transforms the original matrix A of the large SLE into two smaller matrices, the compacted matrix $\overline{A}$ and the pointer matrix P. The $\overline{A}$ matrix stores all non-zero entries while the P matrix saves their original positions in the matrix A. The two smaller matrices require less storage area exponentially than the original matrix as the system increases in size. Moreover, the SMA does not store any zero results, which also saves storage area. (Appendix B)

2. Shorter duration of computer processing time --
    Since the SMA operates selectively, it has fewer operating steps to be performed in solving the large SLE. This algorithm requires considerably less computer processing time.

A brief description of the SMA is shown in Appendix C, and a detailed description is found in the first suggested reference (pp. 1 - 40).

Testing

Simple computer simulated tests were done to compare the Gaussian Elimination Algorithm and the Sparse Matrix Algorithm. Besides needing less computer processing time, the SMA also needs less computer storage area in solving large SLE (Appendix D). A detailed testing report is found in Numerical Analysis Memo 815.

Implementation

The SMA can be implemented as a subroutine in our computer system by developing one of our own with the assistance of our Computer System Group. Moreover, the subroutine can be bought from HP, IBM or Honeywell for a reasonable price (Appendix E). In any case, we can have higher efficiency and considerably less computer processing expenditure because more than fifty percent of the designs of the Electronic Circuit Design Group involve large systems of linear equations. We expend twenty dollars for each computer processing minute (Computer System Memo 15).

Conclusion

The Sparse Matrix Algorithm is efficient in solving large systems of linear equations. It can be implemented as a subroutine in our computer system without much difficulty. Its implementation is highly recommended. At the same time, the present Gaussian Elimination Subroutine should be retained as it is still useful in solving small systems of linear equations.

American
Electronics
Incorporated

To: Mr. John Nicol
            Chief Optoelectronics Engineer
            Optoelectronics Department

From: Khalil Najafi
            Assistant Optoelectronics Engineer
            Optoelectronics Department

Date: October 30, 1979

Subject: Liquid Crystal Displays: Analysis of Failure and Recommendations for
            Solution

Dist: Mr. Edward Jones, President
            American Electronics Incorporated

Foreword

Although becoming fairly well established as displays in small instruments,
Liquid Crystal Displays (LCDs) are still regarded with some skepticism by many
designers. Since American Electronics Incorporated (AEI) plans to use LCDs instead of Light Emitting Diodes (LEDs) for its future products, I was assigned by Mr.
John Nicol to investigate the LCDs modes of failure, to provide solutions to prevent
these failures, and to propose different kinds of tests to detect these modes of failure.

The following report describes the results of my investigation: it explains the
shortcomings of LCDs and proposes means for overcoming and detecting these.

Summary

Investigations show that LCDs can be used as good display devices if the
problems stated below are eliminated through the adoption of the solutions suggested. LCDs were found to have problems in two major areas:

1) problems arising as a result of the presence of moisture, oxygen, and
    a high temperature environment, which can cause very severe dam-
    age to the LCDs;

2) problems due to the loss of contact between the connector and the
    indium oxide leads on the LCD.

These problems can be avoided by following recommended procedures:

1) use of good insulating materials and of protective seals against mois-
    ture and temperature;

2) replacement of the current method of creating a contact between the
    connector and the indium oxide leads on the LCD by a display that can
    be soldered directly to the circuit board.

Finally, any occurrence of these problems can be detected by the following recommended tests:

a) test for tolerance against temperature,

b) test for moisture content, and

c) test for high electric fields.

All of these tests should be carried out by the quality control or inspection department.

Discussion

The discussion provides details on the following topics:

1) Failure due to temperature and moisture, and proposed prevention measures.

2) Failure due to loss of contact, and proposed prevention measures.

3) Recommended quality control procedures.

1. FAILURE DUE TO TEMPERATURE AND MOISTURE and Proposed Prevention Measures

One of the main problems facing LCDs is their strong sensitivity to temperature and moisture. These factors can change the behavior of LCDs rather drastically and result in gradual degradation of the device. Two basic categories can be recognized in this case: damages due to combination of high temperature and high humidity and damages due to high temperature only.

1.1 Damages Due to a Combination of Temperature and Moisture

The most damaging environment that an LCD is likely to experience is produced by the combination of high temperature and high humidity. Depending on how the display is constructed, three failure modes can occur under this condition: failure of the polarizers, failure of the adhesive on the polarizer and reflector, and, finally, failure of the liquid crystal mixture.

1.1.1 Failure of the Polarizers

The most common failure occurs with the polarizers. Regular polarizing material will begin to degrade after a short period of time (4 days) when subjected to a temperature of 50°C and 95% relative humidity. Once moisture penetrates the plastic backing of the polarizing material, it affects the polarizing properties. The digit area of the display will start to lose contrast, gradually turning from black to brown until it finally disappears completely.

1.1.1.1 Preventing Failure of the Polarizers

To provide protection against the effects of high temperature and humidity, the use of a polarizing material known as

"K-sheet," introduced by Polaroid Corporation, is recommended. This material has polarization properties which are relatively unaffected by moisture.

1.1.2 Failure of the Adhesive

Moisture can also attack the adhesive on the polarizer and reflector and cause it to begin peeling away from the glass. If the alignment is lost, dark patches will appear in the viewing area like those as already described.

1.1.2.1 Preventing Failure of the Adhesive

To prevent this kind of failure, the use of a truly hermetic seal is recommended to protect the adhesive.

1.1.3 Failure of the Liquid Crystal Mixture

If any of the components in the liquid crystal mixture (such as Schiff-base compounds) are attacked by moisture, the liquid crystal molecules will start to break down, thereby lowering the clearing point of the mixture and the effective upper operating temperature of the display. Due to the ionic by-products of the chemical reactions, the electrical currents of the display will also show a very significant increase.

1.1.3.1 Preventing Failure of the Liquid Crystal Mixture

The procedure described in part 1.1.2.1 should be adopted.

1.2 Failure Due to High Temperature Only

Failure can occur when an LCD is exposed to high temperatures only. The twisted nematic display requires the glass plates to be specially treated to impress a uniform alignment on the liquid crystal molecules next to the glass surface. The quality of this alignment can be degraded by high temperature because of decomposition of the interface layer.

1.2.1 Preventing Failure Due to High Temperature

The procedure used in part 1.1.2.1 should be adopted.

2. FAILURE DUE TO LOSS OF CONTACT

Problems can arise due to loss of contact between the connector and the indium oxide leads on the LCD. The user will often blame the LCD for an electrical failure such as an open segment when in fact it is the connector which is responsible.

2.1 Preventing Failure Due to Loss of Contact

This problem was solved by the introduction of the DIL displays with substrate chips by Liquid Crystal Displays in 1975. They offer a display that can be soldered directly in a circuit board to reduce the possibility of any disconnections.

3. <u>RECOMMENDED QUALITY CONTROL PROCEDURES</u>

It is recommended that AEI provide its quality control or inspection department with the equipment to measure the following characteristics or to make at least the following tests:

1. The electrical current through the cell at room temperature, AC square wave at approximately 32 Hz variable in range 5 to 12 volts to match the range cells. (Normal operating temperature range $-10°C$ to $+50°C$.)

2. The clearing point of the fluid, measured by raising the temperature until the viewing area turns black. This requires a dry-environment oven.

3. Electrical operation of the display at the specified upper operating temperature for 30 days, followed by tests (1) and (2) above to detect any changes.

4. Excursions of temperature outside the operating range. Both low and high temperatures less than $-40°C$ and more than $+50°C$.

5. Humidity testing 95% at 50°C.

6. Response time.

OCEAN UTILIZATION INC.
2714 Johnson
Corvallis, OR 97330
(903) 555-8527

Date: 9 November 1981

To: Mr. Edward Blye, Fleet Co-ordinator
               Del Mar Fish Company

From: Jeffrey D. Paduan, Consultant

Subject: Breakwater Design Recommendations for the new Inlet Beach Harbour Facility

Dist.:   Jack Doner, Construction Supervisor

FOREWORD

On August 12, 1981, Ocean Utilization Inc. was contacted by the Del Mar Fish Company regarding the relocation of the Del Mar Fishing Fleet at Inlet Beach, Florida. This relocation may require the construction of some form of breakwater to protect the fleet from heavy wave action in the nearshore area. I was assigned to investigate the conditions at Inlet Beach to determine if a breakwater was actually needed and, if so, to recommend the most economical design. The purpose of this report is to provide the results of my investigation.

SUMMARY

Because the direction of the most intense waves and the predominant waves are not the same at Inlet Beach, I am recommending that two parallel breakwaters be constructed. This is based on the determination that some form of breakwater construction is clearly necessary to protect the Del Mar fleet given the wave conditions that exist. The breakwaters should be about 30 feet wide in order to sufficiently damp out the incoming wave energy and should extend about 300 feet from shore in order to sufficiently clear the dangerous nearshore area. The breakwaters can be constructed of readily available stones, and the cost of one breakwater appears to be about $500,000. for a total of $1,000,000. for the two breakwaters.

1.)   BACKGROUND

1.1) Nearshore Hazards

The area that is most hazardous for ships to navigate is the area near the shore. There are two reasons for this:
1) Increased chance of the ship running aground in shallow water
2) More severe and treacherous wave conditions near the shore

This report is concerned with the second of these reasons. Although it seems paradoxical that waves would be more treacherous for ships near the shore than for ships far from the shore, it can be explained using a simplified theory of water waves. It can be shown that waves in shallow water have speeds that are propor-

tional to the depth of the water. Because waves approaching the shore find themselves in water that is becoming progressively less deep, the speed of the waves
decreases and the height is forced to increase in order to maintain the same energy level.

1.2) What Is the Surf Zone?

For the purpose of this report, we shall define the "surf zone" to be the area extending from the shore out to a depth where the waves are no longer affected by
the bottom. This depth is defined by wave theory as $L/2$, or one half the wavelength of the incoming wave. Our investigation will yield an average wavelength,
L, and therefore a characteristic surf zone width, d, depending on the depth at our
location.

1.3) What Is a Breakwater?

A breakwater is a physical device that blocks or damps wave action. Breakwaters
are used to provide an alley of safe passage for ships and boats moving, across the
surf zone, from the harbour to the open water. The definition of a breakwater is
general. It includes everything from a simple pile of stones to an ordered pile of
concrete blocks to a solid wall of cement. Logically, these different breakwater designs, utilizing different materials, have varying degrees of effectiveness at shunt-
ing wave action. The particular design chosen for any one location is a function of
wave conditions at that location and of the degree of protection required by the
ships that will utilize the breakwater.

1.3a) Breakwater Width

The width of a breakwater depends on two things: the materials of which it is constructed and the size of the waves attacking it. If the breakwater is solid, it need
only be wide enough to be stable. Cement breakwaters usually are ten to twenty
feet wide at the bottom and converge at the top to only a few feet. These relatively
narrow breakwaters are 100% efficient at damping out wave energy because they
are solid and let no water pass through. Stone breakwaters, on the other hand, or
other designs using loose-fitting, porous materials allow some of the water and
wave energy to pass through, and so in order to be effective they must be wide
enough to damp out the required amount of wave energy. It has been shown that
for most applications porous breakwaters must be at least as wide as one third of
the wavelength. This width can range, typically, from ten to fifty feet.

1.3b) Breakwater Length

Breakwaters should be long enough to extend through the surf zone. This length
will depend on the wavelength and the water depth at a particular location.

1.4) Critical Parameters for Describing Nearshore Waves

1.4a) The Predominant Wave Height

The predominant wave height is defined as the average height of waves approaching from the predominant wave direction.

1.4b) The Predominant Wave Direction

The predominant wave direction is defined as the direction from which waves are
most likely to approach the beach. Far offshore this direction is usually the same

for large stretches of coastline, but because the topography or depths of a particular beach turn waves due to refraction (see section 1.5) the predominant wave direction will be much different in the surf zone than offshore and slightly different for adjacent stretches of beach.

1.4c) The Direction of Most Intense Waves

The direction of most intense waves is defined as the direction from which the largest waves approach the beach. The direction will depend on which direction the strongest storms typically come from and on the topography of the beach. Every beach has a unique topography. Although many flat, gently sloping beaches may be nearly identical, irregularities such as an underwater trough or mound can focus wave energy and create very large waves.

1.5) Wave Refraction

The dependence of wave speeds in shallow water on depth leads to a phenomenon known as refraction. Refraction means the bending of waves. For example, if a wave approaches a beach at an angle the part of the wave closest to the shore will reach the shallow water first and be slowed down while the offshore portion of the wave is still traveling relatively fast. The effect of this is to line up waves more nearly parallel to the beach and to decrease the angle at which they approach the beach. If deep holes or tall mounds are present in the beach topography, refraction can cause wave energy to be intensified on a mound and dispersed over a hole or trough. This phenomenon is illustrated, graphically, in the Appendix. Oceanographers use lines to represent wave fronts. The gaps between successive lines contain equal amounts of energy; therefore the lines that are farther apart represent less intense, smaller waves and the lines that are close together represent more intense, larger waves.

1.5a) The Wave Refraction Program

Oceanographers at O.U.I. have developed a computer simulation program that uses wave theory and water depths to model refraction phenomena.[1] The user must fill in the depth grid and specify the spacing; then he must indicate the direction from which the waves are approaching far offshore and the wavelength of the waves. The program then provides a graphic output depicting the waves and how they are refracted as they approach the beach.

2.) DATA

2.1) Offshore Wave History

In order to determine the relative frequencies of the directions from which waves approach the Inlet Beach site, the Coast Guard records from ships traveling across the Gulf of Mexico were consulted.[2] We also examined the records of wind and wave data from the National Weather Service's Tampa, Florida, station.[3]

2.2) Surf Zone Topography

In order to determine the topography in the surf zone we took depth measurements with a small boat and a commercial depth gauge made by Shakespeare (Model 1021). The measurements were taken every 50 feet out to a distance of 1000 feet from shore, and then the process was repeated every 50 feet along the shore for a one half mile stretch of beach centered at the Inlet Beach site. This gave us a depth grid with a fifty-foot resolution and over one thousand data points for use in the wave refraction program.

3.)   ANALYSIS AND RESULTS

3.1) The Predominant Wave Height Indicates a Need for a Breakwater

The predominant wave height at the Inlet Beach facility is of sufficient height to warrant the construction of a breakwater. This was determined by using the refraction program and the depth data described above. As waves are bent by the decreasing depths in the nearshore zone, they are crowded together. This crowding shows up in the refraction program and allows us to calculate how much the wave height increases from offshore to nearshore. Using the offshore wave data, the predominant offshore wave height was determined to be about 4 feet. When this height was corrected for refraction, the predominant wave height was calculated to be about 6 feet. Six feet is 30% higher than the maximum average wave height allowable for the Del Mar fishing fleet according to the fleet design criteria you sent us.[4]

3.2) The Predominant Wave Direction Indicates a Need for an Eastern Breakwater

The predominant wave direction is from the southeast. This direction suggests that the east side of the ship tract through the surf zone should be protected by a breakwater. The predominant wave direction was determined by first examining the offshore wave history and determining that the predominant offshore wave direction is southeast at precisely 45 degrees to the shore west of the site (see Appendix). Waves approach from this direction 91% of the time. In order to compensate for wave refraction in the surf zone, the surf zone depth grid, the predominant offshore wave direction, and the average wavelength (see section 4.2) were fed into the refraction program. The refraction program predicted the predominant wave direction to be from the southeast at 20 degrees to the shore west of the site.

3.3) Direction of Most Intense Waves Indicates a Possible Need for an Eastern Breakwater

The most intense waves approach the Inlet Beach site from the southwest. This suggests that it may be necessary to construct a breakwater on the west side of the ship tract through the surf zone. The direction of the most intense waves was determined in two stages. First, the wave history was consulted to find out what direction the most intense offshore waves come from and how often they approach from this direction. It was found that the most intense offshore waves approach from the south and that waves approach from this direction only 1% of the time. Because any breakwater is going to be ineffective on waves approaching straight on as these waves from the south will and because waves approach from this direction only 1% of the time, we decided to ignore the contribution to the direction of most intense waves by these waves approaching from the south. Second, we used the wave refraction program to test all possible offshore wave directions to see what direction produces the most intense waves in the surf zone. It turned out that offshore waves approaching from the southwest at 45 degrees to the shore east of the site produce the largest waves in the surf zone (see Appendix). These waves in the surf zone average 12 feet high and approach the site at an angle of about 55 degrees to the shore east of the site. Waves approaching the site from an offshore direction of 45 degrees to the shore east of the site were taken as the most intense waves and their direction as the direction of the most intense waves. It happens that waves approach from this direction 6% of the time.

3.4) Surf Zone Profile Indicates an Optimum Breakwater Angle

It turns out that a breakwater built at the Inlet Beach site would be more effective if it were inclined at an angle toward the west shore. The optimum angle is about 85 degrees to the shore west of the site (see Appendix). This angle was determined by first realizing that inclining the breakwater to the west would more directly confront the waves approaching from the predominant wave direction and that if the angle was not less than 70 degrees the breakwater would still effectively combat the waves from the direction of the most intense waves. Next we see that the determining factor becomes the surf zone width. We don't want to make the breakwater any longer than we have to to get across the surf zone. We looked at the depth data in the surf zone to determine the angle that would give the best protection but not increase the cost too much. The surf zone width is pretty regular straight offshore from the site, and so it became a somewhat subjective choice as to the best angle. We chose an angle of about 85 degrees because we felt that any smaller angle would increase the length, and therefore the cost, of the breakwater by an amount that is not justified by the increased effectiveness.

4.) CONSTRUCTION SPECIFICATIONS

4.1) Breakwater Materials

A breakwater constructed of loosely packed stones would be an effective yet relatively inexpensive breakwater for the Inlet Beach site. Stones were determined to be a sufficient building material based on a report published by the Army Corps of Engineers' Coastal Engineering Research Center.[5] At the Center they did extensive testing on different materials for breakwaters. The report compares three different materials: stones, rhombohedrally shaped concrete blocks, and solid concrete walls. The strength and endurance of the three materials were given in terms of the maximum predominant wave height that each could routinely withstand, the maximum wave height that each could withstand for short periods of time (less than 5% of the time), and the percent of initial cost that would be required each ten years to maintain the structure. The figures showed the stones to be least effective, followed by the blocks and then the concrete wall with the highest effectiveness. However, the predominant wave conditions at Inlet Beach are some 26% lower than the maximum given for a stone breakwater. Also, the stone breakwater is much cheaper than either of the other two, averaging about 60% cheaper to construct, and the maintenance cost is comparable at 10% per 10 years, or 1% of the initial cost per year.

4.2) Breakwater Width

Because a stone breakwater is a porous breakwater, the width of the breakwater will depend on the average wavelength, $\underline{L}$, of the incoming waves. Wave theory suggests that the breakwater be at least $\underline{L}/3$ feet wide to insure that the breakwater is not narrow enough to be a nodal point in the wave field and therefore ineffective. The average wavelength at Inlet Beach is about 60 feet; therefore a breakwater that is 30 feet wide would adequately damp out the wave energy and provide a safety margin of some 50% over the theoretically suggested width.

4.3) Breakwater Length

The length is determined by the distance across the surf zone. With an average wavelength of 60 feet, the edge of the surf zone is defined at a depth of 30 feet. From the depth data in the surf zone, it is evident that it would require a breakwater 300 feet long inclined 85 degrees to the shore west of the site to reach a depth of 30 feet and thereby clear the surf zone.

4.4) Breakwater Separation

If two breakwaters are used, the question of their separation distance becomes critical. Clearly, they must be at least twice as far apart as the ships that will travel between them are wide, but it is also critical that they are not too far apart. Model studies have shown that it is better if the insert distance, $\underline{ID}$, is not greater than one wavelength, $\underline{L}$.[6] This suggests a separation distance, $\underline{SD}$, of about 140 feet (see Appendix).

5.) PRELIMINARY COST ESTIMATES

5.1) Cost per Breakwater

We have obtained estimates of production cost from the Brown Construction Co. of Miami and the Watertech Co. of Corpus Christi, totalling $500,000.[7] and $600,000.[8] respectively. Both of these estimates are for stone breakwaters 30 feet wide and include material costs. The Watertech Co. estimate comes from a formula based on the average water depth, which is about 20 feet. For an average depth of 20 feet they quote a price of $2000. per running foot, totalling $600,000. The Brown Co. estimate is based on a more complicated formula: $(\$1000)(\underline{Q})(\underline{dX})$, where $\underline{X}$ is the distance offshore and $\underline{Q} = 1$ for a depth that is less than or equal to 10 feet and $\underline{Q} = .1(\underline{y} - 10) + 1$ for $\underline{y}$ = depth greater than 10 feet. This formula requires an integration of the beach slope across the surf zone. If one assumes a smooth, straight-line slope of 1/10, which is indicated by the depth data, the formula yields a total cost of $500,000. per breakwater.

5.2) Cost Analysis of Two Breakwaters

In eleven years the cost of the second breakwater would be recovered from revenues on fishing days that would otherwise be lost. Using the gross income figures you supplied, I determined that a loss of 6% of your fishing days due to severe waves offshore of the site would result in a revenue loss of $51,000 per year. The additional cost of some $555,000. ($500,000. initial plus $55,000 maintenance) for the second breakwater would therefore be regained in eleven years at these revenue rates.

6.) RECOMMENDATIONS

I am recommending that you construct twin stone breakwaters, 30 feet wide, 300 feet long, 140 feet apart, and inclined 85 degrees to the shore west of your facility at Inlet Beach, Florida, (see Appendix). This breakwater configuration should adequately protect your vessels from both the average waves and the most intense waves approaching the site. The cost for such a breakwater system appears to be about $1,000,000. according to an estimate given by the Brown Construction Co. of Miami. I recommend, however, that you investigate other marine construction firms before making any final decision, and I can suggest several others if you would like me to.

REFERENCES

1)   O.U.I. Wave Refraction Program, based on Airy linear wave theory (1957), adapted at O.U.I. 1976

2)   United States Coast Guard, Reports of Ships of Opportunity, Gulf of Mexico, 1946-1979

3)   National Weather Service, Climatological Weather Bulletin 1950-1959, 1960-1969, 1970-1979, Tampa, Fla., Station

4)   Del Mar Fleet Specifications, personal communication, Edward Blye, Aug. 12, 1981

5)   Coastal Engineering Research Center, Technical Report #78-1333465-2, 1978

6)   "Optimum Breakwater Spacing, A Model Study," Univ. of Michigan Technical Report, #79-004356-1, 1979

7)   Brown Construction Company, 2600 Rickenbacker Causeway, Miami, Fla., production bid for Del Mar Fish Company-Inlet Beach Facility Ship Protection System, via Jason Clark, Sept. 18, 1981

8)   Watertech Inc., 18000 Gorman, Corpus Christi, Tex., Del Mar Construction Bid, via George Leilin, Sept. 23, 1981

Bell, Swartzel, and Thomas
Consulting Engineers
Ann Arbor, Michigan

Date: April 20, 1982

To: G. B. Williams, President
D. E. Hammer, Project Manager
Technical Specialties Company
Ann Arbor, Michigan

From: Douglas J. Thomas

Subject: Design Proposal for Scrubbing Section of Dual-Alkali Flue Gas Desulfurization System

Foreword

In a letter from D. E. Hammer and G. B. Williams dated January 27, 1981,
our company was retained by Technical Specialties to design the scrubbing and
flue gas disposal sections of a dual-alkali flue gas desulfurization (FGD) system
being built for a client in West Virginia. The design has been completed. This report includes:

1. A description of the process
2. A description of the process control scheme
3. Specifications for all required equipment
4. A discussion of design judgments
5. Summaries of capital and operating costs

Summary

The design was started by determining the rate of mass transfer in the system. Using the data, the scrubber was sized and the final material balance was
closed. The flue gas disposal system was then designed, considering the desire to
eliminate a condensing vapor plume. The scrubbing liquor recirculation system
was sized using the liquid rate dictated by mass transfer and assuming 30 minutes
of residence time.

The best scrubber configuration was found to be a single vertical spray
tower equipped with an internal demister. The spray tower is a vertical cylinder
30 feet in diameter and 85 feet high. It is equipped with 230 spray nozzles, 46 in
each of 5 levels, delivering a total of 2300 gallons per minute of scrubbing liquor.
The liquor recirculation system consists of one large recirculation tank, a pump,
and a soda ash/water reaction tank. In addition to the existing stack, a fan and a
fired heater are required for flue gas disposal.

Minimum sulfur dioxide removal in this system should be 73 percent, a safe margin above the required 62.4 percent. Capital cost of the installed system is $3.5 million, and annual utility costs are $360,000.

Introduction

Technical Specialties required a scrubber design for a demonstration dual-alkali FGD system the company had contracted to build. There were no commercial-scale scrubber designs available for this chemistry, and so it was necessary to design a scrubber from first principles using pilot plant scrubbing liquor concentration data. The system was designed to clean the flue gas resulting from combustion of coal of known composition. The flue gas cleaned by this process must comply with a sulfur emission regulation.

The scrubber works by absorbing sulfur dioxide from the flue gas into the scrubbing liquor, an aqueous solution of sodium bisulfite, sodium sulfite, and sodium sulfate. The sulfur dioxide thus absorbed subsequently reacts with sodium sulfite and water to form sodium bisulfite. Sulfur is eliminated from the system in the form of calcium sulfite and calcium sulfate by the scrubbing liquor regeneration system, the design of which was left to Technical Specialties.

The main technical problems encountered in this design were selecting a scrubber type, quantifying mass transfer of sulfur dioxide from the gas to the scrubbing liquor, and devising a method for flue gas disposal.

1. Synopsis of Process

The incoming flue gas, stream F1 on the Process and Instrumentation Flow Diagram, Figure 1, contains 0.146 mole percent sulfur dioxide when the coal is burned with 15 percent excess air. At least 62.4 percent of this sulfur dioxide must be removed before the flue gas is discharged to the atmosphere in stream F3. This is accomplished by mass transfer of sulfur dioxide from the gas to the liquid phase in the countercurrent spray column. The sulfur dioxide, after dissolving in the scrubbing liquor, undergoes an essentially instantaneous irreversible reaction with sodium sulfite and water to form sodium bisulfite. The scrubbing liquor returns to the recirculation tank, where it is mixed with regenerated liquor, stream F2, and the sodium hydroxide from the soda ash/water reactor. A portion of the contents of the recirculation tank is withdrawn to the regeneration loop, stream F4, where sulfur is removed in the form of calcium sulfite and calcium sulfate. The material balance for this process appears in the appendix [page 391].

A sketch of the spray tower appears in the appendix [page 392]. Particulate-free flue gas enters through a duct near the base of the tower. The gas rises through a distributor intended to flatten out the gas velocity profile in the column in order to reduce entrainment of scrubbing liquor droplets. Flue gas rises at approximately 5.4 feet/sec. through 50 feet of gas-liquid contacting area. At 10-foot intervals in the contacting section, banks of 46 spray nozzles spray 10 gpm each in a pattern designed to cover the entire cross section of the tower. The liquor flow

FIGURE 1: Process and Instrumentation Flow Diagram (Flue Gas Scrubbing)

to three of these spray banks may be shut off when it is desired to run the system at less than design capacity. Immediately above the last level of spray nozzles in the tower, a wire mesh demister is installed. The demister was designed within the column in order to save the cost of a separate vessel and to keep gas velocity through the demister low enough that re-entrainment would not be significant. A portion of the make-up water to the system is used as wash water for the demister.

The scrubbed flue gas exits the tower through ducts to a fan, where the pressure is increased from −5.5 inches of water to atmospheric pressure. The gas enters the desulfurization system at −0.5 inches of water and undergoes a total pressure drop of 5 inches of water in the scrubber, demister, and ductwork.

In order to eliminate a condensing vapor plume in the stack gas, the flue gas is reheated in a direct-fired heater. The flue gas is then drawn out of the system by the existing stack.

Scrubbing liquor is circulated by the pump downstream of the recirculation tank. Liquor returns to the tank from the tower by gravity flow.

2. Process Control Scheme

Scrubbing liquor flow to the tower is maintained in constant proportion to the flue gas flow rate. A ratio control system is used to compare flow rates of gas and liquor and to adjust the flow of liquor by means of a pneumatically operated valve on the liquor feed line. All controls in this system are pneumatic for reliability in continuous service in a corrosive environment exposed to the elements.

Pressure in the tower is kept within specifications by a pressure control loop on the tower outlet duct. It was deemed preferable to control pressure at this location rather than to restrict flow of flue gas from the boiler to the scrubber.

A liquid-level control loop is specified for the liquor drain in the tower base. It is intended to maintain five feet of liquid to prevent flue gas bypass of the scrubbing section or intake of atmospheric air through this drain.

A sample tap is provided in the exiting flue gas duct so that samples for evaluating the system performance may be withdrawn. Flue gas analysis facilities are to be supplied by the client.

3. Equipment Specifications

Specifications for the scrubbing liquor pump, flue gas fan, scrubbing liquor recirculation tank, soda ash/water reaction tank, and spray tower are in the appendix [pages 392–396].

Material selection was often dictated by consideration of corrosion resistance. The materials of construction of all major pieces of equipment are listed in the following table.

Materials of Construction

Equipment Item	Material
Scrubber	
Shell	316L Stainless Steel
Piping	"
Demister	"
Distributor	"
Nozzles	Brass
Pump	
Casing	Natural Rubber-Lined Carbon Steel
Impeller	Hard Nickel Alloy
Fan	
Casing	Natural Rubber-Lined Carbon Steel
Impeller	Inconel 625
Recirculation Tank	A36 Carbon Steel with Fiberglass Reinforced Polyester Lining
Fired Heater	Carbon Steel
Soda Ash/Water Reaction Tank	316L Stainless Steel

4. Discussion of Design

Several types of scrubber were considered for this system, but the vertical spray tower was decided to be the best option. Packed towers and tray towers were eliminated from consideration because of their tendencies for fouling and corrosion with the particulate-loaded, corrosive liquor used by this process. Venturi scrubbers were rejected because of difficulty in modelling mass transfer in scaled-up units. It was also difficult to quantify mass transfer in fluidized bed scrubbers, removing them from consideration. Horizontal spray scrubbers were seen to offer no advantage over vertical towers, but caused concern that flue gas could channel through an unsprayed portion of the cross section.

The rate of mass transfer depends upon the solubility of sulfur dioxide in the scrubbing liquor. However, no better value for this solubility than the Henry's Law constant for sulfur dioxide in water was found. The salts in the liquor would be expected to reduce sulfur dioxide solubility. The sensitivity of the mass transfer rate to variation in solubility was tested by decreasing the solubility value by a factor of 1000. This caused a mere 3 percent change in the required liquid rate, and so an exact value for solubility was not considered necessary.

The partial pressure of oxygen in the scrubber was too low and the concentration of sulfite ions in the liquor too high for oxygen absorption to be significant. The effect of dissolution and reaction of oxygen was not considered.

The minimum liquid rate required for mass transfer was multiplied by a factor to account for entrainment, drop size distribution, drop evaporation, and drop coalescence to a film on the tower walls. This factor was arbitrarily set equal to 5 after a study of the data presented by Pigford and Pyle (Ref. 6)

The flue gas exiting the spray tower was assumed to approach saturation at a temperature approaching the temperature of the scrubbing liquor. This fixed the make-up water flow rate. In order to eliminate the condensing vapor plume which this gas would produce on contact with the atmosphere, the design calls for heating it 100 F° above its dew point. Flexibility should be allowed for in the heat duty of the heater should 100-degree heating prove to be insufficient or excessive.

More information is needed regarding the quantity of particulates in the flue gas. If solid matter accumulates in the recirculation tank, it may be necessary to provide agitation for this vessel.

5. Costs

Capital costs were calculated using the method of Guthrie (ref. 2) and adjusting for inflation using Marshall and Swift's index. The capital cost breakdown for the sections of the scrubbing system that we designed are given in the table that appears on the following page.

Capital Cost

Item	Installed Cost (1981)
Scrubber	$1,950,000
Pump	62,693
Fan	395,861
Tank	129,555
Heater	361,819
Reactor Tank	76,981
	2,976,909
Contingency and Contractor Fee, 18%	535,844
Total	$3,512,753

Utility costs are given in the following table. Cost data are from the <u>Chemical Engineers' Handbook</u> (ref. 5), corrected for inflation using the Marshall and Swift Index.

Utility Costs

Electricity	$108,031/year
Water	11,043
Natural Gas	244,045
Total	$363,119/year

Conclusions

The sulfur emission regulations for this plant can be met by using the dual-alkali scrubbing system detailed in this report. A spray tower 30 feet in diameter with 50 feet of gas-liquid contacting should provide more than adequate surface for mass transfer. A liquor rate of 2300 gpm is required for assurance of adequate absorption. A fan, a direct-fired heater, and the existing stack are required for disposal of the scrubbed flue gas. The instrumentation and control system is designed to allow turndown of the system capacity. Liquor feed rate, spray tower pressure, and liquid level in the tower are controlled. Provision is made for flue gas sampling, but facilities for testing are left to the client. The system's present cost is $3.5 million, installed. It will cost $360,000 per year for utilities to run.

Details of the particulate removal system and the scrubbing liquor regeneration loop are left to the principal contractor, Technical Specialties.

References

1. HS Fogler, "The Elements of Chemical Kinetics and Reactor Calculations," Prentice Hall Inc., Englewood Cliffs, New Jersey, 1974.

2. KM Guthrie, "Capital Cost Estimating," Chemical Engineering, March 24, 1969.

3. L Kroop, Master's Thesis, Chemical Engineering Department, University of Michigan, 1981.

4. JJ Matley, "Cost File," Chemical Engineering, January 26, 1981.

5. RH Perry and CH Chilton editors, Chemical Engineers' Handbook, 5th Edition, McGraw-Hill, New York, 1973.

6. RL Pigford and C Pyle, "Performance Characteristics of Spray Type Absorption Equipment," Industrial and Engineering Chemistry, July 1951.

7. MR Tek and JO Wilkes, "Fluid Flow and Heat Transfer," Department of Chemical Engineering, University of Michigan, 1974.

8. RC Weast, editor, CRC Handbook of Chemistry and Physics, 59th Edition, CRC Press, Boca Raton, Florida, 1978.

9. CJ West, editor, "The International Critical Tables," Vol. 3, McGraw-Hill, New York, 1933.

10. RN Wimpress, "Rating Fired Heaters," Hydrocarbon Processing and Petroleum Refiner, October 1963.

MASS BALANCE

Stream	Total	CO_2	H_2O	SO_2	N_2	O_2	$NaHSO_3$	Na_2SO_3	NA_2SO_4	NaCl	$NaNO_3$	Na_2CO_3	Inserts
						Flow Rate lb moles/hr							
F1	24,994.3	3645.4	1539.8	36.5	19,113.5	659.1	-	-	-	-	-	-	4258 lb/hr
F2	2,450.0	-	2388.7	-	-	-	-	22.8	35.5	2.6	0.36	-	-
F3	31,397.9	3645.4	7970.0	9.92	19,113.5	659.1	-	-	-	-	-	-	-
F4	2,299.5	-	2227.4	-	-	-	12.9	17.3	36.9	4.6	0.36	-	1281 lb/hr
F5	6,268.7	-	-	-	-	-	-	-	-	-	-	-	-
F6	*	-	-	-	-	-	-	-	-	-	-	*	-

*Soda ash flow rate was not determined because of inconsistencies in the pilot plant data.

SPRAY TOWER DESIGN

11" DIAMETER

WASH WATER TO DEMISTER

2'

DEMISTER

85'

SPRAY LEVELS AT 10' INTERVALS

6"

50' CONTACT LENGTH

GAS DISTRIBUTOR

GAS INLET 11" SQUARE

12"

5'

LIQ. LEVEL

30' DIAMETER

ELEVATION

STYLE 721 (97.77 FREE VOLUME) P. 18-86 PERRYS
ALLOWABLE VELOCITY = 13.4^{v}/SEC = $.35\sqrt{\dfrac{78.3}{.05376}}$

PRESSURE DROP

P. 20-98	TOWER	.5 to 1.5" H_2O
P. 18-86	DEMISTER	.5" H_2O
TABLE 5-9	DUCTS*	3.0" H_2O
		5.0" H_2O

* ASSUMES 165' OF 8.5'/ DIAM. DUCTS

P = 406.79 - 5 = 402" H_2O
= 14.53 psia

4'-4½"

4'-0¾" TYPICAL

A A

12"

1" = 1' SCALE

10

460 GPM LEVEL
46 NOZZLES/LEVEL
10 GPM/NOZZLE
EACH NOZZLE PRODUCING 800 AVE DIAMETER DROPS, AT 60 PSI PRESSURE DROP. SIMILAR TO SPRAYCO # 3/4 HD4. ENTIRE TOWER CROSS SECTION IS SPRAYED 41 ¼ INCHES BELOW NOZZLES

REDUCING TEE 6"

2½"

3¾"

TEE NOZZLE

SECTION A A

52½"

41¼"

52½"

PUMPS

Page 1 of 1
By   D.J. Thomas
Date: 2-16-81

Item No.   P-1
Name: Scrubbing Liquor Pump
No. Req'd   1

Function:

Type:
   Pump CENTRIFUGAL
   Drive _____

Materials Handled:
   Name of Material: SCRUBBING LIQUOR
   Pumping temp., °F 135
   Specific gravity at t°F 1.26
   Viscosity at t°F 1.35×10^{-3} lb/ft^2
   Normal Flow: lb/hr 1,444,000   GPM 2280
   Specified capacity, GPM 2300
   Suction vessel (origin) RECIRCULATION TANK
   Discharge vessel (destination) SCRUBBER TOWER

Head Data:

	Suction Side		Discharge Side	
	psia	ft.	psia	ft.
Press. over liquid	15		75	110.34
Static head				95.00
Friction drop				2414 combined
Velocity head				1.32
Total head			114	209.1

Driver Power, HP:
   Normal calcd.   20L
Maximum Calcd.   250   @ 20% excess capacity
Specified   _____

COMPRESSORS

By: D.J. Thomas

Item No.:   C-1
Name: FLUE GAS FAN
Number Required: 1

Function:

Type:
  Compressor:
  Driver:

Process Requirements:
  Material Compressed:

Component	Wt.%	Mol%	MW
CO_2			44
H_2O			18
SO_2			64
N_2			28
O_2			32

  Flow rate:
    lb/hr.
    Mol/hr.   25,000
    SCFM (60°F, 14.7 psia).   158,152
    Mean spec. ht. ratio, c_p/c_v.
    Mean MW.   ~30

	Suction	Discharge
Pressure, psia.	14.5	14.7
Temperature, °F.	~150	~150
ACFM.	185,083	185,525

Compressor Design:

No. of Cylinders		Volumetric efficiency, %		
Compression ratio		CFM at suction conditions	185,083	
Stroke, in.		Ideal gas power, HP	148	
Cylinder diam., in.		Compression eff., %	55	
Speed, RPM		Shaft power, HP	265	
Displacement, CFM		Calc. Driver Power, HP		
Clearance, %		Spec. Driver Power, HP	325	@ 20%

excess capacity

By: D.J. Thomas
Date: 2-16-81

Item No.:   T-1

Name: SCRUBBING LIQUOR RECLAMATION TANK

Number Required: 1

Function:

Type:

Process Data:
    Material handled: SCRUBBING LIQUOR
    Composition:

Component	Wt. %	Mol %
H_2O	79.67	
$NaHSO_3$	7.67	
Na_2SO_3	14.32	
Na_2SO_4	10.42	
NaCl	.32	
$NaNO_3$	.06	
Inerts	2.54	

Temperature, °F: Operating pressure, psig.:
    Normal   135   Normal   0
    Maximum   140   Maximum   0

    Sp. G. at t °F   1.26

Flow rate:
    GPM   2386
    lb./hr.   1,505,000

Size Data:
    Residence time ~30 min
    Volumetric capacity 74,592 gal

Design Specifications:
    Material of Construction A36 CS, FIBERGLASS REINFORCED POLYESTER
                     LINING
    Position VERTICAL
    Design temp., °F. 135
    Design, pressure, psig. 0
    Diameter, ft. 23
    Length, ft. 24
    Wall thickness, in. ⅛"
    Corrosion allowance, in. NONE
    Insulation req'd (Yes or No) No
        For what

SODA ASH/WATER REACTION TANK

Material Handled:
    Soda Ash
    Water   155gpm

Temperature
    Normal   70°F
    Maximum   100

Size Data
        Residence Time   30 minutes
        Volumetric Capacity   4700 gallons

Design Specifications
        Material of Construction   316L stainless steel
        Position   Vertical
        Design Temp.   70°F
        Design Pressure   0 psig
        Diameter   10 feet
        Height   8 feet
        Wall thickness   .125 inch
        Corrosion Allowance   none
        Insulation   none

LONG FORMAL REPORT

FUNCTIONAL

SPECIFICATION

OLDSMOBILE DIVISION
GENERAL MOTORS CORPORATION

FASCIA SYSTEM

Prepared by
DIGITAL EQUIPMENT CORPORATION
Software Services

Great Lakes District
21333 Haggerty Road
Novi, Michigan 48050

November 15, 1981
Version 1.0

Summary Table of Contents

APPENDICES

FIGURES